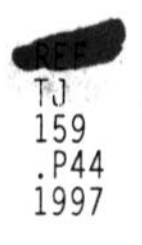

Mechanical Engineering Problems and Solutions

Mechanical Engineering Problems and Solutions

Richard K. Pefley
Santa Clara University

6th Edition

Engineering Press Austin, Texas

ISBN 1-57645-0008-2

Engineering Press P.O. Box 200129 Austin, Texas 78720-0129

Contents

Mechanical Engineering License Review

Exam Files

Professors around the country have opened their exam files and revealed their examination problems and solutions. These are actural exam problems with the complete solutions prepared by the same professors who wrote the problems. Exam Files are currently available for these topics:

Calculus I
Calculus II
Calculus III
Circuit Analysis
College Algebra
Differential Equations
Dynamics
Engineering Economic Analysis
Fluid Mechanics
Linear Algebra
Materials Science
Mechanics of Materials
Organic Chemistry
Physics I Mechanics
Physics III Electricity and Magnetism
Probability and Statistics
Statics
Thermodynamics

For a description of all available **Exam Files**, or to order them, ask at your college or technical bookstore, or call **1-800-800-1651** or write to:

Engineering Press
P.O. Box 1
San Jose, CA 95103-0001

Introduction

BECOMING A PROFESSIONAL ENGINEER

To achieve registration as a Professional Engineer there are four distinct steps: education, fundamentals of engineering (engineer-in-training) exam, professional experience, and finally, the professional engineer exam. These steps are described in the following sections.

Education

The obvious appropriate education is a B.S. degree in mechanical engineering from an accredited college or university. This is not an absolute requirement. Alternative, but less acceptable, education is a B.S. degree in something other than mechanical engineering, or from a non-accredited institution, or four years of education but no degree.

Fundamentals of Engineering (FE/EIT) Exam

Most people are required to take and pass this eight-hour multiple-choice examination. Different states call it by different names (Fundamentals of Engineering, E.I.T., or Intern Engineer) but the exam is the same in all states. It is prepared and graded by the National Council of Examiners for Engineering and Surveying (NCEES). Review materials for this exam are found in other books like Newnan: *Engineer-In-Training License Review.*

Experience

Typically one must have four years of acceptable experience before being permitted to take the Professional Engineer exam, but this requirement may vary from state to state. Both the length and character of the experience will be examined. It may, of course, take more than four years to acquire four years of acceptable experience.

Professional Engineer Exam

The second national exam is called Principles and Practice of Engineering by NCEES, but probably everyone else calls it the Professional Engineer or P.E. exam. All states, plus Guam, the District of Columbia, and Puerto Rico use the same NCEES exam.

MECHANICAL ENGINEERING PROFESSIONAL ENGINEER EXAM

Background

The reason for passing laws regulating the practice of mechanical engineering is to protect the public from incompetent practioners. Beginning about 1907 the individual states began passing *title* acts regulating who could call themselves a mechanical engineer. As the laws were strengthened, the *practice* of certain aspects of mechanical engineering was limited to those who were registered mechanical engineers, or working under the supervision of a registered mechanical engineer. There is no national registration law; registration is based on individual state laws and is administered by boards of registration in each of the states. A listing of the State Boards is in Table 1.1.

Examination Development

Initially the states wrote their own examinations, but beginning in 1966 the NCEES took over the task for some of the states. Now the NCEES exams are used by all states. This greatly eases the ability of a mechanical engineer to move from one state to another and achieve registration in the new state. About 6000 mechanical engineers take the exam each year. As a result about 23% of all mechanical engineers are registered professional engineers.

The development of the mechanical engineering exam is the responsibility of the NCEES Committee on Examinations for Professional Engineers. The committee is composed of people from industry, consulting, and education, plus consultants and subject matter experts. The starting point for the exam is a mechanical engineering task analysis survey that NCEES does at roughly five to ten year intervals. People in industry, consulting and education are surveyed to determine what mechanical engineers do and what knowledge is needed. From this NCEES develops what they call a "matrix of knowledges" that form the basis for the mechanical engineering exam structure described in the next section.

The actual exam questions are prepared by the NCEES committee members, subject matter experts, and other volunteers. All people participating must hold professional registration. Using workshop meetings and correspondence by mail, the questions are written and circulated for review. The problems relate to current professional situations. They are structured to quickly orient one to the requirements, so the examinee can judge whether he or she can successfully solve it. While based on an understanding of engineering fundamentals, the problems require the application of practical professional judgement and insight. Although four hours are allowed for four problems, probably any problem can be solved in 20 minutes by a specialist in the field. A professionally competent applicant can solve the problem in no more than 45 minutes. Multi-part questions are arranged so the solution of each succeeding part does not depend on the correct solution of a prior part. Each part will have a single answer that is reasonable.

Examination Structure

The ten problems in the morning four-hour session are regular computation "essay" problems. In the afternoon four-hour session all ten problems are multiple choice. In each category (Machine Design, Stress Analysis, Power Plant Systems, and so on) about half of the problems will be in the morning session and half in the afternoon. Engineering economics may appear as a component within one or two of the problems.

1. MACHINE DESIGN - Two Problems.
   Fasteners, gears, brakes, belts, clutches, wire rope, bearings, conveyors

2. STRESS ANALYSIS/STRUCTURAL DESIGN - Two Problems.
   Stress analysis of machines, tools or structures including theories of failure, static and fatique life

3. KINEMATICS AND DYNAMICS - One Problem.
   Motion of machines and vehicles and associated forces/energy

4. POWER PLANT SYSTEMS - Two Problems
   Power plant cycles, compressors and turbines, thermal and mechanical efficiency, co-generation

5. POWER PLANT PROCESSES - One Problem.
   Fuels and combustion, combustion stoichiometry, products of combustion; efficiency

6. POWER PLANT COMPONENTS - One Problem.
   Boilers and pressure vessels, expansion tanks, piping for gases and liquids

7. HVAC/R SYSTEMS - Two Problems.
   Heating/cooling/ventilation load calculations, psychrometrics, evaporative cooling, refrigeration specifications

8. HVAC/R COMPONENTS - One Problem.
   Components used in the operation and control of heating, ventilating, cooling and refrigeration equipment

9. CONTROL SYSTEMS - One Problem
   Analysis and performance of general mechanical systems (fluid, thermal, etc.)
   May include (but is not limited to) root locus, stability and/or control diagram components, first and second order systems

10. INSTRUMENTATION/MEASUREMENTS - One Problem
    Specifications of measuring systems, static and dynamic measurement of temperature, pressure flow, etc.

11. VIBRATIONS - One Problem.
    One and two degree of greedom systems, forced vibrations, transmissibility, isolation

12. HEAT TRANSFER - One Problem.
    Condustion, convection and radiation in practical applications

13. THERMODYNAMICS - One Problem
    Work, energy, compressible gases in various processes such as compressors, engines and nozzles

14. HYDRAULICS/PNEUMATICS - One Problem.
    Hydraulic equipment, hydraulic power and control diagrams, pumps and piping head loss calculations

15. MANAGEMENT - One Problem.
    Estimation of production for various products such as sheet metal parts, selection of alternative manufacturing methods based on economics and life cycle studies

16. FIRE PROTECTION - One Problem.
    System specifications, sprinkler systems, mobile systems, control valves, flow measurement and control. NFPA 13, 14, 20 and 291.

Note: The examination is developed with problems that will require a variety of approaches and methodologies including design, analysis, application, economic aspects, and operations.

Taking The Exam

Exam Dates

The National Council of Examiners for Engineering and Surveying (NCEES) prepares Mechanical Engineering Professional Engineer exams for use on a Friday in April and October each year. Some state boards administer the exam twice a year in their state, while others offer the exam once a year. The scheduled exam dates are:

	April	October
1995	7	27
1996	19	25
1997	18	31
1998	24	30

People seeking to take a particular exam must apply to their state board several months in advance.

Exam Procedure

Before the morning four-hour session begins, the proctors will pass out an exam booklet and solutions pamphlet to each examinee. There are likely to be civil, chemical, and electrical engineers taking their own exams at the same time. You must solve four of the ten mechanical engineering problems.

The solution pamphlet contains grid sheets on right-hand pages. Only work on these grid sheets will be graded. The left-hand pages are blank and are for scratch paper. The scratchwork will not be considered in the scoring.

If you finish more than 30 minutes early, you may turn in the booklets and leave. In the last 30 minutes, however, you must remain to the end to insure a quiet environment for all those still working, and to insure an orderly collection of materials.

The afternoon session will begin following a one-hour lunch break. The afternoon exam booklet will be distributed along with an answer sheet. The booklet will have ten 10-part multiple choice questions. You must select and solve four of them. An HB or #2 pencil is to be used to record your answers on the scoring sheet.

Exam-Taking Suggestions

People familiar with the psychology of exam-taking have several suggestions for people as they prepare to take an exam.

1. Exam taking is really two skills. One is the skill of illustrating knowledge that you know. The other is the skill of exam-taking. The first may be enhanced by a systematic review of the technical material. Exam-taking skills, on the other hand, may be improved by practice with similar problems presented in the exam format.

2. Since there is no deduction for guessing on the multiple choice problems, an answer should be given for all ten parts of the four selected problems. Even when one is going to guess, a logical approach is to attempt to first eliminate one or two of the five alternatives. If this can be done, the chance of selecting a correct answer obviously improves from 1 in 5 to, say, 1 in 3.

3. Plan ahead with a strategy. Which is your strongest area? Can you expect to see one or two problems in this area? What about your second strongest area? What will you do if you still must find problems in other areas?

4. Have a time plan. How much time are you going to allow yourself to initially go through the entire ten problems and grade them in difficulty *for you to solve them*? Consider assigning a letter, like A, B, C and D, to each problem. If you allow 15 minutes for grading the problems, you might divide the remaining time into *five* parts of 45 minutes each. Thus 45 minutes would be scheduled for the first - and easiest - problem to be solved. Three additional 45 minute periods could be planned for the remaining three problems. Finally, the last 45 minutes would be in reserve. It could be used to switch to a substitute problem in case one of the selected problems proves too difficult. If that is unnecessary, the time can be used to check over the solutions of the four selected problems. A time plan is very important. It gives you the confidence of being in control, and at the same time keeps you from making the serious mistake of misallocation of time in the exam.

5. Read all five multiple choice answers before making a selection. The first answer in a multiple choice question is sometimes a plausible decoy - not the best answer.

6. Do not change an answer unless you are absolutely certain you have made a mistake. Your first reaction is likely to be correct.

7. Do not sit next to a friend, a window, or other potential distractions.

Exam Day Preparations

There is no doubt that the exam will be a stressful and tiring day. This will be no day to have unpleasant surprises. For this reason we suggest that an advance visit be made to the examination site. Try to determine such items as:

1. How much time should I allow for travel to the exam on that day? Plan to arrive about 15 minutes early. That way you will have ample time, but not too much time. Arriving too early, and mingling with others who also are anxious, will increase your anxiety and nervousness.

2. Where will I park?

3. How does the exam site look? Will I have ample workspace? Where will I stack my reference materials? Will it be overly bright (sunglasses) or cold (sweater), or noisy (earplugs)? Would a cushion make the chair more comfortable?

4. Where is the drinking fountain, lavoratory facilities, payphone?

5. What about food? Should I take something along for energy in the exam? A bag lunch during the break probably makes sense.

What To Take To The Exam

The NCEES guidelines say you may bring the following reference materials and aids into the examination room for your personal use only:

1. Handbooks and textbooks
2. Bound reference materials, provided the materials are and remain bound during the entire examination. The NCEES defines "bound" as books or materials fastened securely in its cover by fasteners which penetrate all papers. Examples are ring binders, spiral binders and notebooks, plastic snap binders, brads, screw posts, and so on.
3. Battery operated, silent non-printing calculators.

At one time NCEES had a rule that did not permit "review publications directed principally toward sample questions and their solutions" in the exam room. This set the stage for restricting some kinds of publications from the exam. State boards may adopt the NCEES guidelines, or adopt either more or less restrictive rules. Thus an important step in preparing for the exam is to know what will - and will not - be permitted. We suggest that if possible you obtain a written copy of your state's policy for the specific exam you will be taking. Recently there has been considerable confusion at individual examination sites, so a copy of the exact applicable policy will not only allow you to carefully and correctly prepare your materials, but also will insure that the exam proctors will allow all proper materials that you bring to the exam.

As a general rule we recommend that you plan well in advance what books and materials you want to take to the exam. Then they should be obtained promptly so you use the same materials in your review that you will have in the exam.

License Review Books

We suggest you select license review books that have a 1994 or more recent copyright. The exam used to be 17 essay questions and 3 multiple-choice problems, including a full scale engineering economics problem. The engineering economics problem is gone (at least as a separate question) and half the remaining 20 problems are now multiple choice.

Textbooks

If you still have your university textbooks, we think they are the ones you should use in the exam, unless they are too out of date. To a great extent the books will be like old friends with familiar notation.

Bound Reference Materials

The NCEES guidelines suggest that you can take any reference materials you wish, so long as you prepare them properly. You could, for example, prepare several volumes of bound reference materials with each volume intended to cover a particular category of problem. Maybe the most efficient way to use this book would be to cut it up and insert portions of it in your individually prepared bound materials. Use tabs so specific material can be located quickly. If you do a careful and systematic review of mechanical engineering, and prepare a lot of well organized materials, you just may find that you are so well prepared that you will not have left anything of value at home.

Other Items

Calculator - NCEES says you may bring a battery operated, silent, non-printing calculator. You need to determine whether or not your state permits pre-programmed calculators. Extra batteries for your calculator are essential, and many people feel that a second calculator is also a very good idea.

Clock - You must have a time plan and a clock or wristwatch.

Pencils - You should consider mechanical pencils that you twist to advance the lead. This is no place to go running around to sharpen a pencil, and you surely do not want to drag along a pencil sharpener.

Eraser - Try a couple to decide what to bring along. You must be able to change answers on the multiple choice answer sheet, and that means a good eraser. Similarly you will want to make corrections in the essay problem calculations.

Exam Assignment Paperwork - Take along the letter assigning you to the exam at the specified location. To prove you are the correct person, also bring something with your name and picture.

Items Suggested By Advance Visit - If you visit the exam site you probably will discover an item or two that you need to add to your list.

Clothes - Plan to wear comfortable clothes. You probably will do better if you are slightly cool.

Box For Everything - You need to be able to carry all your materials to the exam and have them conveniently organized at your side. Probably a cardboard box is the answer.

Exam Scoring

Essay Questions

The exam booklets are returned to Clemson, SC. There the four essay question solutions are removed from the morning workbook. Each problem is sent to one of many scorers throughout the country.

For each question an item specific scoring plan is created with six possible scores: 0, 2, 4, 6, 8, and 10 points. For each score the scoring plan defines the level of knowledge exhibited by the applicant. An applicant who is minimally qualified in the topic is assigned a score of 6 points. The scoring plan shows exactly what is required to achieve the 6 point score. Similar detailed scoring criteria are developed for the two levels of superior performance (8 and 10 points) and the three levels of inferior performance (0, 2, and 4 points). Every essay problem submitted for grading receives one of these six scores. The scoring criteria may be based on positive factors, like identifying the correct computation approach, or negative factors, like improper assumptions or calculation errors, or a mixture of both positive and negative factors. After scoring, the graded materials are returned to NCEES, which reassembles the applicants work and tabulates the scores.

Multiple Choice Questions

Each of the four multiple choice problems is 10 points, with each of the ten questions of the problem worth one point. The questions are machine scored by scanning. The input data are evaluated by computer programs to do error checking. Marking two answers to a question, for example, would be detected and no credit given. In addition, the programs identify those questions with statistically unlikely results. There is, of course, a possibility that one or more of the questions is in some way faulty. In that case a decision will be made by subject matter experts on how the situation should be handled.

Passing The Exam

In the exam you must answer eight problems, each worth 10 points, for a total raw score of 80 points. Since the minimally qualified applicant is assumed to average six points per problem, a raw score of 48 points is set equal to a converted passing score of 70. Stated bluntly, you must get 48 of the 80 possible points to pass. The converted scores are reported to the individual state boards in about two months, along with the recommended pass or fail status of each applicant. The state board is the final authority of whether an applicant has passed or failed the exam.

Although there is some variation from exam to exam, the following gives the approximate passing rates:

Applicant's Degree	Passing Exam
Engineering from accredited school	62%
Engineering from non-accredited school	50
Engineering Technology from accredited school	42
Engineering Technology from non-accredited school	33
Non-Graduates	36
All Applicants	56

Although you want to pass the exam on your first attempt, you should recognize that if necessary you can always apply and take it again.

THIS BOOK

The task of preparing for the day-long mechanical engineer examination is a formidable one. The logical approach is to undertake a systematic review of the categories within the mechanical engineering exam, and to do it at about the level of understanding required to successfully pass the exam. This book is designed to help with that task. The problems have been selected to represent an appropriate topic and level of difficulty. Many are actual exam problems, but are not from the NCEES exams, as they do not release their problems for publication.

The names of the contributors to this book appear on the first page of each chapter. Thanks to people who have studied the prior editions of this book, a number of errors have been identifies and removed. If you note any errors in this edition, a letter, written to the Engineering Press address, would be greatly appreaciated.

Table 1.1. State Boards of Registration for Engineers

State	Mail Address	Phone
AL	301 Interstate Park Drive, Montgomery 36109	205-242-5568
AK	P.O. Box 110806, Juneau 99811	907-465-2540
AZ	1951 W. Camelback Rd, Suite 250, Phoenix 85015	602-255-4053
AR	P.O. Box 2541, Little Rock 72203	501-324-9085
CA	2535 Capitol Oaks Dr, #300, Sacramento 95833-2926	916-920-7466
CO	1560 Broadway, Ste. 1370, Denver 80202	303-894-7788
CT	165 Capitol Ave., Rm G-3A, Hartford 06106	203-566-3386
DE	2005 Concord Pike, Wilmington 19803	302-577-6500
DC	614 H Street NW, Rm 923, Washington 20001	202-727-7454
FL	1940 N. Monroe St., Tallahassee 32399	904-488-9912
GA	166 Pryor Street SW,Rm 504, Atlanta 30303	404-656-3926
GU	P.O. Box 2950, Agana, Guam 96910	671-646-1079
HI	P.O. Box 3469, Honolulu 96801	808-586-2702
ID	600 S. Orchard, Ste. A, Boise 83705	208-334-3860
IL	320 W. Washington St, 3/FL,Springfield 62786	217-785-0820
IN	100 N. Senate Ave, Rm 1021, Indianapolis 46204	317-232-2980
IA	1918 S.E. Hulsizer, Ankeny 50021	515-281-5602
KS	900 Jackson, Ste 507, Topeka 66612-1214	913-296-3053
KY	160 Democrat Drive, Frankfort 40601	502-564-2680
LA	1055 St. Charles Ave, Ste 415, New Orleans 70130	504-568-8450
ME	State House, Sta. 92, Augusta 04333	207-287-3236
MD	501 St. Paul Pl, Rm 902, Baltimore 21202	410-333-6322
MA	100 Cambridge St, Rm 1512, Boston 02202	617-727-9956
MI	P.O. Box 30018, Lansing 48909	517-335-1669
MN	133 E. Seventh St, 3/Fl, St. Paul 55101	612-296-2388
MS	P.O. Box 3, Jackson 39205	601-359-6160
MO	P.O. Box 184, Jefferson City 65102	314-751-0047
MT	111 N. Jackson Arcade Bldg, Helena 59620-0407	406-444-4285
NE	P.O. Box 94751, Lincoln 68509	402-471-2021
NV	1755 E. Plumb Lane, Ste 135, Reno 89502	702-688-1231
NH	57 Regional Dr., Concord 03301	603-271-2219

NJ	P.O. Box 45015, Newark 07101	201-504-6460
NM	440 Cerrillos Rd., Ste A, Santa Fe 87501	505-827-7316
NY	Madison Ave, Cult Educ Ctr., Albany 12230	518-474-3846
NC	3620 Six Forks Rd., Raleigh 27609	919-781-9499
ND	P.O. Box 1357, Bismarck 58502	701-258-0786
MP	P.O. Box 2078, Siapan, No Mariana Is. 96950	670-234-5897
OH	77 S. High St. 16/Fl, Columbus 43266-0314	614-466-3650
OK	201 NE 27th St, Rm 120, Oklahoma City 73105	405-521-2874
OR	750 Front St, NE, Ste 240, Salem 97310	503-378-4180
PA	P.O. Box 2649, Harrisburg 17105-2649	717-783-7049
PR	P.O. Box 3271, San Juan 00904	809-722-2122
RI	10 Orms St, Ste 324, Providence 02904	401-277-2565
SC	P.O. Drawer 50408, Columbia 29250	803-734-9166
SD	2040 W. Main St, Ste 304, Rapid City 57702	605-394-2510
TN	Volunteer Plaza, 3/Fl, Nashville 37243	615-741-3221
TX	P.O. Drawer 18329, Austin 78760	512-440-7723
UT	P.O. Box 45805, Salt Lake City 84145	801-530-6628
VT	109 State St., Montpelier 05609-1106	802-828-2875
VI	No. 1 Sub Base, Rm 205, St. Thomas 00802	809-774-3130
VA	3600 W. Broad St., Richmond 23230-4917	804-367-8514
WA	P.O. Box 9649, Olympia 98507	206-753-2548
WV	608 Union Bldg., Charleston 25301	304-348-3554
WI	P.O. Box 8935, Madison 53708-8935	608-266-1397
WY	Herschler Bldg., Rm 4135E, Cheyenne 82002	307-777-6155

1

Mathematical Fundamentals

PETER A. SZEGO

A survey of the entirety of applied mathematics is beyond the scope of this review. Instead we focus attention on a number of common stumbling blocks. Although all of these are of an elementary character, experience has shown that they are at the root of a high proportion of errors in engineering mathematics.

FORMULA SUBSTITUTION

As simple a matter as substituting in formulas sometimes causes difficulties. Consider for example the quadratic equation, along with its solution, as expressed by the formulas:

$$ax^2 + bx + c = 0, \qquad x = \frac{-b \pm \sqrt{b^2 - 4ac}}{2a}$$

A numerical example, say $2x^2 - 3x + 5 = 0$, is handled in a straightforward manner. We either make the substitutions mentally or (to be quite certain) we make up a little equivalence table, thus:

$$a = 2,\ b = -3,\ c = 5 \quad \therefore x = \frac{-(-3) \pm \sqrt{(-3)^2 - 4(2)(5)}}{2(2)}$$

$$= \frac{3 \pm \sqrt{9-40}}{4} = \frac{3 \pm i\sqrt{31}}{4}$$

Sometimes, however, we must deal with a quadratic equation written as a formula (i.e., with symbol coefficients) and, moreover, some of the symbols may coincide with those in the "standard" form. For example, suppose we wish to solve for z in $2z^2 + az + c = 0$. To avoid confusion we need to distinguish the "a" and "c" of our problem from the like symbols in our reference formula. One way to do this is to use some special mark on the symbols of the reference formula, thus:

$$\bar{a}\bar{x}^2 + \bar{b}\bar{x} + \bar{c} = 0, \qquad \bar{x} = \frac{-\bar{b} \pm \sqrt{\bar{b}^2 - 4\bar{a}\bar{c}}}{2\bar{a}}$$

Once more, we set up an equivalence table:

$$\bar{a} = 2,\ \bar{b} = a,\ \bar{c} = c,\ \bar{x} = z$$

so that

$$z = \frac{-a \pm \sqrt{a^2 - 4(2)c}}{2(2)}$$

The discussion of the quadratic equation is only by way of example. The problem of symbol substitution arises in all areas of mathematics. Texts and reference works typically use a unique set of symbols for certain categories of problems. For example, in calculus one usually sees "x" used as the independent variable and "y" used as the dependent variable. If in a particular problem some other symbols appear, we must carry out the appropriate translations in the manner illustrated above.

QUADRATIC EQUATION

In the previous section, we gave the "standard" formula for the solutions of the quadratic equation. If the discriminant $(b^2 - 4ac)$ is positive, the roots are real numbers; if $(b^2 - 4ac)$ is negative, the roots are complex; and if the discriminant is zero, the roots are identical and real (double root). In applications, our understanding of the physical situation usually will indicate whether the roots must be real or complex, and this should agree with the predictions of the formula. In other words, we have here an example of the use of physical understanding as a check and verification on our mathematics.

A common source of confusion is the distinction between the pairings real/complex and positive/negative. As stated above, the question of whether the roots are real or complex depends on the sign of the discriminant. The question of the sign of the roots, themselves, can arise only if the roots are real. The sign of a complex root is not a meaningful concept.

Again, physical understanding can serve to help interpret mathematical results. Suppose, for example, that the unknown in a quadratic equation represents a length, and that both roots are real, with one positive and one negative. Since a negative length generally is not possible, we reject the negative root as being extraneous and take the positive root as the desired result. But what if both roots had been negative (or worse yet, complex)? Then we have no acceptable solution and something must be wrong with the derivation!

LOGARITHMS

The log symbol

$$\log_b N = x \quad \text{means} \quad b^x = N$$

In this definition, the symbols b, x, N are all real and b and N must be positive. In particular, note that the log of a negative number is undefined (actually, in advanced applications this case can be defined as a complex number). Thus we have another check point. If our calculations lead to, say $\log_{10}(-5.3)$, an error probably has occurred.

When the logarithm appears in calculus, the base b usually will be the number e; sometimes such a log is written as ln. Conversion formulas from one base to another are available in references. We have to make sure that conversions are carried out in the correct direction. Suppose, for example, we have

$$\log_{10} 2 = 0.301$$

and we wish to convert to the base e (i.e., to find $\log_e 2$). A fool proof method is to rewrite the $\log_{10}$ expression according to the definition of the log,

$$10^{0.301} = 2$$

and then to take the $\log_e$ of both sides. Thus,

$$\log_e 10^{0.301} = \log_e 2, \quad \text{or} \quad \log_e 2 = 0.301\ (\log_e 10)$$

where the quantity in the parenthesis is a standard conversion number. Fortunately, most hand calculators handle $\log_{10}$ and $\log_e$ computations directly so this tedious approach can be avoided.

TRIGONOMETRIC FUNCTIONS

One source of confusion is how to handle angles not falling in the first quadrant. Say we seek sin 120°. Sketch a unit circle in an

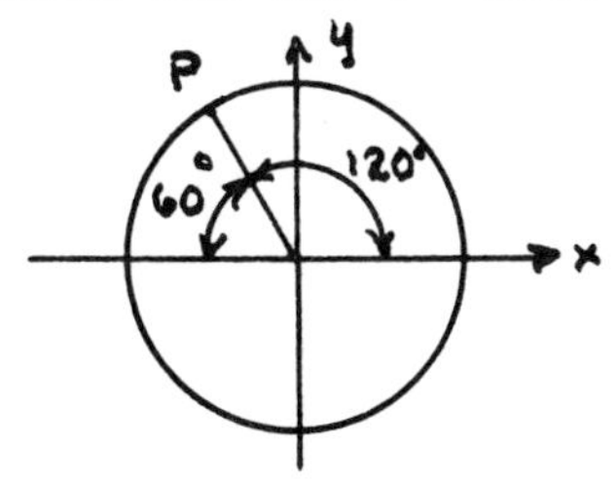

xy-plane with the angle 120° as shown. The sine function is defined as the y coordinate of P. From the sketch we see that this is the same as sin 60° or $\sqrt{3}/2$. Similarly, cos 120° = x(P) = - cos 60° = $-\frac{1}{2}$. Note the appearance of the minus sign because of the negative value of x(P).

Many problems require only a knowledge of the trig functions for the three angles 30°, 45°, 60°. These cases can be remembered easily by constructing two right triangles as shown:

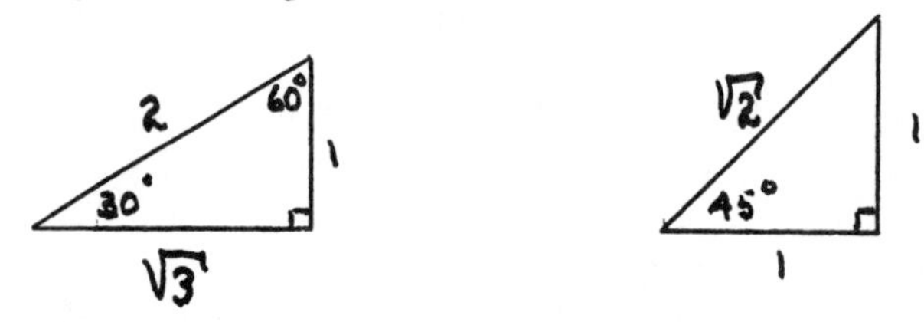

For example, $\sin 60° = \sqrt{3}/2$ can be read off the sketch at once. A common error, however, is to mix up the role of the 30° and 60°. In dealing with a particular trigonometric function, such as the sine, it may help to say to oneself "sine is the opposite over the hypotenuse" as one points to the appropriate sides of the triangle.

A frequently occurring situation is that the value of a trigonometric function is given and one seeks the value of another trig function. Suppose, for example, one knows that $\sin x = 0.2$ and that one seeks the value of $\cos x$. Several methods can be used. One possibility is to obtain the value of x from a slide rule or table and then to look up $\cos x$. A second, and perhaps preferable approach, is to sketch a right triangle illustrating the given data, to calculate the third side by the Pythagorean formula, and then to read off the

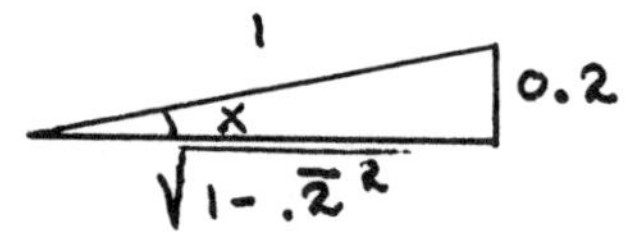

desired result. This second approach remains valid if our starting point is a formula (e.g., $\sin x = q$, then $\cos x = \sqrt{1 - q^2}$). This method amounts to the same thing as using appropriate trigonometric identities ($\sin^2 x + \cos^2 x = 1$, in this case). But the triangle method may be easier to remember than the identities.

In calculus applications, angles generally will be expressed in radian measure. Care must be taken to convert properly between radians and degrees when this is necessary. A simple way to remember the proper relation is to consider the whole circle, for which the angle 360° is equivalent to 2π radians.

DERIVATIVES AND INTEGRALS

The most commonly used derivative and integral formulas are set out below.

$$\frac{dx^n}{dx} = nx^{n-1}, \qquad \frac{d}{dx}\{f[z(x)]\} = \frac{df}{dz}, \frac{dz}{dx} \quad \text{(chain rule)}$$

$$\frac{d(u \cdot v)}{dx} = u\frac{dv}{dx} + v\frac{du}{dx}, \quad \frac{d(u/v)}{dx} = \frac{v\,u' - u\,v'}{v^2},$$

$$\frac{d(\log_e u)}{dx} = \frac{1}{u}\frac{du}{dx}, \quad \frac{d(\sin u)}{dx} = \cos u\frac{du}{dx}, \quad \frac{d(\cos u)}{dx} = -\sin u\frac{du}{dx},$$

$$\int dx = x + C, \quad \int x^n\,dx = \frac{x^{n+1}}{n+1} + C \;(n \neq -1), \quad \int \frac{dx}{x} = \log_e x + C,$$

$$\int \cos x\,dx = \sin x + C, \quad \int \sin x\,dx = -\cos x + C$$

The principal difficulty experienced by engineers in applying these rules arises from the problem of proper substitution. Suppose, for example, we seek

$$\frac{d}{dt}[t^a \sin(5t^2)]$$

First, we apply the product rule for $(u \cdot v)'$ with $u = t^a$, $v = \sin(5t^2)$ (note that t now replaces the role of x). Thus, our desired derivative becomes

$$t^a \frac{d}{dt}[\sin(5t^2)] + \sin(5t^2)\frac{d}{dt}(t^a)$$

Next we take care of $\sin(5t^2)$ by using the chain rule together with the special rules for x^n and $\sin x$. Thus:

$$\frac{d}{dt}[\sin(5t^2)] = \cos(5t^2)\frac{d}{dt}(5t^2) = 10t \cos 5t^2$$

Finally, the sought for derivative becomes

$$10t^{a+1}\cos(5t^2) + at^{a-1}\sin(5t^2)$$

Other points to watch are:

1. The appearance of the "open" constant of integration in the indefinite integral. This constant may be needed in order to meet all the conditions of a particular problem.

2. Look out for the way the algebraic signs appear in the derivatives and integrals of the trigonometric functions. This can be confusing. If in doubt, check with a table of integrals.

3. In the integration of x^n be sure that $n \neq -1$. Otherwise, use the special formula for this case.

In addition to the above there is an occasional mix-up between the formulas

$$\frac{dx^n}{dx} = nx^{n-1} \quad \text{and} \quad \frac{dn^x}{dx} = n^x \log_e n$$

However, the need to find the derivative of n^x is relatively rare.

ACCURACY

Even with a hand calculator accuracy can be a problem in rare situations. This might occur if intermediate values in the computation are not recorded with sufficient accuracy. This arises in the subtraction of two quantities of almost equal magnitude. Typically, this can occur in solving the quadratic equation when both roots are real and b and $\sqrt{b^2 - 4ac}$ are almost of the same magnitude.

Let us look at an example. We wish to find

$$-10 + \sqrt{101}.$$

Now $\sqrt{101} = 10.05$, so that the overall result is approximately 0.05. However, if we took $\sqrt{101}$ to only three significant figures, we would have 10.0 and a final result of zero instead of 0.05. Problems of this nature which involve a square root can be handled by the approximate formula (based on the binomial expansion):

$$\sqrt{1+\varepsilon} \cong 1 + \frac{1}{2}\varepsilon \quad |\varepsilon| << |$$

where ε can be either positive or negative, but must be small in magnitude compared with one. We transform to the standard form as follows:

$$\sqrt{101} = \sqrt{100 + 1} = \sqrt{100(1 + .01)}$$

$$= 10\sqrt{1 + .01} \cong 10\ (1 + .005) = 10.05$$

Other commonly used approximation formulas are

$$\sin x \cong x, \quad \cos x \cong 1 - \frac{1}{2}x^2$$

where x must be expressed in radian measure and must be of small magnitude compared with one.

DIFFERENTIAL EQUATIONS

We restrict our attention to two special differential equations which occur often in mechanical engineering.

The d.e. (differential equation)

$$\frac{d^2y}{dx^2} = f(x)$$

is directly solvable by successive integration. Thus if

$$\frac{d^2x}{dt^2} = 6t^2 - 2a$$

we have

$$\frac{dx}{dt} = 2t^3 - 2at + C_1, \quad x = \frac{1}{2} t^4 - at^2 + C_1 t + C_2$$

Note the appearance of the integration constants C_1 and C_2 and note also the symbol variation between the standard form of the differential equation and the example.

A more difficult differential equation, which occurs frequently in vibration problems, has the form

$$\frac{d^2x}{dt^2} + p^2 x = \delta$$

where p and δ are constants. The method of solution of this differential equation is set out in standard texts. Here we give only the result:

$$x(t) = \frac{\delta}{p^2} + C_1 \cos pt + C_2 \sin pt$$

or

$$x(t) = \frac{\delta}{p^2} + A \cos (pt-\alpha)$$

The constants C_1, C_2 and A, α are integration constants. The two forms listed above are equivalent, as can be shown by using appropriate trigonometric identities.

Of frequent occurrence is the special case $\delta = 0$, where the solution specializes by the omission of the first term (involving δ).

The presence of trigonometric terms in the above solution leads to the characteristic to-and-fro motion found in vibrating systems. Now suppose that as the result of applying physical principles we arrive at the somewhat different differential equation

$$\frac{d^2x}{dt^2} - p^2 x = \delta$$

Although this differential equation looks almost the same as the previous one its solution has quite a different behavior. In fact, it does not oscillate at all. Thus, if we anticipate oscillatory behavior and end up with the second differential equation, something has gone wrong in the analysis. Here again is a check point for our mathematics.

REVIEW PROBLEMS

Following are 28 simple problems with answers to serve as a partial review and extension of the subject matter outlined above. Try to work the problems before looking at the answers. For each problem try to discern whether some special point is being brought out.

Following the 28 review problems, there are six problems from past professional examinations with answers outlined. Again, try to work the problems before looking at the answers. Note that in many cases there will be more than one correct way to attack a problem.

1. $\log \frac{P}{N} (a - b)^{1/n}$ is equal to:

   (a) $\log \frac{P}{N} + n \log (a - b)$

   (b) $\log P - \log N + \log (a - b) + \log \frac{1}{n}$

   (c) $\log P - \log N + (\frac{1}{n}) \log (a - b)$

2. $\int_0^{10} (x^2 + 1)\, dx$ is equal to:

   (a) 20 (b) 21 (c) 101 (d) 1001 (e) None of these

3. sin 930° is equal to:

   (a) $-\frac{1}{2}$ (b) $+\frac{1}{2}$ (c) $\frac{2}{\sqrt{3}}$ (d) $-\frac{1}{\sqrt{2}}$ (e) None of these

4. $\log \frac{(a + b)}{c}$ is equal to:

   (a) $\log a + \log b - \log c$

   (b) $\frac{\log a + \log b}{\log c}$

   (c) $\log (a + b) - \log c$

   (d) $e^{\frac{(a + b)}{c}}$

   (e) $\log \frac{a}{c} + \log \frac{b}{c}$

5. The first derivative of $\frac{1}{x}$ with respect to x is:

   (a) $-\frac{1}{x^2}$ (b) $-\frac{2}{x^2}$ (c) $\log_e x$ (d) 1 (e) $\frac{dx}{dy}$

6. Find the slope of the line which is tangent to the parabola $y = 8 x^2 + 4$ at the point where $x = 2$

7. tan (arc sin 0.5) is equal to:

   (a) $\frac{1}{4}$ (b) $\frac{1}{2}$ (c) $\frac{1}{3^{1/2}}$ (d) 1 (e) $3^{1/2}$

8. $\log_{10}(100)^2 - \ln_e 2.718$ is approximately equal to:

   (a) 2 (b) 3 (c) 4 (d) 5 (e) 6

9. For the position-time function, $x = 3t^2 + 2t$, the velocity in the x direction at $t = 1$ is:

   (a) 4 (b) 5 (c) 6 (d) 7 (e) 8

10. Derive the equation of the largest circle that is tangent to both coordinate axes and has its center on the line,

$$2x + y - 6 = 0$$

11. In the equation, $y = \dfrac{-x^3 + 3x + 2}{x^2 + 2x + 1}$, the limit of y as x approaches a value of minus one is:

   (a) zero (b) 1 (c) 2 (d) 3 (e) infinity

12. The cosine of 120° is equal to:

   (a) $\frac{3^{1/2}}{2}$ (b) $\frac{1}{3^{1/2}}$ (c) $\frac{1}{2}$ (d) $-\frac{1}{2}$ (d) $\frac{-3^{1/2}}{2}$

13. If the sine of angle "A" is given as K, the tangent of angle "A" would be:

   (a) $1 - K$ (b) $\frac{1}{K}$ (c) $(1 - K^2)^{1/2}$ (d) $\frac{1}{(1 - K^2)^{1/2}}$

   (e) $\frac{K}{(1 - K^2)^{1/2}}$

14. $\log_{10}(1000)^3$ is equal to:

   (a) 3 (b) 5 (c) 6 (d) 9 (e) 12

15. $\ln_e (2.718)^{xy}$ is approximately equal to:

   (a) xy (b) exy (c) 2.718 xy (d) x + y (e) 0.434 xy

16. Each interior angle of a regular polygon with five sides is:

   (a) 90 degrees (b) 100 degrees (c) 108 degrees

   (d) 120 degrees (e) 130 degrees

17. The number below that has four significant figures is:

   (a) 1414.0 (b) 1.4140 (c) 0.141 (d) 0.01414

   (e) 0.0014

18. The slope of the curve $y = x^3 - 4x$ as it passes through the origin ($x = 0$, $y = 0$) is equal to:

   (a) + 4 (b) + 2 (c) 0 (d) - 2 (e) - 4

19. Find the area bounded by the parabola, $y^2 = 2x$, and the line, $x = 8$.

20. If $u = e^{3y} \cos 2x$, what is $\frac{du}{dt}$ if both x and y are functions of t?

21. The logarithm of the number, - 0.09, is:

(a) positive
(b) negative
(c) zero
(d) a complex number
(e) none of these

22. The value of sin (A + B) where sin A = 1/3 and cos B = 1/4 is:

(a) 7/12
(b) 1/12
(c) $(1 + 30^{1/2})/12$
(d) $(1 + 2\sqrt{30})/12$
(e) $(8^{1/2} - 1)/12$

23. The derivative of $\frac{x}{(x - 1)}$ is:

(a) $\frac{-1}{(x - 1)^2}$
(b) $\frac{1}{(x - 1)^2}$
(c) $\frac{x}{(x - 1)^2}$
(d) $\frac{1}{(x - 1)}$
(e) $\frac{-x}{(x - 1)^2}$

24. The value of $\int (1 - x)^{1/2}\, dx$ is:

(a) no value
(b) $(2/3)(1 - x)^{3/2}$
(c) $(-1/3)(1 - x)^{3/2} + C$
(d) $(-2/3)(1 - x)^{3/2} + C$
(e) $(-3/2)(1 - x)^{3/2} + C$

25. $\int \frac{x\, dx}{x^2 + 1}$ equals:

(a) $\frac{1}{2} \ln (x^2 + 1)$
(b) $\ln (x^2 + 1) + C$
(c) $\ln (x^2 + 1)$
(d) $\frac{(x^2 + 1)^2}{2} + C$
(e) $\frac{1}{2} \ln (x^2 + 1) + C$

26. The integral, $\int_1^2 (x^2 - 1)\,dx$, equals:

    (a) 0 (b) 2 (c) 3/4 (d) 4/3 (e) 5/3

27. The area bounded by the x axis, the lines, x = 1 and x = 3, and by $y = x^3$ is equal to:

    (a) 81/4 (b) 20 (c) 10 (d) - 20 (e) 41/2

28. The expression $A^3 + B\,X$ written in Fortran is:

    (a) A*3 + BX
    (b) A*3 + B*X
    (c) A**3 + B*X
    (d) A**3 + BX
    (e) A*3 + B**X

ANSWERS TO PROBLEMS

1. c
2. e
3. a
4. c
5. a
6. 32
7. c
8. b
9. e
10. $(x - 6)^2 + (y + 6)^2 = 36$
11. d
12. d
13. e
14. d
15. a
16. c
17. d
18. e
19. 128/3
20. $\frac{du}{dt} = 3\,e^{3y} \cos 2x \frac{dy}{dt} - 2\,e^{3y} \sin 2x \frac{dx}{dt}$
21. d
22. d
23. a
24. d
25. e
26. d
27. b
28. c

MATH FUNDAMENTALS 1

A cylindrical water tank which is 35 ft. in diameter and 105 ft. in length is placed temporarily on an 18.5° slope as shown in the figure. The filler opening is located flush with the top of the tank at the mid-point.

What is the maximum volume of water which can be placed in the tank?

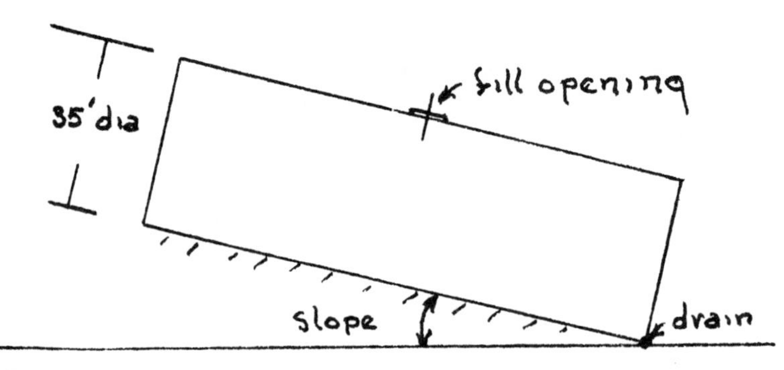

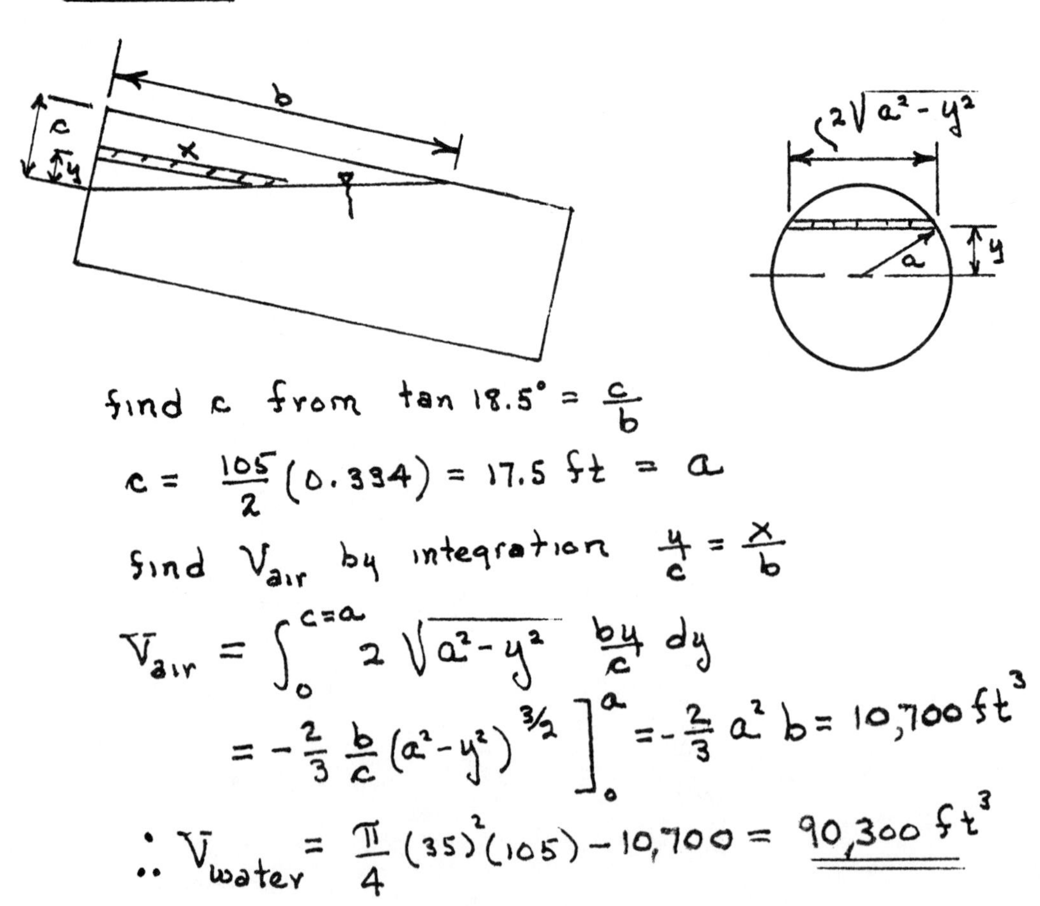

MATH FUNDAMENTALS 2

A bag contains four white balls and six black balls. Three balls are drawn in succession without replacing any ball after it is drawn. Find the probability (chances in a hundred) that the first and second balls will be white and the third ball black.

Solution

1st drawing	4W, 6B	$P(W) = 4/10$
2nd "	3W, 6B	$P(W) = 3/9$
3rd "	2W, 6B	$P(B) = 6/8$

$$P(WWB) = \frac{4}{10} \times \frac{3}{9} \times \frac{6}{8} = \frac{1}{10}$$

or 10 chances out of 100

MATH FUNDAMENTALS 3

A solid steel cylinder is three inches in diameter and eighteen inches long. It has been proposed that a concentric hole two inches in diameter be bored into the cylinder from one end and filled with lead so as to move the center of gravity of the cylinder 1/2 inch toward the bored end. Steel weighs 490 pounds per cubic foot and lead weighs 705 pounds per cubic foot. Is this proposal a reasonable one? Explain, verifying by analysis and calculations.

Solution

Without any lead insert, the c.g. is at the center of the large cylinder. Imagine the lead plug growing in length. At first, the c.g. moves toward the bored end. Eventually, the c.g. coincides with the end face of the plug (see figure). Further lengthening of the plug, now moves the c.g. back towards its original position.

Maximum c.g. excursion is for the sketched configuration

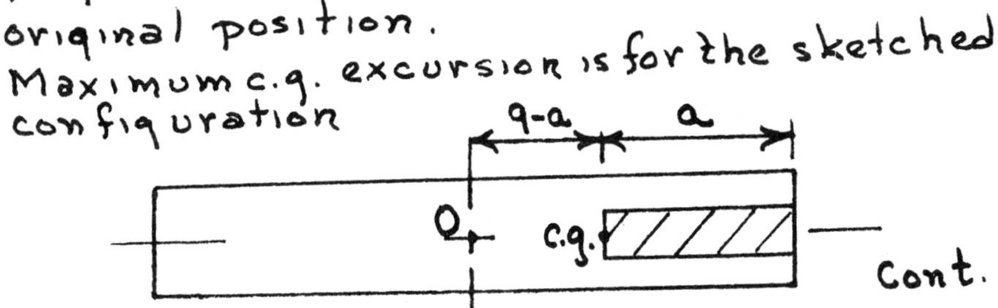

Cont.

(3) Cont

$$\Sigma M_o: (9-a)\left(\frac{\pi}{4}\cdot 3^2\cdot 18\cdot\frac{490}{1728}+\frac{\pi}{4}\cdot 2^2\cdot a\cdot\frac{705-490}{1728}\right)$$

$$= 0+\left(9-a+\frac{1}{2}a\right)\left(\frac{\pi}{4}\cdot 2^2\cdot a\right)\frac{705-490}{1728}$$

simplifying

$$a^2+\frac{81\times 98}{43}a-\frac{9\times 81\times 98}{43}=0$$

$$\therefore\ a=-\frac{81\times 49}{43}\pm\sqrt{\left(\frac{81\times 49}{43}\right)^2+\left(\frac{9\times 81\times 98}{43}\right)}$$

for physical meaning we must take the plus sign.

Then: $a=-\frac{81\times 49}{43}\left[-1+\sqrt{1+\frac{43\times 2}{49\times 9}}\right]$

$$=9-\frac{43}{98}+\frac{2}{81}\left(\frac{43}{98}\right)^2-\cdots$$

$$=9-0.438+0.005-\cdots\doteq 9-0.433$$

Maximum displacement of c.g. is 0.433 in so proposal is not reasonable

MATH FUNDAMENTALS 4

If the probability of an event occurring is 1/3, what is the probability that it will occur exactly twice (2) in five (5) times?

Solution

Binomial distribution (independent events with two possible outcomes)

n = no. of trials p = probability of success on a single trial

x = no. of "successes"

$P(x)$ = probability of exactly x successes

$$P(x)=\binom{n}{x}p^x(1-p)^{n-x}=\binom{5}{2}\left(\frac{1}{3}\right)^2\left(1-\frac{1}{3}\right)^3=\frac{5!}{2!3!}\cdot\frac{1}{9}\cdot\frac{8}{27}=0.329$$

MATH FUNDAMENTALS 5

A system is made up of three black boxes in series having the following reliability:

(a) 0.99 (b) 0.95 (c) 0.70

If the system reliability requirement is 0.85 and redundancy is permitted for one box, can the requirement be achieved?

Solution

Reliability of series $R_s = 0.99 \times 0.95 \times 0.70 = 0.658$

$\therefore$ required R not attainable in series. So, we use redundancy for weakest box (0.70).

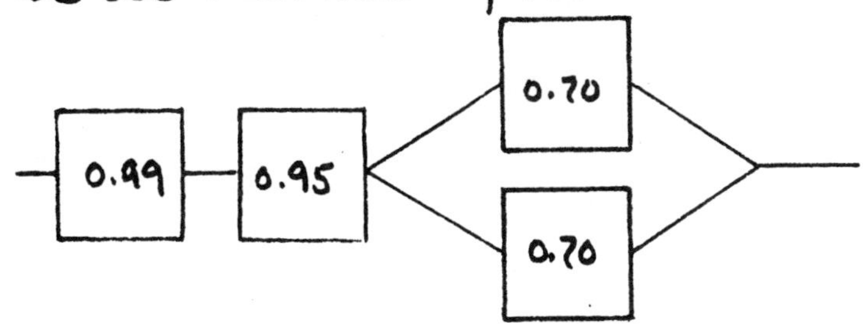

equivalent r for redundant boxes:

$R_{.70/.70}$ = prob. (at least one box ok)

= 1 - prob. (both boxes bad)

$= 1 - 0.30 \times 0.30 = 0.91$

$R_s = 0.99 \times 0.95 \times 0.91 = \underline{0.856 > 0.85}$

requirement can be achieved

MATH FUNDAMENTALS 6

A sample of two hubs are sent to a laboratory for testing. The hubs were selected from a lot of six. What is the probability that at least one will be found defective if two of the original six are defective?

Solution

Lot contains 2 bad and 4 good hubs (2B, 4G)

let P_n = prob. of drawing n bad hubs.

hen: $P_0 + P_1 + P_2 = 1 \quad \therefore P_1 + P_2 = 1 - P_0$

$= 1 - \text{prob.}(GG) = 1 - \frac{4}{6} \cdot \frac{3}{5} = \underline{0.6}$

2 Force and Stress Analysis

HAROLD M. TAPAY

STATICS

Statics is that part of engineering mechanics in which a study is made of force systems, equivalent force systems and the external effects that these forces produce on bodies which are at rest or moving with constant velocity. These external effects are usually referred to as reactive forces or just reactions; and the forces that cause them are called active forces or just loads. Some examples of these loads on a vehicle are the weight of the vehicle, the weight of the occupants, and the air resistance. Bodies which are at rest or moving with constant velocity have their active and reactive force systems in equilibrium which means no resultant force or no resultant moment. In solving for the reactive forces a very useful visual aid is a diagram of the structure or a member of the structure in which all the forces applied by contacting bodies plus gravity forces (weight) are shown; this diagram is called a "free body diagram".

The principles used in statics are as listed below:

1. Parallelogram Law -

   Forces F_1 and F_2 can be replaced by F_3, i.e., diagonal of parallelogram through point of concurrency of F_1 and F_2.

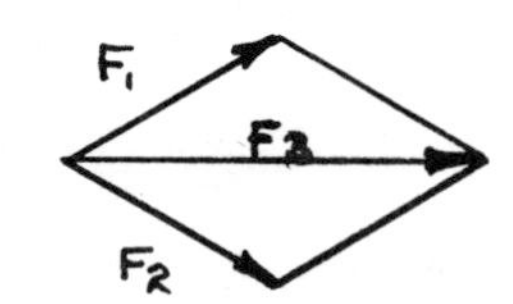

2. Principle of Transmissibility -

   External effects, i.e., reactions R_L and R_R are not changed by moving Force F from A to B or to any point on its line of action.

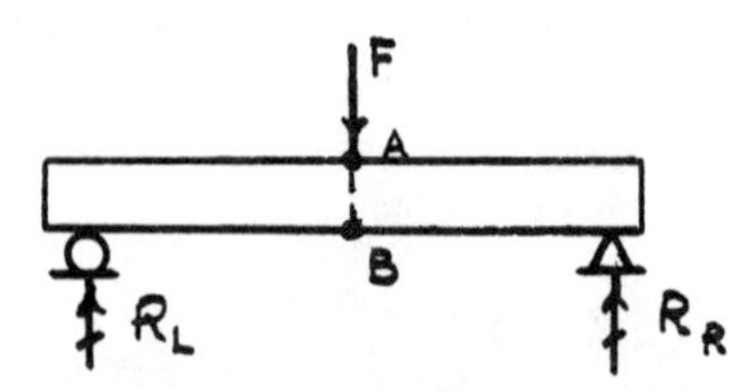

3. Newton's 1st Law -

   Contains principle of equilibrium of forces.

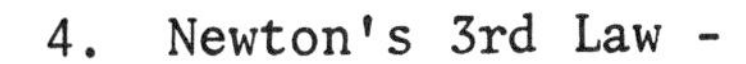

4. Newton's 3rd Law -

   Active and reactive forces are equal in magnitude, opposite in direction and collinear, e.g., $|R_A| = |W|$

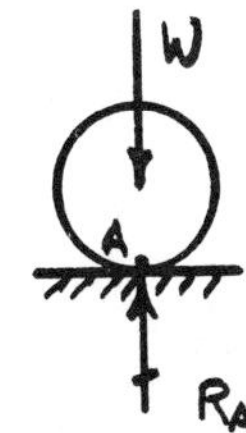

5. Varignon's Theorem - or Principle of Moments -

   Moment of a force about any point is equal to the sum of the moments of its components about the same point, i.e., $r \times F = r \times F_1 + r \times F_2 + r \times F_3$

   $$\text{where } F = \underbrace{F_1 + F_2 + F_3}_{\text{components of } F}$$

For <u>force systems not</u> in <u>equilibrium</u>:

Resultant Force $R = F_1 + F_2 + F_3 + \ldots$

or $R_x = \Sigma F_x$; $R_y = \Sigma F_y$; $R_z = \Sigma F_z$

<u>Equations of equilibrium</u> for a force system are:

$$\Sigma F_x = 0; \quad \Sigma F_y = 0; \quad \Sigma F_z = 0$$

$$\Sigma M_x = 0; \quad \Sigma M_y = 0; \quad \Sigma M_z = 0$$

<u>Very important force system</u> - <u>couple</u>; its resultant is a moment

M_z whose magnitude = Fd

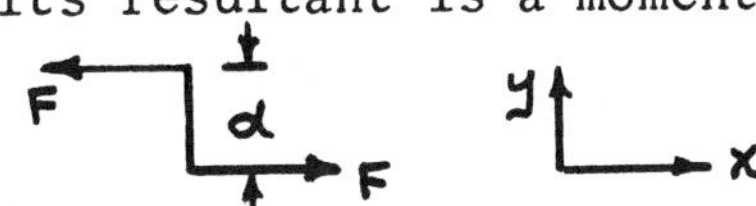

<u>Very important equivalent force system</u>:

<u>Special reactive forces</u>:

Maximum static friction - $(F_s)_{max}$

Body on verge of moving -

$(F_s)_{max} = \mu_s N$

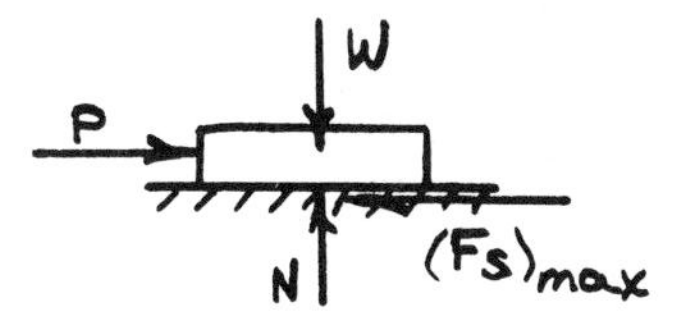

Kinetic friction - F_k

Body moving $F_k = \mu_k N$

μ_s - coeff. of static friction

μ_k - coeff. of kinetic friction

$\mu_s > \mu_k$

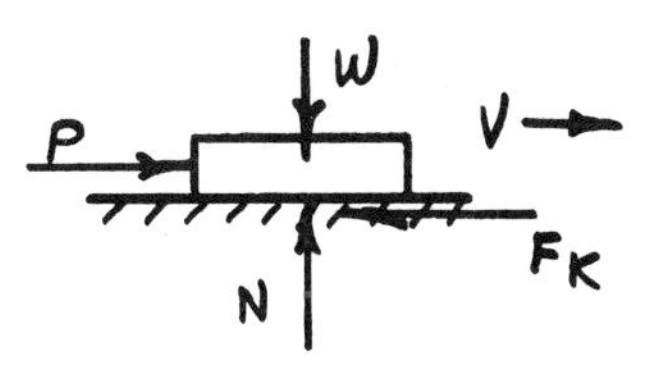

STRENGTH OF MATERIALS

Strength of materials is that part of engineering mechanics in which the strength, deflection and stability of structural members are investigated under different force systems. This investigation results in formulas for stresses, strains, beam deflections, angles of twist, buckling loads, etc. Some of these formulas contain mechanical properties of the material from which the member is made; these properties having been determined previously in a materials testing laboratory. The derivation of the above formulas requires the use of some of the following:

1. Principles of statics.
2. Materials properties, e.g., μ, E, G, etc.
3. Assumed strain distributions.
4. Mathematical properties of members, cross section, e.g., A, I, J.
5. Stress-strain law.

Types of strain
- Normal strain $\varepsilon \frac{in}{in}$ - tensile and compressive
- Shearing strain $\gamma \frac{in}{in}$

For axial loading -

$\Delta x = P\ell/EA$

$\epsilon_x = \Delta x/\ell$

$\epsilon_y = \epsilon_z = -\mu\epsilon_x$

For bending -

$M \quad \epsilon_x \propto y : \epsilon_x = y/\rho$

For torsion

γ_{max}, M_t, ℓ, r

$\gamma \propto r$

ϕ = angle of twist = $M_t\ell/GJ$

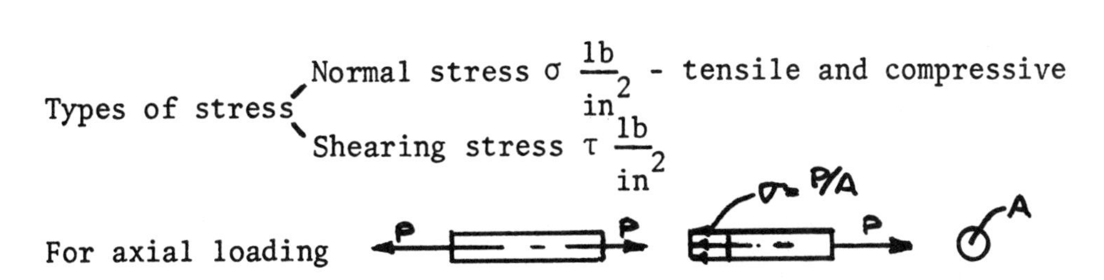

For bending of initially straight beams whose material obeys Hooke's Law

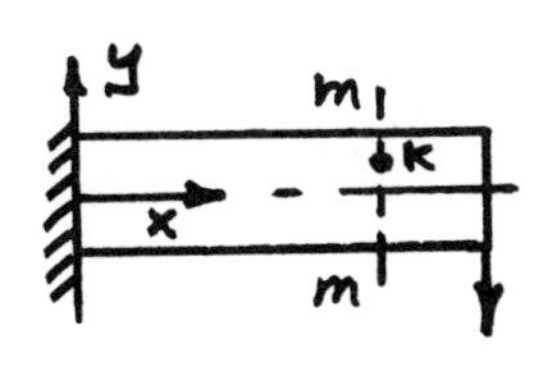

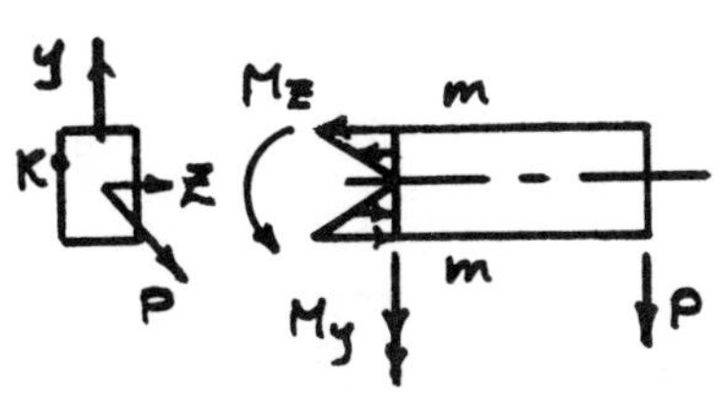

$$\sigma = \pm \frac{M_{zz}\, y}{I_{zz}} \pm \frac{M_{yy}\, z}{I_{yy}}$$

For fiber K both flexure Stresses are tensile.

For torsion of straight members of circular cross section whose material obeys Hooke's Law

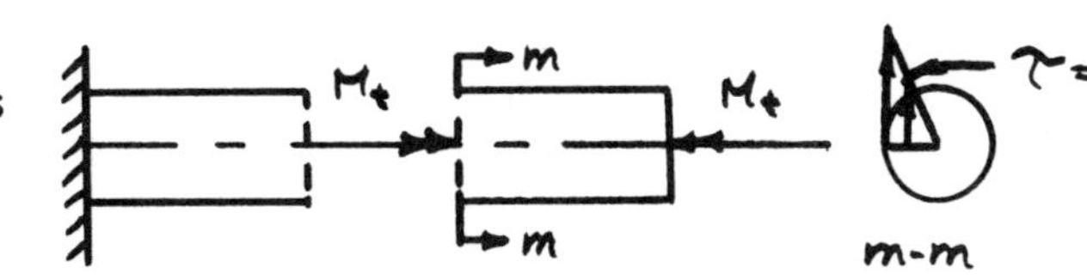

For bending of initially straight beam in one plane

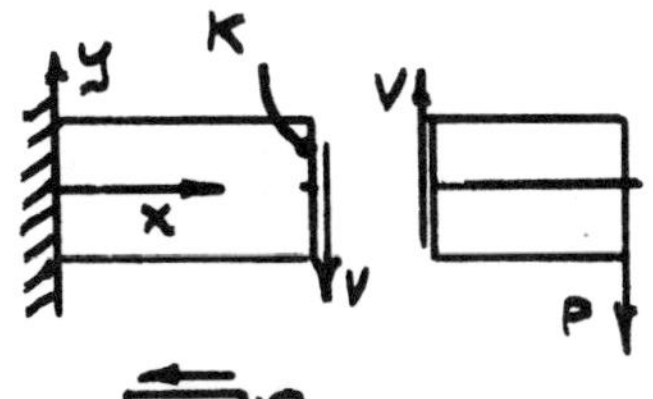

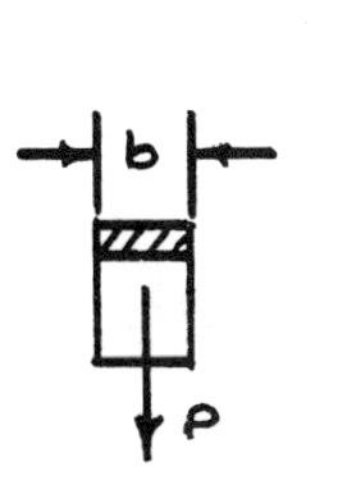

$$\tau_{xy} = \frac{VQ}{bI}$$

$Q = \int y\, dA$ for Shaded Area

Hooke's Generalized Law

$$\varepsilon_x = \frac{\sigma_x}{E} - \mu \frac{\sigma_y}{E} - \mu \frac{\sigma_z}{E}$$

$$\varepsilon_y = -\mu \frac{\sigma_x}{E} + \frac{\sigma_y}{E} - \mu \frac{\sigma_z}{E}$$

$$\varepsilon_z = -\mu \frac{\sigma_x}{E} - \frac{\sigma_y}{E} + \frac{\sigma_z}{E}$$

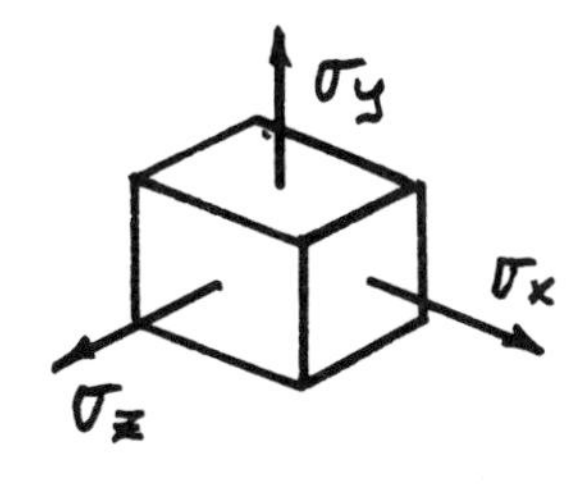

Hooke's Law for shearing stress and shearing strain

$\tau_{xy} = G\, \gamma_{xy}$ $\qquad \tau_{xy} = \tau_{yx}$

$\tau_{yz} = G\, \gamma_{yz}$ $\qquad \tau_{yz} = \tau_{zy}$

$\tau_{zx} = G\, \gamma_{zx}$ $\qquad \tau_{zx} = \tau_{xz}$

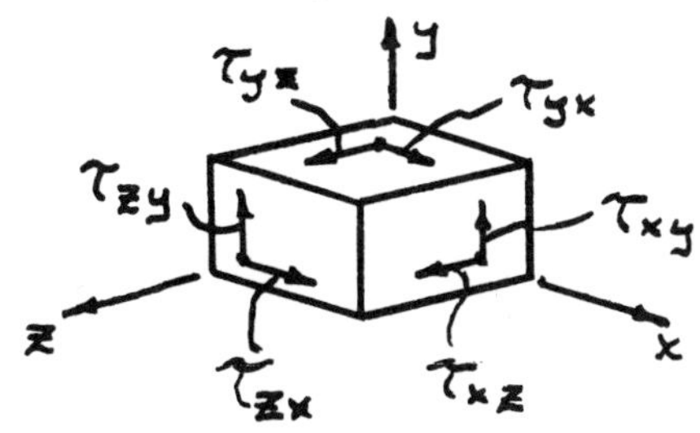

Maximum and Minimum Principal Stresses

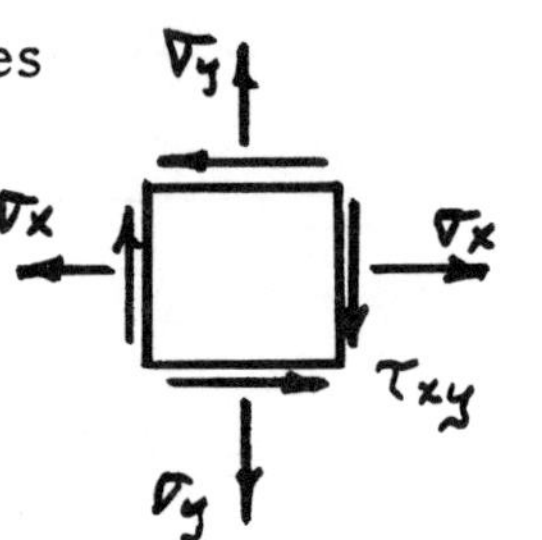

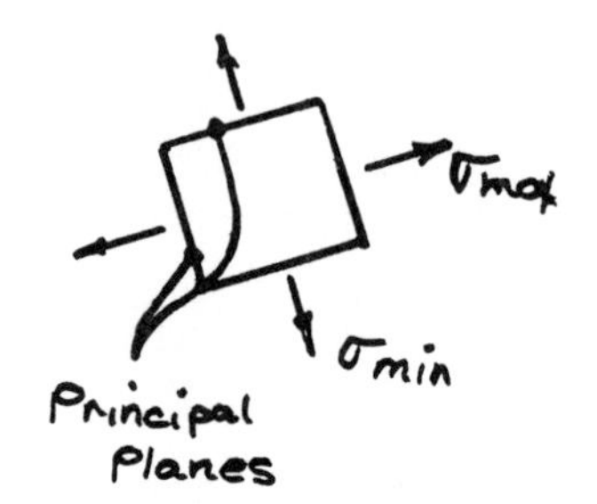

$$\sigma_{\substack{max\\min}} = \frac{\sigma_x+\sigma_y}{2} \pm \sqrt{\left(\frac{\sigma_x-\sigma_y}{2}\right)^2 + \tau_{xy}^{\;2}}$$

Maximum & Minimum Shearing Stresses

$$\tau_{\substack{max\\min}} = \pm\sqrt{\left(\frac{\sigma_x-\sigma_y}{2}\right)^2 + \tau_{xy}^{\;2}}$$

Beam Deflections - y for initially straight beams

Deflection Curve

Algebraic equation for deflection curve can be obtained by solving the differential equation of the deflection curve; namely,

$$EI\,\frac{d^2y}{dx^2} = M$$

Elastic Buckling Loads for Long Columns

$P_{crit.}$

ℓ

$P_{crit.}$

$$P_{crit.} = \frac{n\pi^2 EI}{\ell^2}$$

n - constant - depends on end conditions

e.g., n = 1 for pinned ends.

n = 4 for built in ends.

FORCE AND STRESS ANALYSIS 1

The first section of a rotary well drilling unit has a 6" outside diameter and a 5" inside diameter. If the weight of the drill string above it is 120,000 lbs and it requires a torque of 30,000 lb inch to rotate it, will the stresses in this section be below the allowables of 60,000 psi compression and 20,000 psi shear?

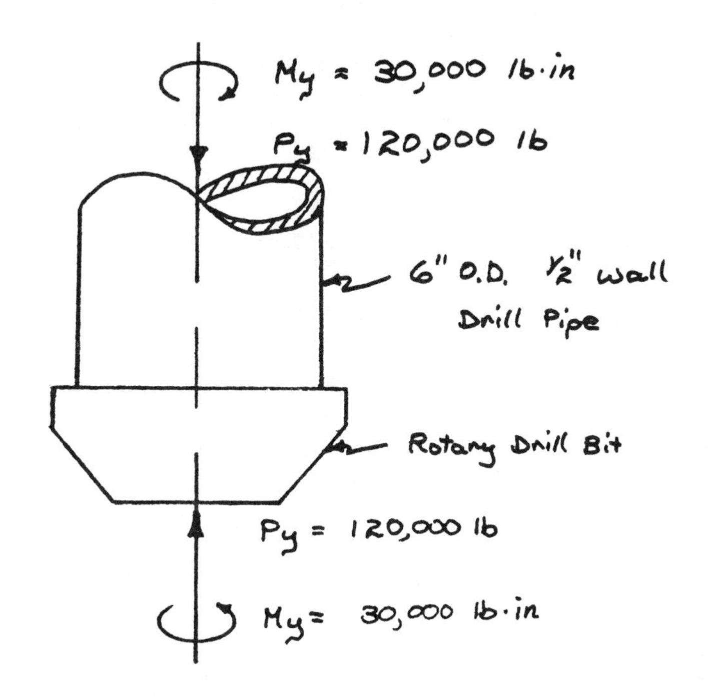

Solution

$$\tau_{xy} = \text{Torsional Shearing Stress} = \frac{(M_y)\,c}{J} = \frac{30{,}000\,(3)}{\frac{\pi}{32}(6^4 - 5^4)}$$

$$= \frac{90{,}000}{67.8} = \underline{\underline{1{,}330\ \text{lb/in}^2}}$$

$$\sigma_y = \text{Compressive Stress} = -\frac{P_y}{A} = \frac{-120{,}000}{\frac{\pi}{4}(6^2 - 5^2)}$$

$$= -\frac{120{,}000}{8.63} = \underline{\underline{-13{,}900\ \text{lb/in}^2}}$$

$$\sigma_{min} = \frac{\sigma_x + \sigma_y}{2} - \sqrt{\left(\frac{\sigma_x - \sigma_y}{2}\right)^2 + \tau_{xy}^2}$$

$$= \frac{-13{,}900}{2} - \sqrt{\left(\frac{13{,}900}{2}\right)^2 + (1{,}330)^2} = -6{,}950 - 7{,}070$$

$$= \underline{\underline{-14{,}020\ \text{lb/in}^2}} \quad \left[\sigma_{allow.} = -60{,}000\ \text{lb/in}^2\right]$$

$$\tau_{max} = \sqrt{\left(\frac{\sigma_x - \sigma_y}{2}\right)^2 + \tau_{xy}^2} = \underline{\underline{7{,}070 \text{ lb/in}^2}}$$

$$\left[\tau_{allow.} = 20{,}000 \text{ lb/in}^2\right]$$

The largest compressive and shearing stresses are below the allowable stresses in compression and shear.

FORCE AND STRESS ANALYSIS 2

A rod is fixed at one end and a linear spring is connected between the free end of the rod and a fixed wall as shown in Fig. (a). The modulus of elasticity E, of the rod material, is nonlinear as indicated in Fig. (b). The rod-spring system is relaxed prior to loading. Determine the deflection of the free end of the rod where a load, P = 20 lbs, is applied. Assume K^2L^2/AE_o = 10 lbs and K = 100 lbs/in.

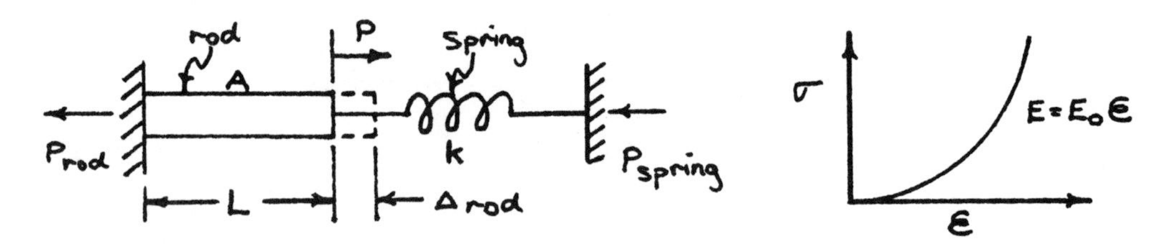

A - Cross sectional area of rod, in^2.
k - Spring constant, lbs/in.

Fig. (a) Spring-Rod System

E - Modulus of elasticity, lbs/in^2.
E_o - Constant.
ϵ - Strain, in/in.

Fig. (b) Stress-Strain Relationship of Rod Material

<u>Solution</u>

1) $\Sigma F_x = 0$ $\qquad P_{rod} = P_{spring} = 20$

2) Since <u>supports</u> are <u>non-yielding</u>, $\Delta_{rod} = \Delta_{spring}$

3) From stress-strain curve for rod material:

$$\frac{d\sigma}{d\epsilon} = E = E_0 \epsilon$$

$$\sigma = \frac{E_0}{2} \epsilon^2$$

$$P_{rod} = \sigma A = (E_0 A/2) \epsilon^2$$

For rod, $\epsilon = \Delta_{rod}/L$

For spring, $\Delta_{spring} = P_{spring}/K$

$$\therefore \quad \epsilon = P_{spring}/KL \quad : \quad \epsilon^2 = (P_{spring})^2/K^2 L^2$$

$$P_{rod} = \frac{E_0 A}{2} \frac{(P_{spring})^2}{K^2 L^2} = \frac{(P_{spring})^2}{2 \frac{K^2 L^2}{A E_0}} = \frac{(P_{spring})^2}{20}$$

Substituting into Eq. (1) gives:

$$\frac{1}{20}(P_{spring})^2 + P_{spring} = 20$$

$$P_{spring}^2 + 20 P_{spring} - 400 = 0$$

$$P_{spring} = \frac{-20 \pm \sqrt{20^2 + 4(400)}}{2} = \frac{-20 \pm 44.8}{2}$$

$$= \underline{\underline{12.4 \text{ lb}}}$$

$$\Delta_{rod} = \Delta_{spring} = \frac{P_{spring}}{K} = \frac{12.4}{100} = 0.124 \text{ in}$$

$$\longrightarrow \quad \approx \underline{\underline{1/8 \text{ in}}}$$

FORCE AND STRESS ANALYSIS 3

The solid circular shaft shown below of diameter 3.0 inches is simply supported by two bearings at A and D. The shaft carrier two pulleys at B and C. Pulley B is 15.0 inches in diameter and carries the vertical loads P_{z1} = 4000 lbs and P_{z2} = 2000 lbs. Pulley C is 12.0 inches in diameter and carries two horizontal loads in the X direction, P_{x1} = 5750 lbs and P_{x2} = 3250 lbs.

REQUIRED: Find the maximum principal stress for element E halfway along the shaft and at the outside fibre: Element E is at X = 1.5, Y = 10.0, Z = 0.0 inches. Neglect the weight of the shaft and pulleys in your calculations.

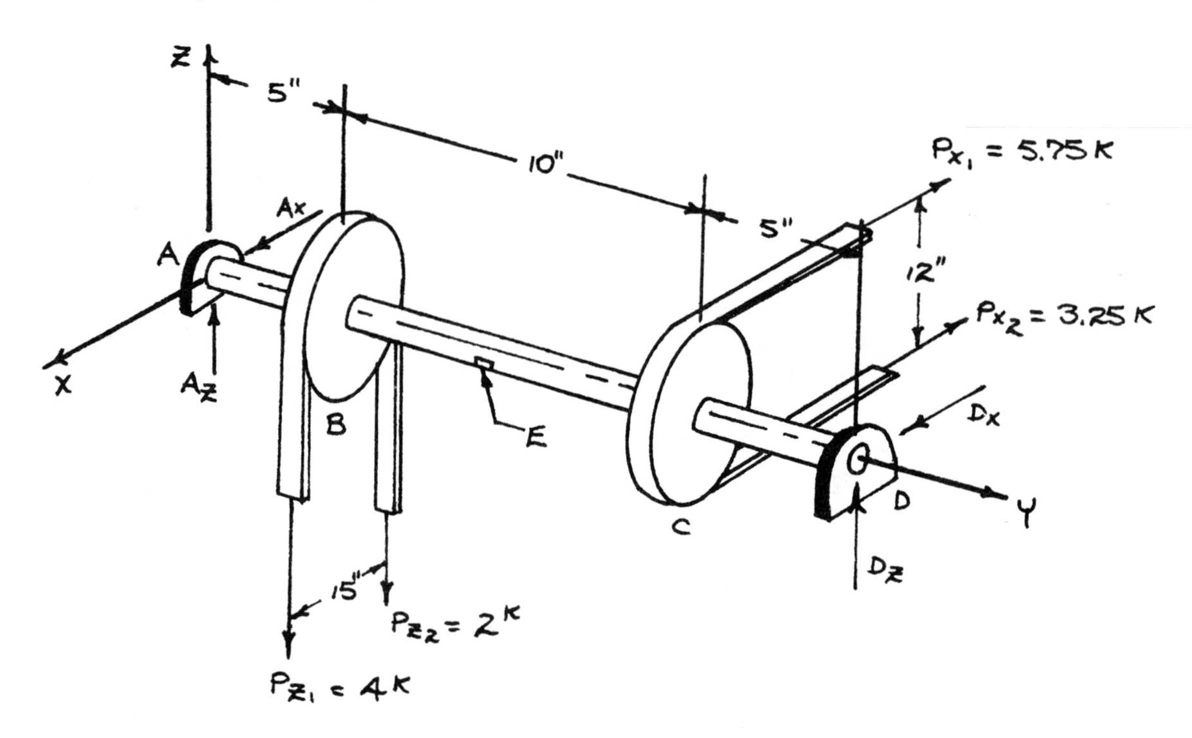

Solution

I. Determination of Bending Moment "M_z", Shear "U_z", and Twisting Moment "M_y" at section containing Element E.

$\Sigma(M_z)_D = 0 \qquad 20\,A_x = 9(5) \qquad A_x = 2.25\,K$

$\Sigma(M_x)_D = 0 \qquad 20\,A_z = 15(6) \qquad A_z = 4.5\,K$

$M_z = 10(2250) = \underline{\underline{22500 \text{ lb. in}}}$

$V_z = 6K - A_z = \underline{\underline{1.5K}}$

$M_y = 4(7.5) - 2(7.5) = 15 \text{ Kip in} = \underline{\underline{15{,}000 \text{ lb.in}}}$

II. For Element E

σ_y = Bending Stress due to $M_z = \dfrac{M_z c}{I} = \dfrac{32\, M_z}{\pi d^3}$

$= \dfrac{32(22{,}500)}{\pi(3)^3} = \underline{\underline{-8500 \text{ psi}}}$

τ_{yz} = Torsional shearing stress due to $M_y = \dfrac{M_y c}{J}$

$= \dfrac{16\, M_y}{\pi d^3} = \dfrac{16(15{,}000)}{\pi(3)^3} = \underline{\underline{2830 \text{ psi}}}$

τ_{yz} = Direct Shearing Stress due to $V_z = \dfrac{Q V_z}{I d}$

$= \dfrac{4}{3}\dfrac{V_z}{A}$; $A = \dfrac{\pi d^2}{4} = 7.08 \text{ in}^2$

$= \dfrac{4}{3}\dfrac{1500}{7.08} = 280 \text{ psi}$

$(\tau_{yz})_{Total} = 2830 + 280 = \underline{\underline{3110 \text{ psi}}}$

σ_{max} = Max. principal stress $= \dfrac{\sigma_y}{2} + \sqrt{\left(\dfrac{\sigma_y}{2}\right)^2 + (\tau_{yz})^2}$

$= -4250 + \sqrt{-4250^2 + 3110^2}$

$= -4250 + 5250$

$= \underline{\underline{1000 \text{ psi}}}$

FORCE AND STRESS ANALYSIS 4

A rotating beam fatigue test was run which yielded the test results shown plotted on semi-log coordinates on the following page. Under pure bending, the specimens failed at 52 KSI and 10^5 cycles. The endurance limit for the material was at 30 KSI (10^6 cycles).

REQUIRED: (a) Determine the number of cycles to failure at 41 KSI.

(b) For a test specimen minimum diameter of 0.3 inches, what is the bending moment carried at the endurance limit?

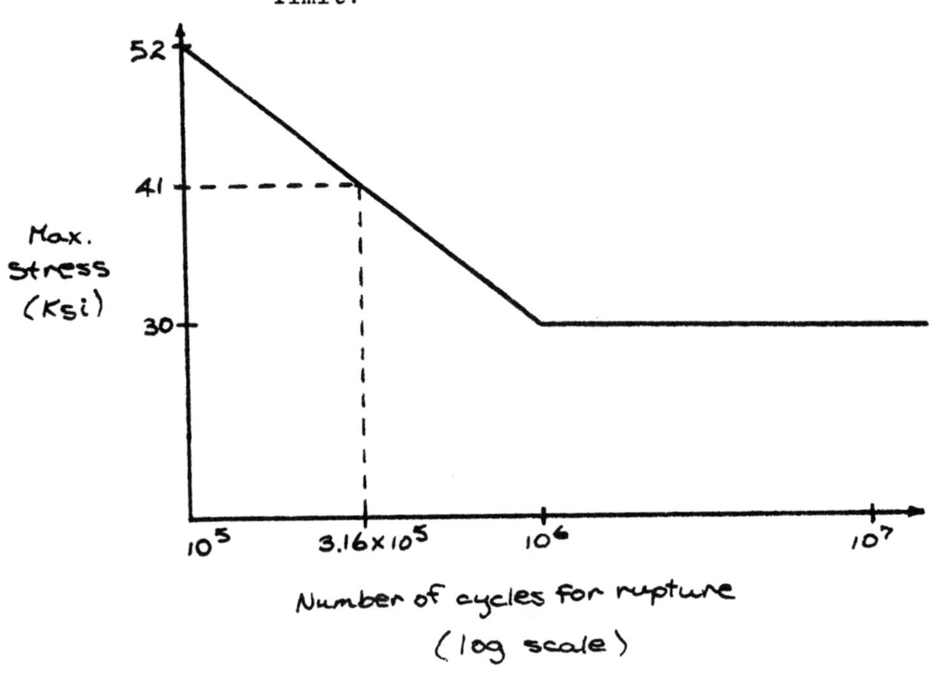

Solution

a) From the above graph for a fatigue strength of 41 Ksi, the number of cycles for rupture is

$$\approx \underline{\underline{320,000}}$$

b) M - Bending Moment carried at Endurance limit

$$M = \frac{(\sigma_{End.\,limit})(I)}{c} = \frac{30,000}{d/2}\left(\frac{\pi d^4}{64}\right) \quad ; \quad d = 0.3 \text{ in.}$$

$$M = \frac{30,000\, \pi d^3}{32} = \frac{30,000\, \pi (0.3)^3}{32} = \underline{\underline{79.3 \text{ lb}\cdot\text{in}}}$$

FORCE AND STRESS ANALYSIS 5

A boat trailer frame is being designed as shown. The support points carry the weight of the boat equally, with the axle attachment directly below the center support. A 4 x 2 in channel (area = 1.57 in^2 and I = 3.83 in^4) has been chosen as the primary member with a round bar held out from the flange by vertical spacers for additional strength. The center of the bar will be 6 in from the flange. If the maximum stress is to be limited to 10,000 psi, what is the diameter and location of the bar? Boat weighs 1500 lbs.

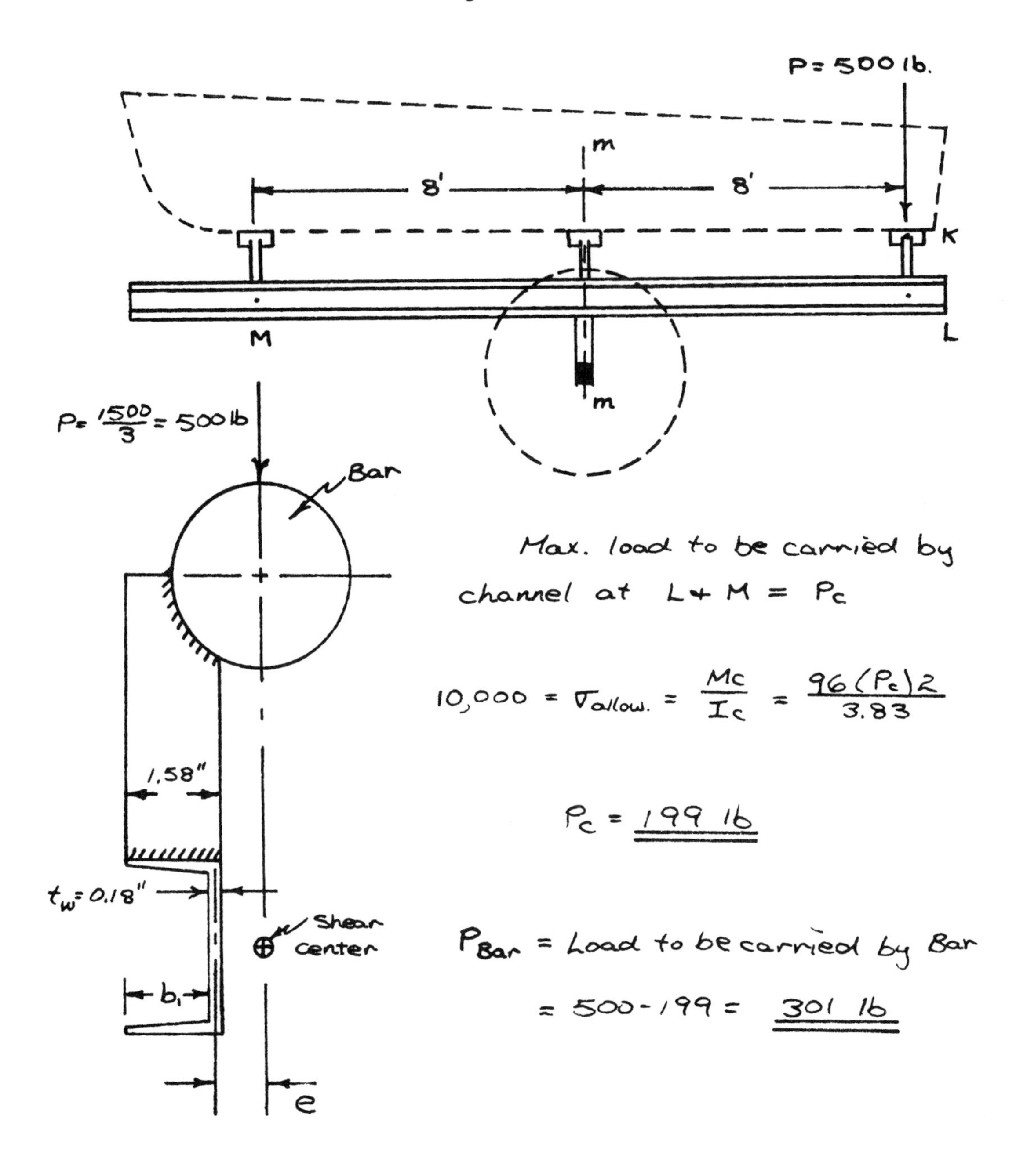

Max. load to be carried by channel at L + M = P_c

$$10{,}000 = \sigma_{allow.} = \frac{Mc}{I_c} = \frac{96(P_c)2}{3.83}$$

$$P_c = \underline{\underline{199 \text{ lb}}}$$

P_{Bar} = Load to be carried by Bar

$$= 500 - 199 = \underline{\underline{301 \text{ lb}}}$$

For no torsional shearing stresses in channel, line of action of load P applied by boat at each support should pass through shear center of channel.

Location of Shear Center

$$e = \frac{\frac{1}{2} b_1}{1 + \frac{1}{6}\frac{A_{web}}{A_{flange}}}$$

$(t_f)_{ave.} = 0.296\ in.$

$b_1 = 1.58 - 0.18 = 1.4\ in.$

$A_{web} \approx (4 - 0.3)(0.18) = .66 in^2$

$A_{flange} \approx 1.4(0.296) = .41\ in^2$

$$e = \frac{\frac{1}{2}(1.4)}{1 + \frac{0.66}{6(0.41)}} = 0.55\ in$$

Solving for required Moment of Inertia of bar.

$$y_K = y_L$$

$$\frac{P_{Bar} X^3}{3E I_{Bar}} = \frac{P_c X^3}{3E I_c}$$

$$I_{Bar} = \frac{\pi d^4}{64} = \frac{P_{bar} I_c}{P_c} = \frac{301(3.88)}{199} = 5.8\ in^4$$

$$d^4 = \frac{64(5.8)}{\pi}\ ; \quad d = 3.3\ in.$$

Longitudinal center line of this bar should be 0.55 in. from center line of web (lateral direction).

A more economical design (much lighter) would be to have a continuous shear web as the vertical spacer. This web would be welded to the channel and to a much lighter round bar or tubing.

FORCE AND STRESS ANALYSIS 6

Compute the minimum web thickness required to prevent shear buckling in the beam shown below.

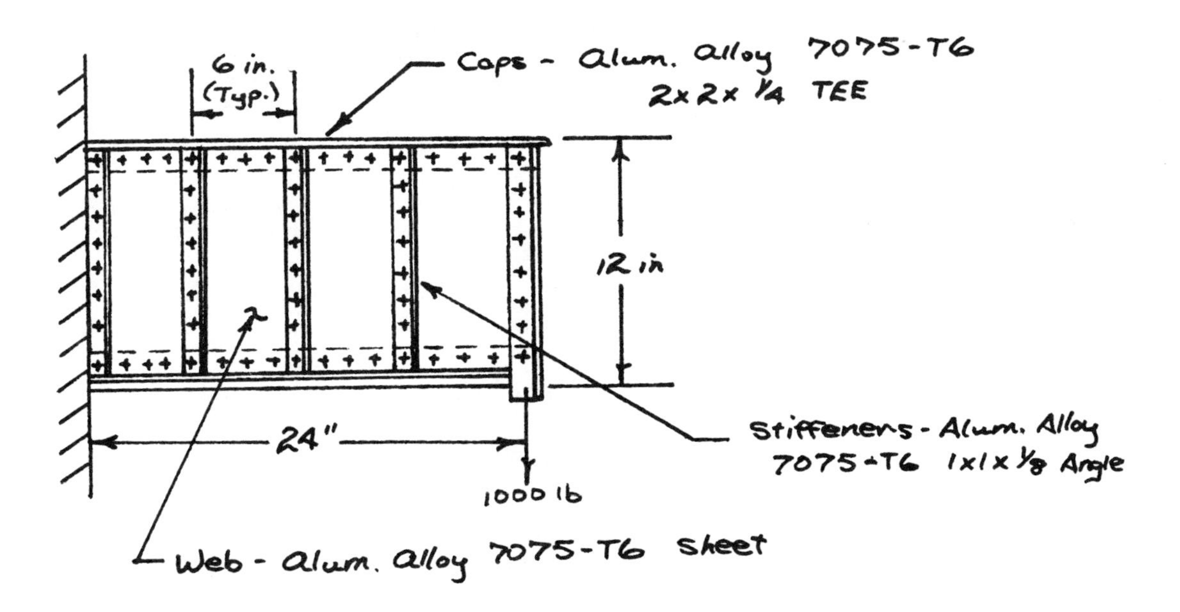

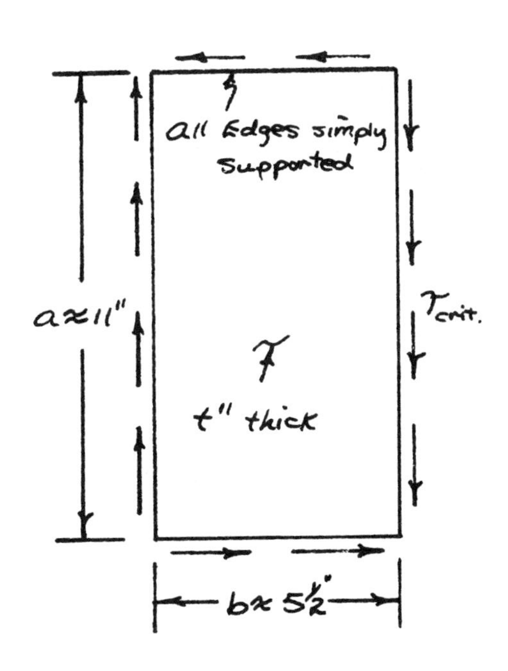

$$\tau_{vert.} \approx \frac{1000}{12t} \quad ; \quad \mu \approx 1/3$$

$$E = 10^7 \text{ lb/in}^2$$

$$\frac{1000}{12t} \approx \tau_{crit.} = \frac{K_s \pi^2 E}{12(1-\mu^2)}\left(\frac{t}{b}\right)^{2*}$$

For $a/b \approx 2$

$$K_s \approx 5.8$$

$$1000(1-\mu^2)(b)^2 = K_s \pi^2 E t^3$$

$$t^3 = \frac{1000(8/9)(5.5)^2}{5.8(\pi^2)(10^7)}$$

$t^3 = 47\ (10^{-6})$

$t = 3.6 \times 10^{-2} = \underline{\underline{0.036\ in}}$ ⟵

Axial load in stiffeners $P_s \approx \frac{Vd}{h}$ * ; $A_s = 0.23\ in^2$

$V = 1000\ lb$; $h = 11\ in.$; $d \approx 6\ in.$

$P_s = \frac{1000(6)}{11} = \underline{\underline{545\ lb.\ c}}$

Axial Stress in Stiffener

$\sigma = \frac{P_s}{A_s} = -\frac{545}{0.23} = \underline{\underline{-2380\ lb/in^2}}$

(low, no buckling)

Max. axial load in Compression Flange $A_c = 1.07\ in^2$

$F_c = \frac{M_{max}}{h} + \frac{V}{2}$ * $\approx \frac{24(1000)}{11} + 500$

$= 2680\ lb.$

$\sigma = \frac{F_c}{A_c} = \frac{2680}{1.07} = \underline{\underline{-2500\ lb/in^2}}$

(low, no buckling)

* Bruhn – Analysis and Design of Airplane Structures

FORCE AND STRESS ANALYSIS 7

The equation governing beam deflection is $\frac{d^2y}{dx^2} = \frac{M}{EI}$.

Using the finite difference method, set up the necessary equations with boundary conditions included, to establish the deflection curve for the cantilever beam shown. Use the increment h = L/4. Do not solve the equations. (Assume EI constant for the beam.)

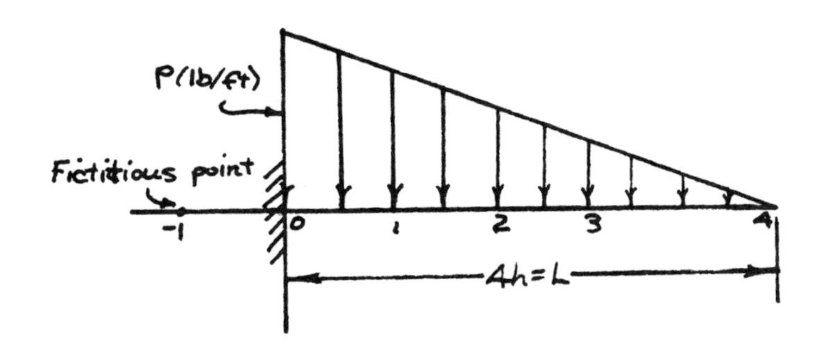

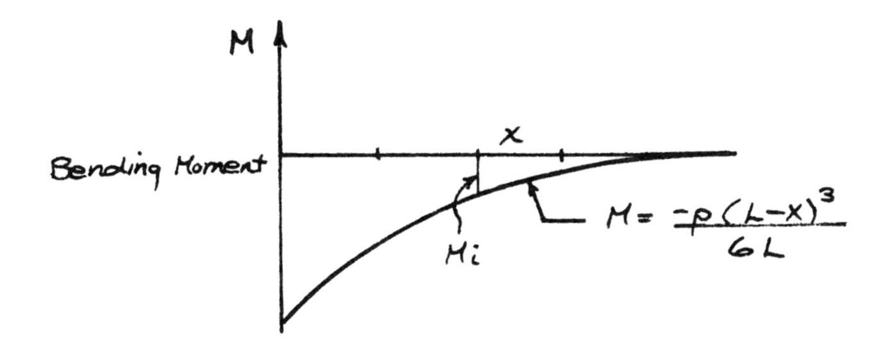

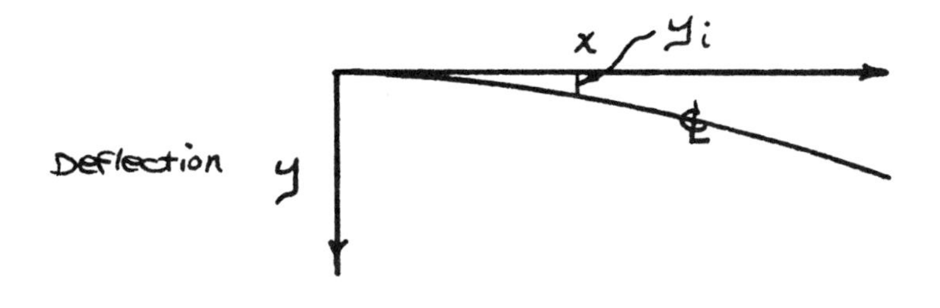

From Finite Difference Method

$$\left(\frac{d^2y}{dx^2}\right)_i \approx \frac{y_{i+1} - 2y_i + y_{i-1}}{h^2}$$

From differential equation of elastic curve

$$\left(\frac{d^2y}{dx^2}\right)_i = -\frac{M_i}{(EI)_i}$$

$$\therefore\ y_{i+1} - 2y_i + y_{i-1} = \frac{-h^2 M_i}{(EI)_i}$$

For $i = 1$:

$$y_2 - 2y_1 + y_0 = \frac{-h^2 M_1}{EI} = -\frac{L^2 M_1}{16EI}$$

$$y_0 = 0$$

$$\therefore\ y_2 - 2y_1 = -\frac{L^2 M_1}{16EI} \longleftarrow (1)$$

Similarly:

$$y_3 - 2y_2 + y_1 = -\frac{L^2 M_2}{16EI} \longleftarrow (2)$$

$$y_4 - 2y_3 + y_2 = -\frac{L^2 M_3}{16EI} \longleftarrow (3)$$

For point 0

$$y_1 - 2y_0 + y_{-1} = \frac{-L^2 M_0}{16EI}$$

and $\left(\frac{dy}{dx}\right)_0 \approx \frac{y_1 - y_{-1}}{2h} = 0 \qquad \therefore\ y_1 = y_{-1}$

$$\therefore\ 2y_1 = -\frac{L^2 M_0}{16EI} \longleftarrow (4)$$

Now have 4 equations and 4 unknowns.

FORCE AND STRESS ANALYSIS 8

A pin-connected tripod frame consisting of three round steel bars AB, AC and AD supports a load of 2000 lb at A. Each bar is 30 inches long and has a cross-sectional area of 0.5 in^2. The radius of the base circle BCD is 12 inches.

If no buckling occurs, find the stress in each of the bars in psi and state whether tension or compression.

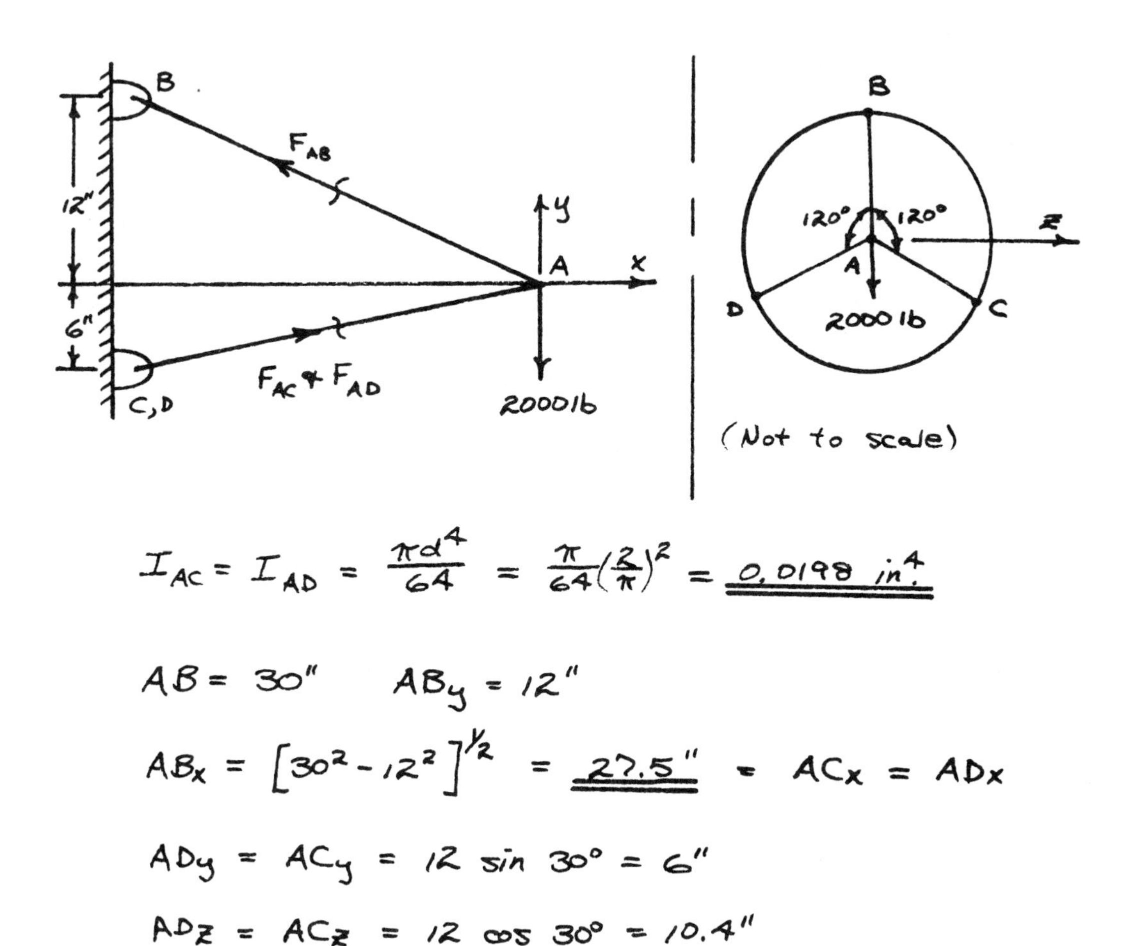

$$I_{AC} = I_{AD} = \frac{\pi d^4}{64} = \frac{\pi}{64}\left(\frac{2}{\pi}\right)^2 = \underline{\underline{0.0198 \text{ in}^4}}$$

$AB = 30''$ $\quad AB_y = 12''$

$$AB_x = \left[30^2 - 12^2\right]^{1/2} = \underline{\underline{27.5''}} = AC_x = AD_x$$

$$AD_y = AC_y = 12 \sin 30° = 6''$$

$$AD_z = AC_z = 12 \cos 30° = 10.4''$$

Force System is concurrent at A

$\underline{\Sigma F_z = 0}$

$$+(F_{AD})_z - (F_{AC})_z = 0$$

$$\frac{10.4}{30}(F_{AD}) - \frac{10.4}{30}(F_{AC}) = 0$$

$F_{AD} = F_{AC}$ ⟵

$\Sigma F_y = 0$

$$(F_{AB})_y + (F_{AC})_y + (F_{AD})_y - 2 = 0$$

$$\frac{12}{30} F_{AB} + \frac{2(6)}{30} F_{AC} - 2 = 0$$

$$12 F_{AB} + 12 F_{AC} = 60$$

$$F_{AB} + F_{AC} = 5 \longleftarrow (1)$$

$\Sigma F_x = 0$

$$(F_{AC})_x + (F_{AD})_x - (F_{AB})_x = 0$$

$$2\left(\frac{27.5}{30}\right) F_{AC} - \frac{27.5}{30} F_{AB} = 0$$

$$2 F_{AC} - F_{AB} = 0 \longleftarrow (2)$$

Add (1) to (2)

$$3 F_{AC} = 5 \qquad F_{AC} = \underline{\underline{1.67\ K}} = F_{AD}$$

$$\sigma_{AC} = \sigma_{AD} = \frac{1.67}{0.5} = \underline{\underline{3.34\ Ksi}} \text{ compr.} \longleftarrow$$

$$F_{AB} = 5 - F_{AC} = 3.33\ K$$

$$\sigma_{AB} = \frac{3.33}{0.5} = \underline{\underline{6.66\ Ksi}} \text{ tensile} \longleftarrow$$

$$(P_{crit.})_{AC \text{ or } AD} = \frac{\pi^2 EI}{\ell^2} = \frac{\pi^2 (3)(10^7)(0.0198)}{900(1000)} = \underline{\underline{6.55K}}$$

No Buckling Occurs, $6.55\ K > 1.67\ K$

FORCE AND STRESS ANALYSIS 9

What will be the vertical deflection of rigid member BC under force P?

AB and CD have length L, area A, modulus of elasticity E, and moment of inertia I.

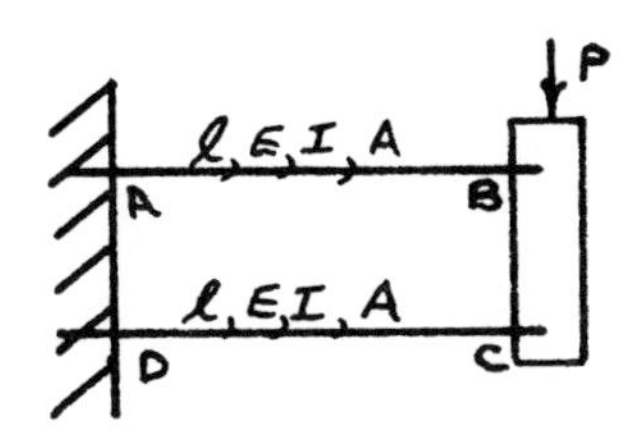

Solution

Deflected Shape

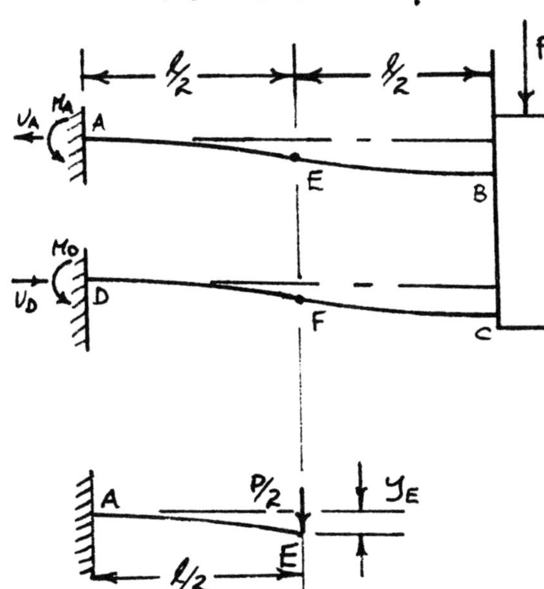

* Points E & F are points of contraflexure. ($M_E = M_F = 0$)

* Since members AB & DC have same properties, length, etc.

$V_A = V_D = P/2$ ←

$M_A = M_D = \frac{P}{2}\frac{l}{2} = \frac{Pl}{4}$ ←

A — $P/2$ — y_E — E — $l/2$

$$y_B = 2y_E = y_C = 2y_F$$

$$= 2\left[\frac{(P/2)(l/2)^3}{3EI}\right] = \frac{Pl^3}{24EI} \leftarrow$$

Vertical Deflection of Rigid Member $= \dfrac{Pl^3}{24EI}$

FORCE AND STRESS ANALYSIS 10

A truck axle is to be designed to withstand a torque load of 2880 ft-lbs and a wheel load of 2000 lbs.

(a) For the configuration shown, what is the axle diameter required to the nearest 1/16 inch?

$$\text{Assume } S_{t\ max} = 100{,}000 \text{ psi}$$

$$S_{s\ max} = 50{,}000 \text{ psi}$$

(b) The minimum factor of safety shall be 2.50 in tension. What is the factor of safety in shear?

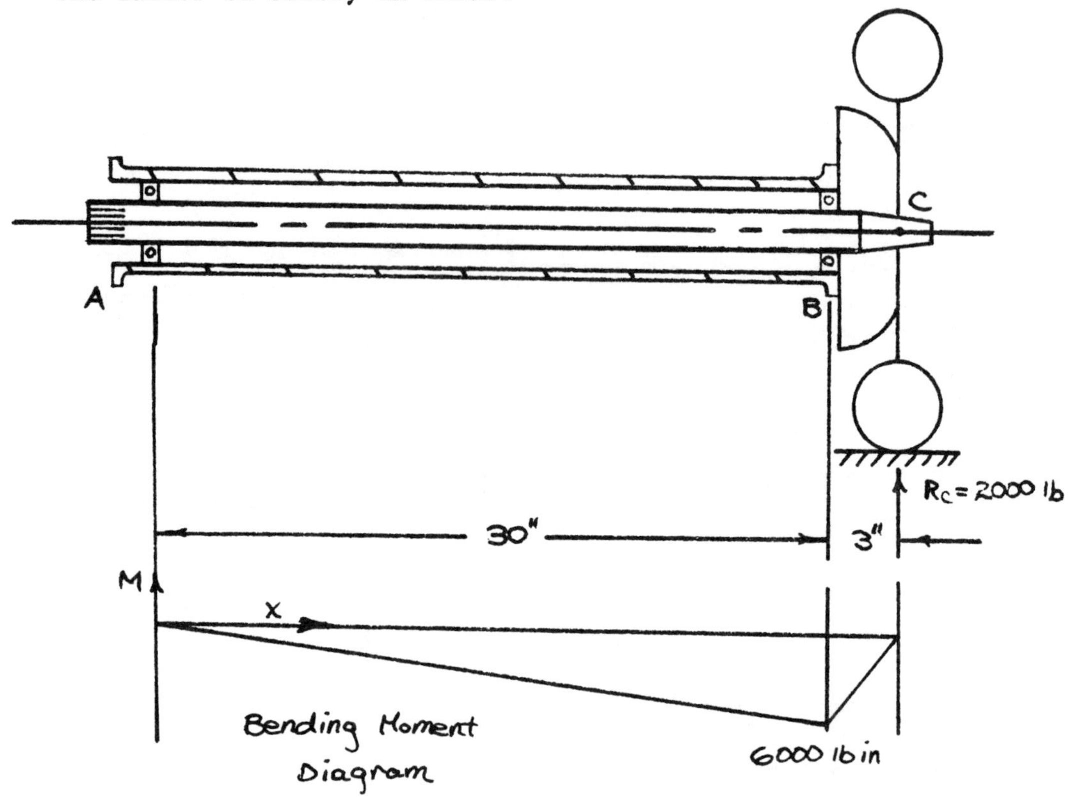

$$\sigma_x = \frac{Mc}{I} = \frac{6000(32)}{\pi d^3} \quad ; \quad \tau_{xy} = \frac{2880(12)(16)}{\pi d^3}$$

$$= \frac{34600(16)}{\pi d^3}$$

$$\text{a) } (S_s)_{max} = 50{,}000 = \sqrt{\left(\frac{\sigma_x}{2}\right)^2 + \tau_{xy}^2}$$

$$= \sqrt{\left[\frac{6000(16)}{\pi d^3}\right]^2 + \left[\frac{34600(16)}{\pi d^3}\right]^2}$$

$$50{,}000 = \frac{16}{\pi d^3}\sqrt{6000^2 + 34{,}600^2} = \frac{16}{\pi d^3}(35{,}100)$$

$$d^3 = \frac{16(35{,}100)}{50{,}000(\pi)} = 3.57$$

$$d = \underline{\underline{1.53 \text{ in}}} \longleftarrow$$

$$(S_t)_{max} = 100{,}000 = \frac{\sigma_x}{2} + \sqrt{\left(\frac{\sigma_x}{2}\right)^2 + \tau_{xy}^2}$$

$$= \frac{6000(16)}{\pi d^3} + \frac{16}{\pi d^3}\sqrt{6000^2 + 34{,}600^2}$$

$$= \frac{16}{\pi d^3}\left[6000 + 35{,}100\right] = \frac{16}{\pi d^3}(41{,}100)$$

$$d^3 = \frac{16(41{,}100)}{\pi(100{,}000)} = 2.09 \text{ in}^3$$

$$d = \underline{\underline{1.28 \text{ in}}} \longleftarrow$$

Required Diameter $= \underline{\underline{1\,9/16''}} \longleftarrow$

---

b) For a factor of Safety of $\underline{\underline{2.5}}$ in tension

$$d^3 = \frac{16(41{,}000)}{\pi\left(\frac{100{,}000}{2.5}\right)} = 5.21 \qquad d = \underline{\underline{1.74 \text{ in}}}$$

$$(S_s)_{max} = \frac{16(35{,}100)}{\pi(1.74)^3} = 33{,}900 \text{ lb/in}^2$$

$$\text{Safety Factor in Shear} = \frac{50{,}000}{33{,}900} = \underline{\underline{1.47}}$$

FORCE AND STRESS ANALYSIS 11

During a static loading test, a 45 degree strain rosette attached to the test object indicated the following strain outputs: $\varepsilon_x = 600$, $\varepsilon_m = -83$, and $\varepsilon_y = 100$ microinches. The material's modulus of elasticity is 30 x 106 psi, while Poisson's ratio is 0.3.

From this data draw Mohr's circle of stress for this point indicating the maximum and minimum stress values and the stresses along the x and y directions. Find also the angle between the x axis and the axis of maximum principal stress.

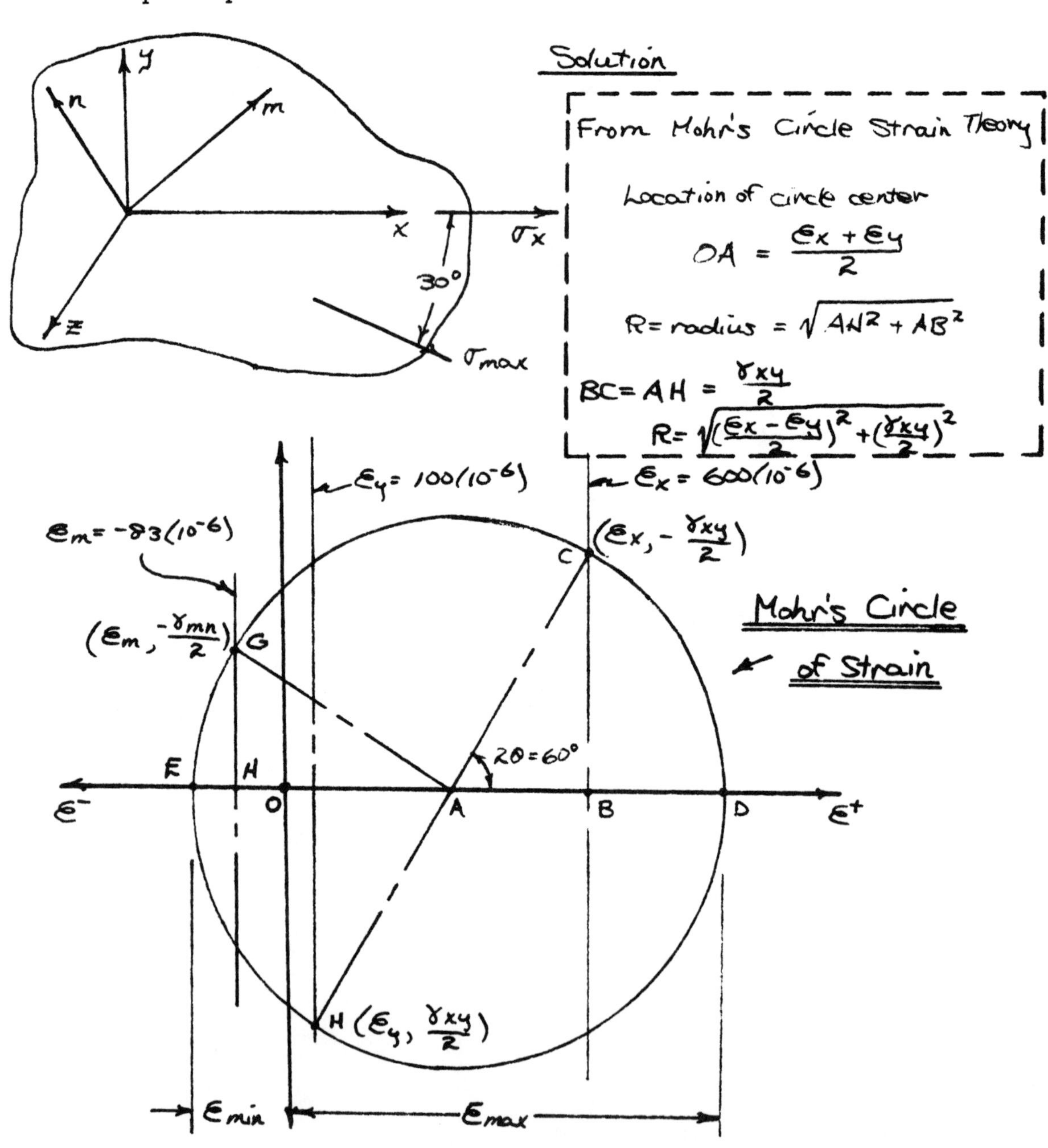

Radius of Mohrs Circle of Strain

$$= AC = (AB^2 + BC^2)^{1/2} = \left[(250\times10^{-6})^2 + (433\times10^{-6})^2\right]^{1/2}$$

$$= 500\times10^{-6} \text{ in/in}$$

$$\varepsilon_{max} = OD = 850\times10^{-6} \text{ in/in}$$

$$\varepsilon_{min.} = OE = -150\times10^{-6} \text{ in/in}$$

$$\sigma_{max} = \frac{E}{1-u^2}\left[\varepsilon_{max} + u\,\varepsilon_{min}\right]$$

$$= \frac{30\times10^6}{1-0.3^2}\left[850\times10^{-6} - 0.3(150)(10^{-6})\right]$$

$$= 33\left[850 - 45\right] = \underline{\underline{26{,}500 \text{ lb/in}^2}} \leftarrow$$

$$\sigma_{min} = \frac{E}{1-u^2}\left[\varepsilon_{min} + u\,\varepsilon_{max}\right]$$

$$= 33\left[-150 + 255\right] = \underline{\underline{3{,}460 \text{ lb/in}^2}} \leftarrow$$

$$\sigma_x = \frac{E}{1-u^2}\left[\varepsilon_x + u\,\varepsilon_y\right] = 33\left[600 + 0.3(100)\right] = \underline{\underline{20{,}800 \text{ lb/in}^2}}$$

$$\sigma_y = \frac{E}{1-u^2}\left[\varepsilon_y + u\,\varepsilon_x\right] = 33\left[100 + 0.3(600)\right] = \underline{\underline{9{,}250 \text{ lb/in}^2}}$$

* Note - above stress are also found from Mohr's circle of stress on the following page.

Mohr's Circle of Stress

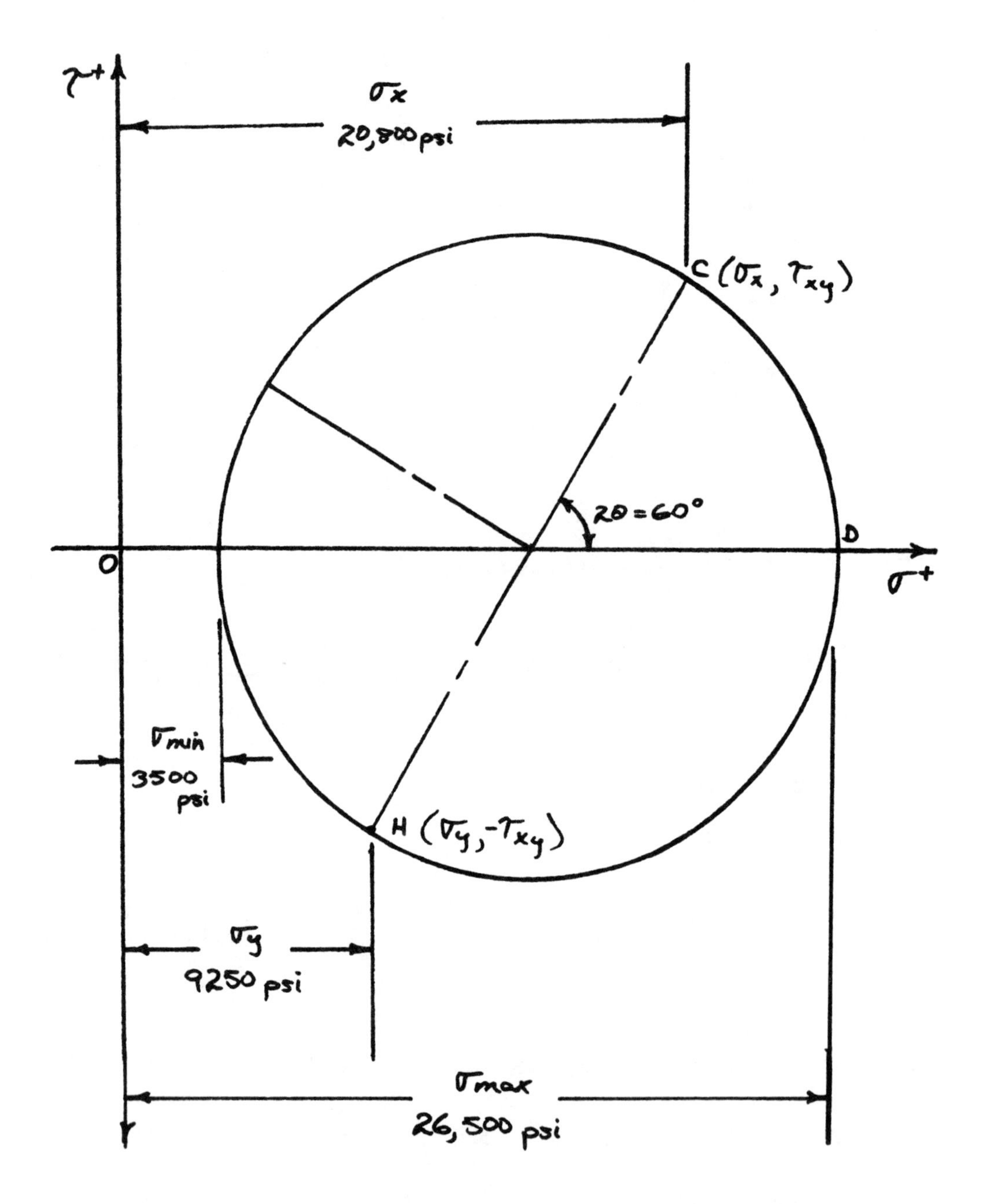

angle between σ_x + σ_{max} = θ = 30° ⟵

FORCE AND STRESS ANALYSIS 12

A uniformly-loaded cantilever beam is supported at the free end, but the support is not at the same level as the fixed end. It was jacked into place when the beam was fully loaded, thus reducing the deflection of the free end by 50%.

What are the reactions at each end? What is the bending moment at the fixed end?

Solution

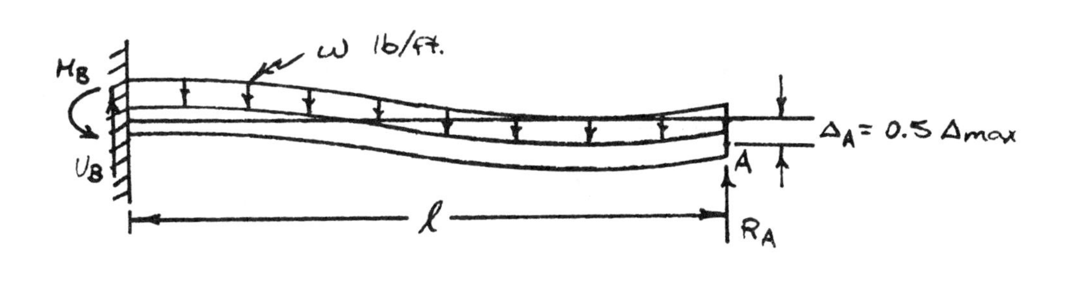

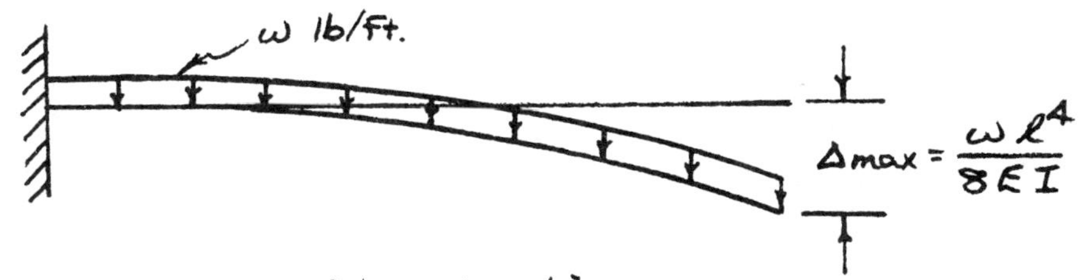

For cantilever with w loading

$$\Delta_{max} = \frac{w\ell^4}{8EI} = \Delta_A$$

For cantilever with R_A loading $\quad \Delta_A = -\frac{R_A \ell^3}{3EI}$

$$\frac{w\ell^4}{8EI} - \frac{R_A \ell^3}{3EI} = \frac{0.5\,w\ell^4}{8EI} = \frac{w\ell^4}{16EI}$$

$$R_A = \frac{3w\ell}{16}$$

$$M_B = -\frac{w\ell^2}{2} + R_A(\ell) = -\frac{w\ell^2}{2} + \frac{3}{16}w\ell(\ell)$$

$$= -\frac{5}{16}w\ell^2$$

FORCE AND STRESS ANALYSIS 13

A load cell in the form of a bent aluminum strip is shown in the diagram. The strip is one-quarter inch wide and one-tenth inch thick. Determine the strain gage reading for a 5 lb load.

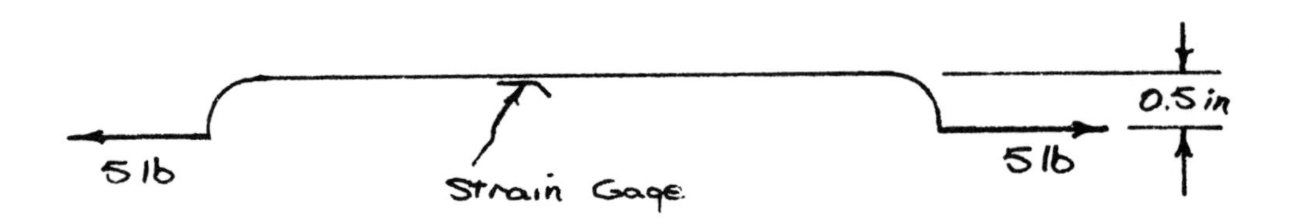

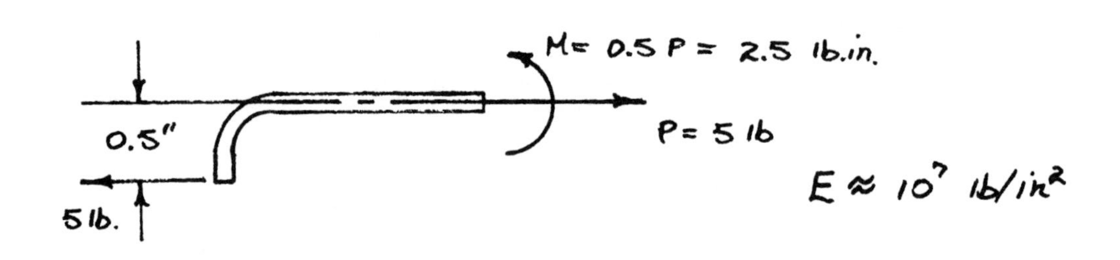

Strain at location of gage $\epsilon_K = \frac{\sigma_K}{E}$

$$\sigma_K = P/A + \frac{Mc}{I}$$

$$= \frac{5}{(1/4)(1/10)} + \frac{2.5\left(\frac{1}{20}\right)}{\frac{1/4\left(1/10\right)^3}{12}}$$

$$= 200 + 6000 = \underline{\underline{6200 \text{ lb/in}^2}}$$

$$\epsilon_K = \frac{6200}{10 \times 10^6} = \underline{\underline{620 \times 10^{-6} \text{ in/in}}} \longleftarrow$$

FORCE AND STRESS ANALYSIS 14

A square steel bar of 1 x 1 inch cross section and 6 feet long is to be used as a column. The ends of the bar are rounded so as to be free to rotate, but are placed in sockets and may not be displaced. If an average unit stress of 30,000 psi is allowable, find the maximum load that this column may support.

Solution

For a long column with rounded ends

Elastic Buckling load $P_{crit.} = \frac{\pi^2 EI}{\ell^2}$; r = gyration radius

For a long column the minimum ℓ/r for above formula to be valid is obtained as follows:

$$\frac{P_{crit.}}{A} = \frac{\pi^2 E \overbrace{(Ar^2)}^{I}}{A\ell^2} = \frac{\pi^2 E}{(\ell/r)^2} = \sigma_{crit} \approx \sigma_{Prop.\,Limit}$$

For elastic Buckling with a $\sigma_{prop.\,limit} \approx 45$ Ksi,

$$\left(\frac{\ell}{r}\right)^2_{min} = \frac{\pi^2 E}{45{,}000} = \frac{\pi^2(30)(10^6)}{45{,}000} = 6570$$

$$\left(\frac{\ell}{r}\right)_{min} = \underline{\underline{81}} \longleftarrow \quad b = h = 1 \text{ in.}$$

For a column with a 1" x 1" cross section + ℓ = 72"

$$r = \left(\frac{I}{A}\right)^{1/2} = \left[\frac{bh^3}{12(bh)}\right]^{1/2} = \frac{h}{\sqrt{12}} = \frac{1}{\sqrt{12}} = 0.28 \text{ in.}$$

$$\frac{\ell}{r} = \frac{72}{0.28} = 257 \longleftarrow$$

The above column is definitely a long column so

the maximum load that the column can carry is not governed by strength but by stiffness

$$P_{max} = P_{critical} = \frac{\pi^2 EI}{\ell^2} = \frac{\pi^2 (3)(10^7)(1^4)}{12(72)^2}$$

$$P_{max} = \underline{\underline{4760 \text{ lb}}} \longleftarrow$$

Assuming a Safety factor of 3,

$$P_{allow} = \frac{4760}{3} \approx \underline{\underline{1600 \text{ lb}}} \longleftarrow$$

3

Dynamics and Vibrations

HAROLD M. TAPAY

DYNAMICS

Dynamics is that part of engineering mechanics in which a study is made of particles and bodies in motion; it is divided into branches--namely, kinematics and kinetics. Kinematics consists essentially of deriving equations relating to displacement, velocity, acceleration and time. Kinetics consists essentially of deriving equations relating to the unbalanced force or the unbalanced moment to the change in motion.

Kinematics:

Types of particle motion -

(1) Rectilinear motion

$$V = \text{absolute velocity} = \frac{ds}{dt} \qquad (1)$$

O — S — Path

$$a = \text{absolute accel.} = \frac{dV}{dt} \qquad (2)$$

$$S = \text{displacement} \qquad VdV = ads \qquad (3)$$

(2) Curvilinear motion

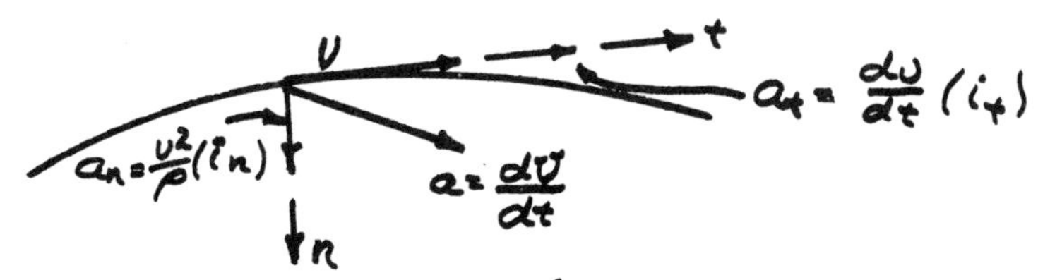

Tangential component of "a" $= a_t = \frac{dv}{dt}(i_t)$

Normal component of "a" $= a_n = \frac{v^2}{P}(i_n)$ $\qquad$ v = speed

Types of Rigid Body Plane Motion:

(1) Rectilinear Translation - all particles move on parallel straight lines and have same displacements, velocities and accelerations.

(2) Curvilinear Translation - same as above except that all particles move on congruent curves.

(3) Fixed Axis Rotation - all particles move on concentric circular paths

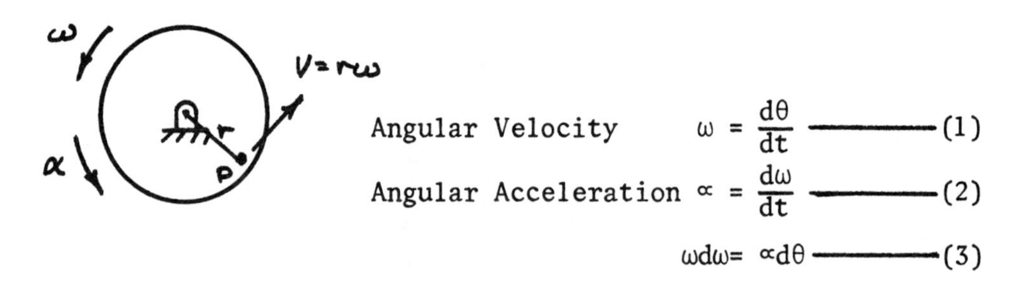

Angular Velocity $\omega = \frac{d\theta}{dt}$ ——— (1)

Angular Acceleration $\alpha = \frac{d\omega}{dt}$ ——— (2)

$\omega d\omega = \alpha d\theta$ ——— (3)

For Particle P -

Tangential acceleration of particle $a_t = \frac{rd\omega}{dt} = r\alpha$

Normal acceleration of particle $a_n = \frac{v^2}{r} = r\omega^2$

(4) General Plane Motion - a motion which can be resolved into translation plus fixed axis rotation

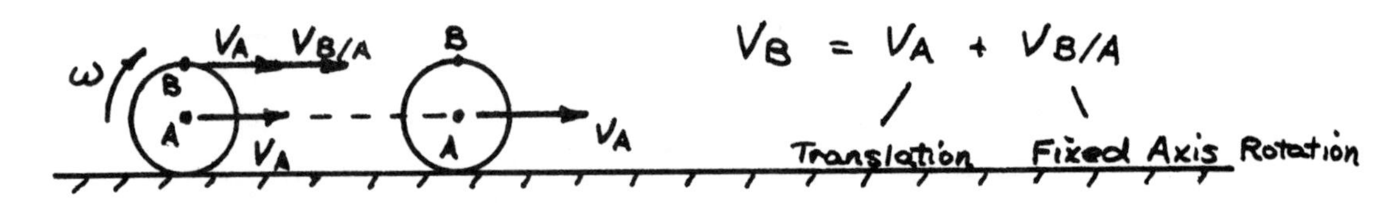

Kinetics:

Newton's 2nd Law of Motion for a particle -

$$F = ma \qquad (1)$$

F = unbalanced force
m = mass of particle
a = absolute acceleration

For English Gravitation System of Units -

F - lb; m - slugs = $\frac{W}{g}$; a - ft/sec^2

For a system of particles (rigid or non-rigid body) -

$$F = (\Sigma m)(\bar{a}) = M\bar{a} \qquad (2)$$

F = unbalanced force
M = mass of system of particles
a = absolute acceleration of center of mass

For rigid-body rotation about fixed axis

$$(M_o)_z = \int (r^2 dm)\alpha \qquad (3)$$

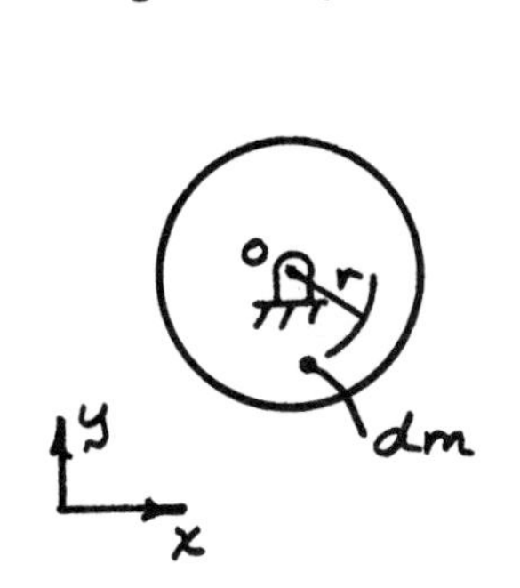

where $\int r^2 dm$ = mass moment of inertia about axis of rotation = $(I_o)_z$

$$(M_o)_z = (I_o)_z \alpha$$

where $\Sigma(M_o)_z$ is unbalanced moment about axis of rotation

Work Energy Principle

By taking Newton's 2nd Law of Motion and using components of "F" and "a" in the tangential direction and multiplying each side of equation by ds and integrating, the work energy principle is obtained:

$$F_t = m \frac{dv}{dt}; \quad F_t ds = m\, dv \frac{ds}{dt}; \quad F_t ds = mv\, dv = dU$$

$$\Delta U_{1\to2} = \int_1^2 dU = m \int_1^2 v\, dv = \frac{1}{2} mv_2^2 - \frac{1}{2} mv_1^2 = \Delta K.E. \qquad (4)$$

where $\frac{1}{2} mv^2$ is kinetic energy of a particle.

$\Delta U_{1\to2}$ = work done by F_t which is tangential component of unbalanced force acting on the particle

Work Done by a Force on a Spring

$\Delta U_{1\to2}$ = work done by a force P slowly applied to a linear elastic spring = 1/2 Px = 1/2 kx^2

k = spring constant = P/x
l = free length of spring

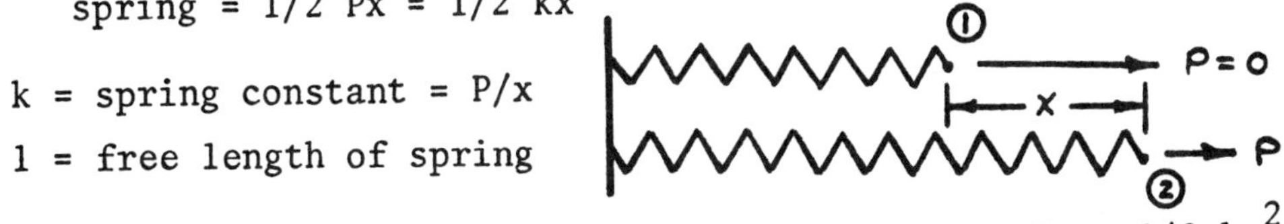

Potential energy of strain for spring in position 2 = V_e = 1/2 kx^2

$$\Delta V_e = (V_e)_2 - (V_e)_1 = 1/2\ kx^2$$

Potential energy of position in gravitational field = V_g

Second form of work energy equation

$$\Delta U_{1\to2} = \Delta K.E. + \Delta V_g + \Delta V_e \text{ ——— (5)}$$

If no work is done then K.E. + V_g + V_e = constant, i.e., mechanical energy is conserved.

Linear Impulse - Linear Momentum Principle

By rearranging Newton's 2nd Law of Motion the linear impulse-linear momentum principle is obtained; G = linear momentum -

$$F = m\frac{dV}{dt} \quad Fdt = d(mv) = dG \text{ ——— (6)}$$

where Fdt is linear impulse and d(mv) is change in linear momentum.

Integrating Eq. (6) for a particle,

$$mV_1 + \int Fdt = mV_2$$

If no unbalanced force acts then linear momentum "G" is conserved, i.e.,

$$G = \text{constant} \text{ ——— (7)}$$

Angular Impulse - Angular Momentum Principle

(Angular Momentum of a Particle)$_z$ = (Moment of Linear Momentum)$_z$

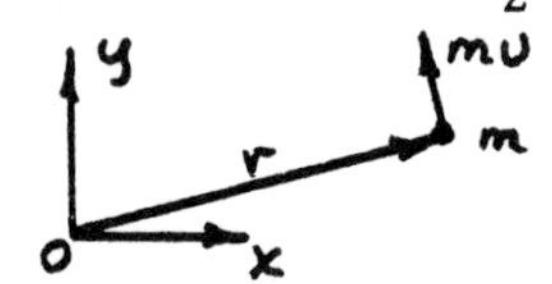

$$H_z = (mV)(r) = \text{Angular Momentum of Particle}$$

For a rotating rigid body

$$H_z = \int (dm)(r\omega)\, r = \omega \int r^2 dm = I_z\omega \text{ ——— (8)}$$

$$\Sigma M_z = I_z\alpha = I_z\frac{d\omega}{dt} = \frac{d}{dt}(I_z\omega) = \frac{dH_z}{dt} \text{ ——— (9)}$$

$$\Sigma M_z\, dt = d(H_z) = d(I_z\omega) \text{ ——— (10)}$$

where ΣM_z dt is angular impulse and $d(I_z\omega)$ is change in angular momentum.

Integrating Eq. (10),

$$I_z\omega_1 + \int_1^2 \Sigma M_z\, dt = I_z\omega_2 \text{ ——— (11)}$$

From Eq. (10) if no unbalanced moment acts then angular momentum H_z is conserved, i.e.,

$$H_z = \text{constant} \text{ ——— (12)}$$

Gyroscopic Motion

When the axis of rotation of a rotating body itself rotates, i.e., precesses, a gyroscopic moment is set up; this moment is obtained from angular momentum-angular impulse principle as follows:

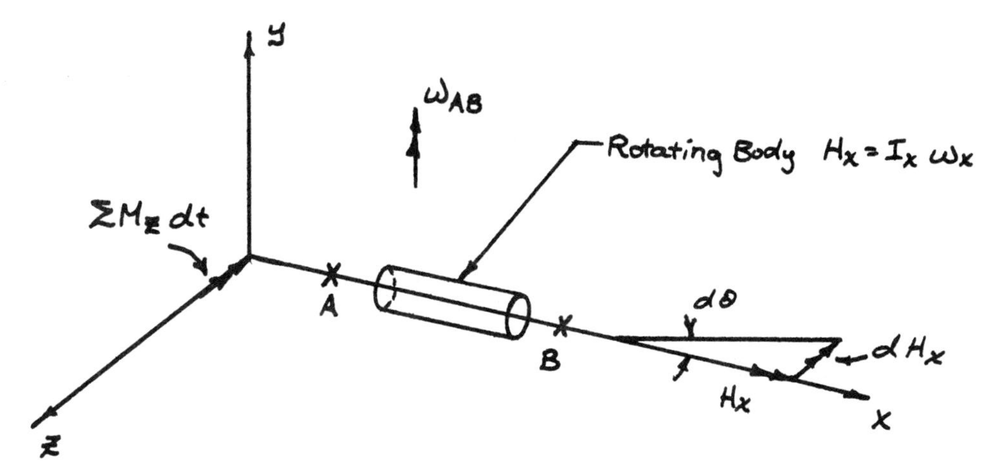

$$\text{Gyroscopic Moment} = \Sigma M_z = \frac{d(H_x)}{dt} = \frac{(H_x d\theta)}{dt} = I_x \left(\frac{d\theta}{dt}\right) \omega_x \text{——}(13)$$

$$\frac{d\theta}{dt} = \omega_{AB} = \text{precession velocity}$$

VIBRATIONS

Natural frequency "f" of free vibrations for systems shown below are obtained by solving equations of motion.

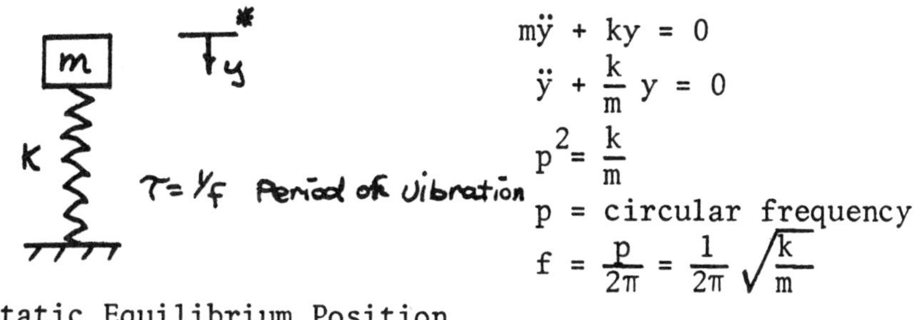

$$m\ddot{y} + ky = 0$$

$$\ddot{y} + \frac{k}{m} y = 0$$

$$p^2 = \frac{k}{m}$$

$$p = \text{circular frequency}$$

$$f = \frac{p}{2\pi} = \frac{1}{2\pi}\sqrt{\frac{k}{m}}$$

*Static Equilibrium Position

$$k = \frac{3EI}{\ell^3}$$

$$m\ddot{y} + ky = 0$$

$$\ddot{y} + \frac{k}{m} y = 0$$

$$f = \frac{p}{2\pi} = \frac{1}{2\pi}\sqrt{\frac{k}{m}}$$

$$\tau = \frac{1}{f}$$

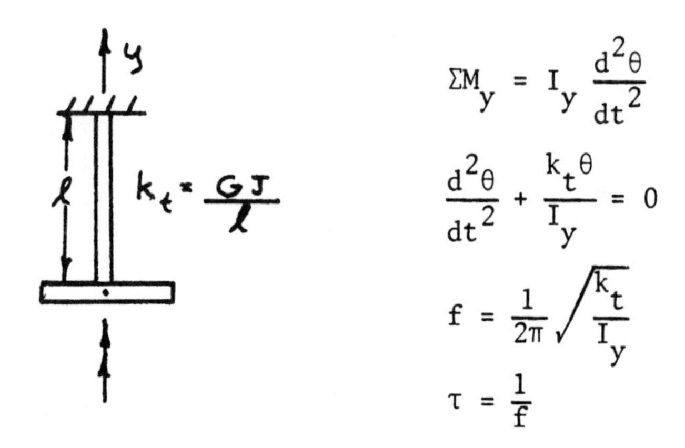

$$\Sigma M_y = I_y \frac{d^2\theta}{dt^2}$$

$$\frac{d^2\theta}{dt^2} + \frac{k_t\theta}{I_y} = 0$$

$$f = \frac{1}{2\pi}\sqrt{\frac{k_t}{I_y}}$$

$$\tau = \frac{1}{f}$$

Natural frequency of damped free vibrations:

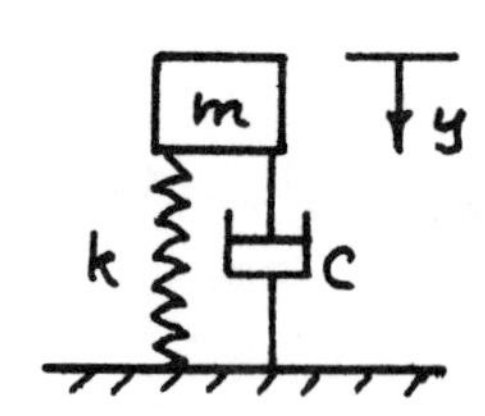

$m\ddot{y} + c\dot{y} + ky = 0$

C = damping coefficient

C_c = critical damping coefficient = 2mp

p = circular frequency for no damping

$f = \frac{\omega}{2\pi}$ where ω is circular frequency of damped free vibrations

$$\omega = p\sqrt{1 - \left(\frac{C}{C_c}\right)^2} \qquad f = \left(\frac{p}{2\pi}\right)\sqrt{1 - \left(\frac{C}{C_c}\right)^2}$$

Damped forced vibrations:

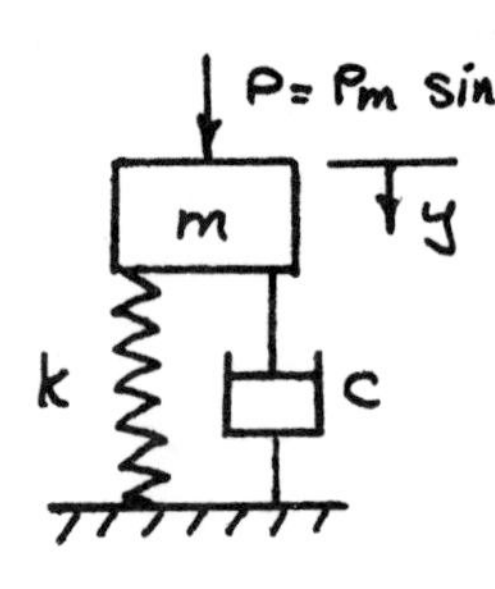

Equation of motion -

$m\ddot{y} + c\dot{y} + ky = P_m \sin \omega t$

Amplitude of steady-state vibration produced by P is

$$y_m = \frac{\frac{P_m}{k}}{\sqrt{\left[1 - \left(\frac{\omega}{p}\right)^2\right]^2 + \left[2\left(\frac{C}{C_c}\right)\left(\frac{\omega}{p}\right)\right]^2}}$$

$$\text{Magnification factor} = \frac{y_m}{\frac{P_m}{k}}$$

For support vibrating - $\delta = \delta_m \sin \omega t$

$$y_m = \frac{\delta_m}{\sqrt{\left[1 - \left(\frac{\omega}{p}\right)^2\right]^2 + \left[2\left(\frac{C}{C_c}\right)\left(\frac{\omega}{p}\right)\right]^2}}$$

$$\text{Magnification factor} = \frac{y_m}{\delta_m}$$

DYNAMICS AND VIBRATIONS 1

In a recent edition of Mechanical Engineering, published by ASME, an inventor described a patented skid he had developed for towing heavy objects. The device is shown below in an idealized representation. A mass M oscillates relative to a wedge shaped form M_o. The wedge is of angle α, the spring constant is K and has unstretched length X_o. There is no friction acting between mass M and the wedge, but the coefficient of friction between the wedge and ground is μ. When the mass M is close to its minimum displacement relative to the wedge ($X_{r\,min}$), it creates a normal force and friction large enough so the skid does not move backwards. When the mass M is close to its maximum displacement relative to the wedge ($X_{r\,max}$), the inertia force overcomes friction; and the wedge and mass go forward a distance X. In this way the skid can tow an object intermittently forward if it is attached to the wedge.

REQUIRED: Analyze the following two unloaded situations:

(a) When X_r is minimum find the minimum coefficient of friction needed to prevent the skid from going backward. $\ddot{X}$ is + 100 ft/sec^2 for that position.

(b) If μ = 0.5, find the $\ddot{X}$ acceleration of the skid forward when X_r is maximum. $\ddot{X}_r$ is - 100 ft/sec^2 for that position.

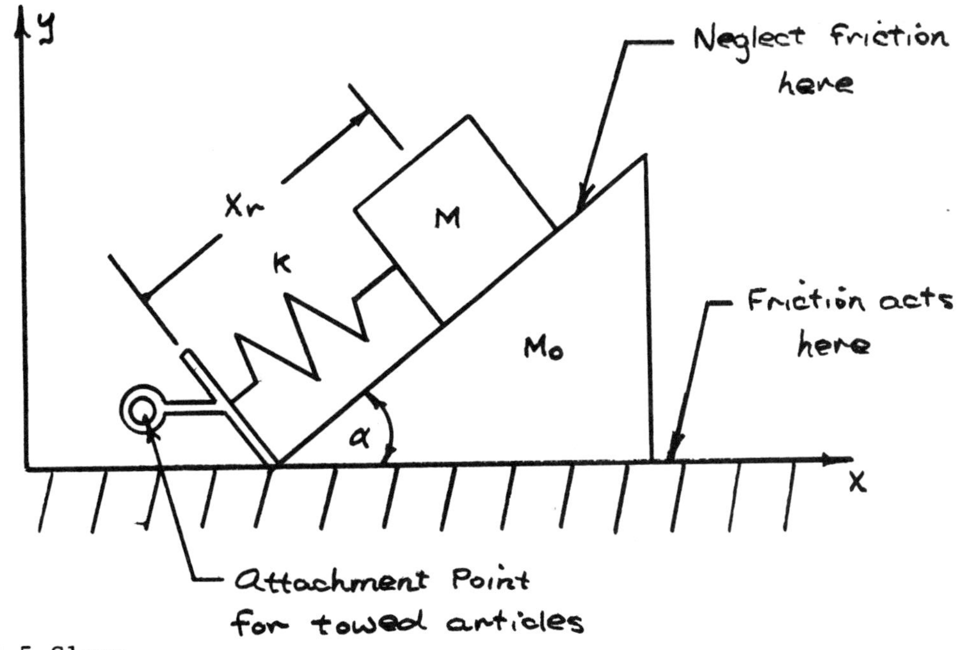

M = 4.5 Slugs
M_o = 9.5 slugs
K = 90 lbs/in
X_o = 12 in
α = 75°

Solution

Freebody Diagram

a)

$\ddot{X}_R = 100$ ft/sec^2

M, M_o, $(X_R)_{min}$, $\alpha = 75°$

$M\ddot{X}_R$, N, W, $K(X_R)_{min}$, 450 lb, 306 lb, 145 lb, $(F_{So})_{max.}$, N_o, y, x

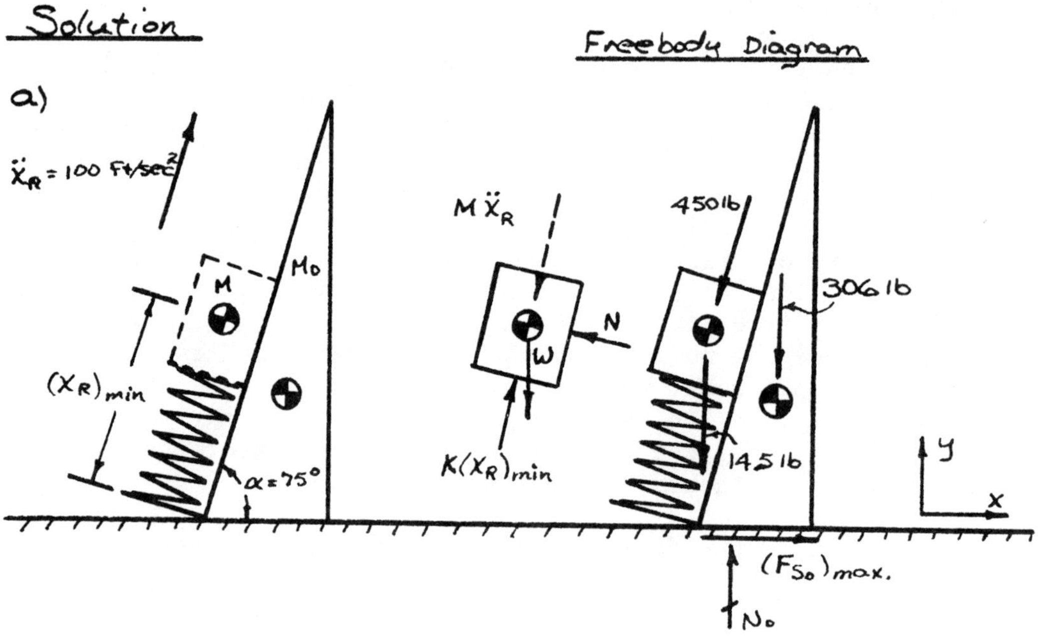

$M = 4.5$ slugs : $W = mg = 4.5(32.2) = \underline{\underline{145 \text{ lb}}}$

$M\ddot{X}_R = 4.5(100) = \underline{\underline{450 \text{ lb}}}$

$M_o = 9.5$ slugs : $W_o = M_o g = \underline{\underline{306 \text{ lb}}}$

From Freebody Diagram :

$\Sigma F_y = 0$ $\qquad N_o = 450 \cos 15° + 145 + 306$

$= 435 + 145 + 306$

$= \underline{\underline{886 \text{ lb}}}$

$\Sigma F_x = 0$ $\qquad (F_{So})_{max.} = 450 \sin 15° = \underline{\underline{116.5 \text{lb}}}$

Minimum coeff. of friction $\mu = \dfrac{(F_{So})_{max}}{N_o} = \dfrac{116.5}{886}$

$\approx \underline{\underline{0.13}} \longleftarrow$

b) $\ddot{X}_R = -100\ \text{ft/sec}^2$: $M\ddot{X}_R = 4.5(100) = \underline{\underline{450\ \text{lb}}}$

Freebody Diagram of Assembly

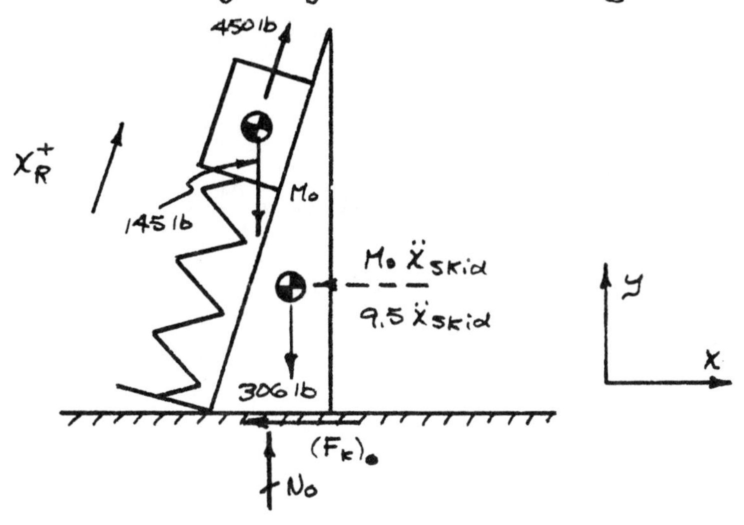

$\Sigma F_y = 0$

$$N_o = 306 + 145 - 450 \cos 15° = 451 - 450(0.966)$$
$$= \underline{16\ \text{lb.}}$$

$$(F_K)_o = \mu_K N_o = 0.5(16) = \underline{\underline{8\ \text{lb}}}$$

$\Sigma F_x = 0$

$$+450 \cos 15° = 9.5\ \ddot{x}_{skid} - 8 = 0$$

$$\ddot{x}_{skid} = \frac{116.5 - 8}{9.5} = \underline{\underline{11.4\ \text{ft/sec}^2}} \longleftarrow$$

Acceleration of skid is 11.4 ft/sec² to the right.

DYNAMICS AND VIBRATIONS 2

(a) Show, by equating "centrifugal" and "gravitational" forces, that speed of a satellite in a circular orbit about the earth is given by:

$$V = R_e\sqrt{\frac{g_e}{R}}$$

where V = orbital speed
R_e = radius of earth
g_e = gravitational acceleration at surface of earth
R = radius of orbit

(Note: Assume the earth is uniform and spherical, and that the gravitational attractive force between the earth and a remote object varies inversely with the square of the distance between their mass centers.)

(b) Use the formula to compute the distance from the earth's surface to a synchronous communication satellite.

<u>Solution</u>

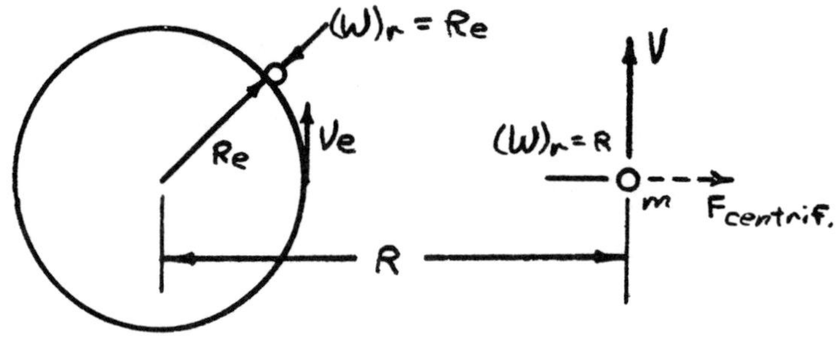

a) $$F_{centrif.} = \frac{m\upsilon^2}{R} = (W)_{r=R} = \frac{m\, m_{earth}\, G}{R^2}$$

$$\upsilon^2 = \frac{G m_e}{R}$$

$$(W)_{r=R_e} = m g_e = \frac{m\, m_e\, G}{R_e^2}$$

$$G m_e = g_e R_e^2$$

$$\upsilon^2 = \frac{g_e R_e^2}{R}$$

$$\upsilon = R_e\sqrt{\frac{g_e}{R}}$$

b) For a synchronus satellite

$$\frac{U_e}{U} = \frac{R_e}{R} \qquad U = \frac{R}{R_e} U_e$$

$$U_e = R_e \omega \qquad \omega = \text{Earth's Angular Velocity}$$

$$= \frac{1(2\pi)}{24(60)(60)} = 7.28(10^{-5})\ \text{sec}^{-1}$$

$$U_e = (3960)(5280)(7.28)(10^{-5}) = \underline{\underline{1520\ \text{ft/sec}}}$$

$$U = \frac{R}{R_e} U_e = R_e \sqrt{\frac{g_e}{R}}$$

$$R^3 U_e^2 = R_e^4 g_e \qquad R^3 = \frac{R_e^4 g_e}{U_e^2}$$

$$R^3 = \frac{(3960)^4 (5280)^4 (32.17)}{(1520)^2} = 2.67 \times 10^{24}$$

$$R = 1.39 \times 10^8\ \text{ft} = \underline{\underline{26{,}400\ \text{miles}}}$$

Distance from Earth's surface to satellite:

$$26{,}400 - 3960 = \underline{\underline{22440\ \text{miles}}}$$

DYNAMICS AND VIBRATIONS 3

A straight piece of 1/8 inch diameter steel rod is clamped at one end in a vise; the other end is free.

Find the length in inches to make the fundamental bending frequency 264 Hz.

Assume steel weighs 0.284 lb/in^3.

Solution

For cantilever beam:

f = fundamental bending freq. = 264 cycles/sec.

$$f \overset{*}{=} \frac{CK(10^5)}{L^2} \qquad K = \text{gyration radius} = \left(\frac{I}{A}\right)^{1/2}$$

$$K = R/2$$

$C = 1.13$ for steel cantilever beam

$$f = \frac{1.13}{L^2}\left(\frac{1}{32}\right)(10^5) = 264$$

$$L^2 = \frac{1.13(10^5)}{264(32)} = 13.4$$

$$L = \underline{\underline{3.67\ in.}} \longleftarrow$$

* This equation comes from

$$f = \frac{a}{2\pi}\left(\frac{1.875}{L}\right)^2 \quad \text{where } a^2 = \frac{EIg}{A\gamma}$$

DYNAMICS AND VIBRATIONS 4

A child on a "pogo stick" can be idealized (for engineering purposes) as a mass on a spring, as shown in the diagram. At the top of a jump the C.G. of the child is above the ground by an amount H_o.

Compute the maximum axial load in the spring (stiffness K, unloaded length L_o) at the bottom of the jump.

Given: Child's weight: 100 lbs
K = spring stiffness: 100 lbs/in
H_o: 2 ft
L_o: 4 in

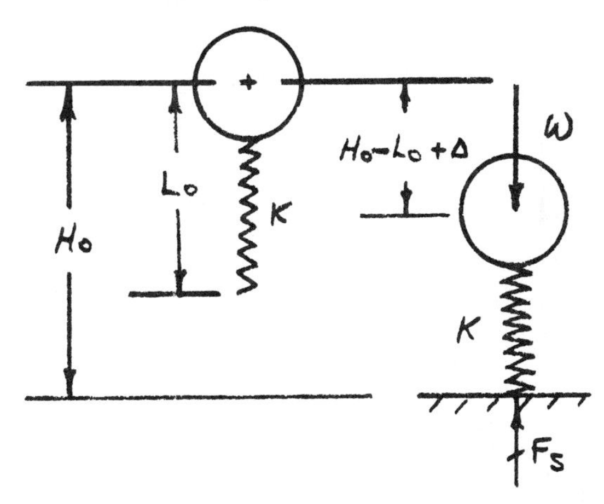

Solution

Let Δ = max. compressive deformation in spring at impact

Using Work-energy principle :

$$\Delta U = \Delta K.E. + \Delta U_g + \Delta U_e$$

$$0 = 0 - (W)(H_o - L_o + \Delta) + \tfrac{1}{2} K \Delta^2$$

$$0 = 0 - 100(24 - 4 + \Delta) + 50 \Delta^2$$

$$0 = -2000 - 100 \Delta + 50 \Delta^2$$

$$\Delta = \frac{100 \pm \sqrt{(-100)^2 + 4(2000)(50)}}{100}$$

$$= \frac{100 \pm}{100} = \underline{\underline{7.4 \text{ in}}}$$

THIS SPRINGS FREE LENGTH Lo IS ONLY 4 IN., ∴ IT CAN'T ACCOMODATE A LINEAR ELASTIC DEFLECTION OF 7.4 IN, SPRING WILL BOTTOM OUT.

DYNAMICS AND VIBRATIONS 5

The exhaust turbine of an aircraft directly drives a small supercharger at a constant speed of 18,000 rpm. The shaft connecting the two is parallel to the aircraft's horizontal center line. The turbine weighs 20 lbs; the blower weighs 10 lbs. Their radii of gyration are 4.0 and 4.5 inches, respectively.

If the bearings supporting the shaft are spaced 10 inches apart, determine the dynamic bearing loads created when the plane makes a 100 ft radius horizontal turn on the ground at a constant speed of 15 mph.

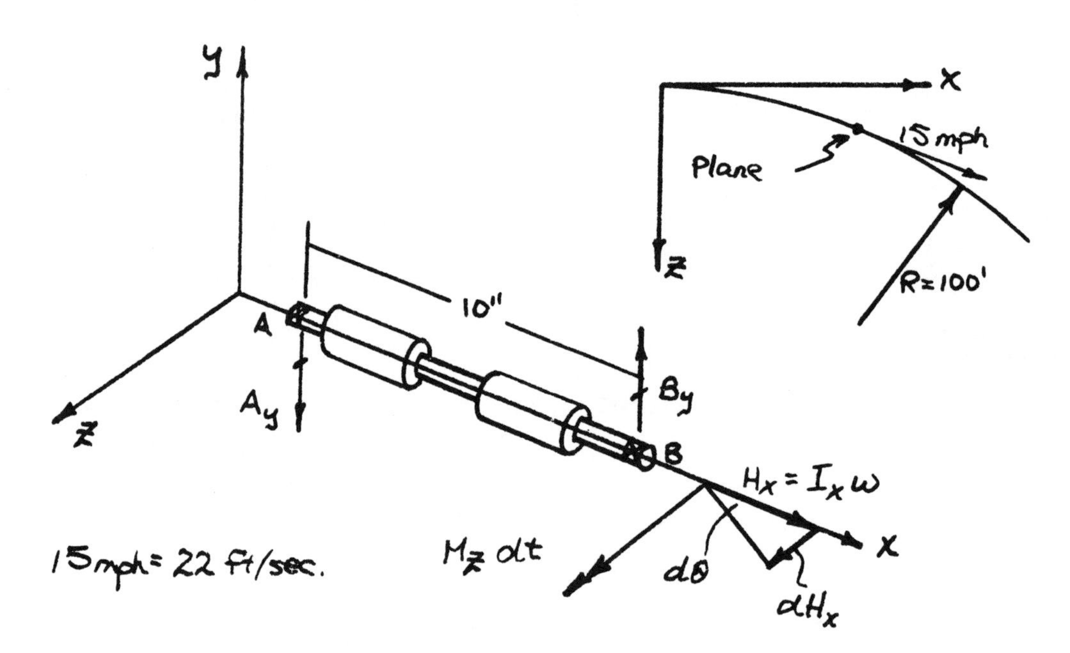

From Angular Impulse - Angular Momentum Principle

Change in angular momentum of Turbine - Supercharger Assembly $= dH_x = M_z\, dt$

$$M_z = \frac{dH_x}{dt} = I_x \omega \left(\frac{d\theta}{dt}\right)$$

$\frac{d\theta}{dt}$ = Precession velocity of shaft axis

$= V/R = 22/100 = 0.22$ rad/sec.

$$I_x = \frac{20}{g}\left(\frac{4}{12}\right)^2 + \frac{10}{g}\left(\frac{4.5}{12}\right)^2$$

$$= \frac{2.22}{g} + \frac{1.41}{g} = \frac{3.63}{g} \text{ slug}\cdot\text{ft}^3$$

$$\omega = \frac{18{,}000}{60}(2\pi) = \underline{\underline{600\,\pi \text{ rad/sec}}}$$

Dynamic Bearing loads $A_y = B_y = \frac{M_z}{10}(12)$

$$= 1.2\,M_z$$

$$M_z = \frac{3.63}{g}(600\pi)(0.22) = 46.7 \text{ lb}\cdot\text{ft}$$

$$A_y = B_y = (1.2)(46.7) = \underline{\underline{56 \text{ lb}}} \leftarrow$$

DYNAMICS AND VIBRATIONS 6

In launching a torpedo from a submarine, the turbulence within a torpedo tube produces a torque on the torpedo's propeller. The propeller is directly attached to the armature of an electric motor, which is free to turn.

Through how many revolutions does the propeller turn if the average torque is 3,000 in-lb and it takes 500 millisec for the torpedo to leave the tube?

Assume the armature can be idealized as a solid copper cylinder 6 inches long and 6 inches in diameter. Neglect the moment of inertia of the propeller.

Solution

$$\bar{I}_{arm} = \tfrac{1}{2} m R^2 \qquad \gamma_{copper} = 556 \text{ lb/ft}^3$$

$$\bar{I}_{arm} = \tfrac{1}{2} \frac{(556)}{g} \frac{\pi}{4}\left(\tfrac{1}{2}\right)^2\left(\tfrac{1}{2}\right) = \underline{\underline{0.85 \text{ slug ft}^2}}$$

$$\text{Angular Acceleration of Armature} = \alpha = \frac{\Sigma \bar{M}}{\bar{I}}$$

$$= \frac{3000}{12(0.85)} = \underline{\underline{294 \text{ rad/sec}^2}}$$

$$\text{Angular Movement of Propeller} = \theta$$

$$\theta = \tfrac{1}{2}\alpha t^2 = \tfrac{1}{2}(294)\left(\tfrac{1}{2}\right)^2 = 36.8 \text{ rad.}$$

$$= \underline{\underline{5.87 \text{ rev.}}}$$

DYNAMICS AND VIBRATIONS 7

The torsion pendulum shown in the figure, a thin rectangular aluminum plate suspended by massless vertical wire and rigid support rods, is used to measure moment of inertia of any given object. The plate has the dimensions 18" x 1" x 24"; the pendulum's period of free oscillation is 1.1 second. When an object was placed on the plate with its center of mass coincident with the plate's center, the observed period of oscillation changed to 1.2 seconds.

What is the moment of inertia of this object?

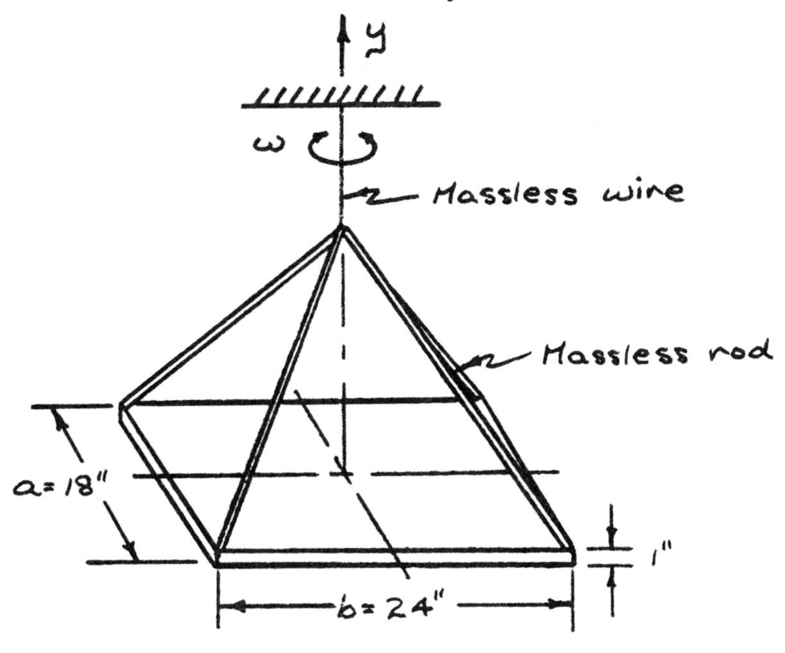

Solution

$$\text{Period} = \tau = 2\pi\sqrt{\frac{I_y}{K}}$$

where K = torsional spring const. of wire

$$I_y = (I_y)_{PL} = \frac{1}{12} m_{PL}(a^2 + b^2) \qquad \gamma_{Al.} = .1 \text{ lb/in}^3$$

$$m_{PL} = \frac{18(24)(1)(0.1)}{g} = 1.34 \text{ slugs}$$

$$\tau^2 = \frac{4\pi^2 (I_y)_{PL}}{K}$$

$$(I_y)_{pt} = \frac{1.34}{12}(1.5^2 + 2^2) = \underline{\underline{0.697 \text{ slug ft}^2}}$$

$$K = \frac{4\pi^2(0.697)}{(1.1)^2} = 22.8 \text{ lb}\cdot\text{ft./rad.}$$

$$\text{New Period} = \tau_{new} = 2\pi\sqrt{\frac{(I_y)_{pt} + (I_y)_{object}}{K}}$$

$$= 2\pi\sqrt{\frac{I_y}{K}}$$

$$K = \frac{4\pi^2(I_y)}{(1.2)^2} = 22.8$$

$$I_y = \frac{(1.2)^2(22.8)}{4\pi^2} = \underline{\underline{0.83 \text{ slug ft}^2}}$$

$$(I_y)_{object} = 0.83 - 0.7 = \underline{\underline{0.13 \text{ slug ft}^2}} \leftarrow$$

DYNAMICS AND VIBRATIONS 8

The figure represents resilient packing of a rigid instrument weighing 15 lbs. The box is dropped 5 ft from rest and strikes a hard surface. Assume that, at the instant of impact, there is no relative velocity between the instrument W and the box, and the force in the packing is zero.

(a) Find the spring constant k, in pounds per inch, required to limit the motion of instrument W to 2 inches relative to the box.

(b) Find the maximum deceleration of the instrument in g's.

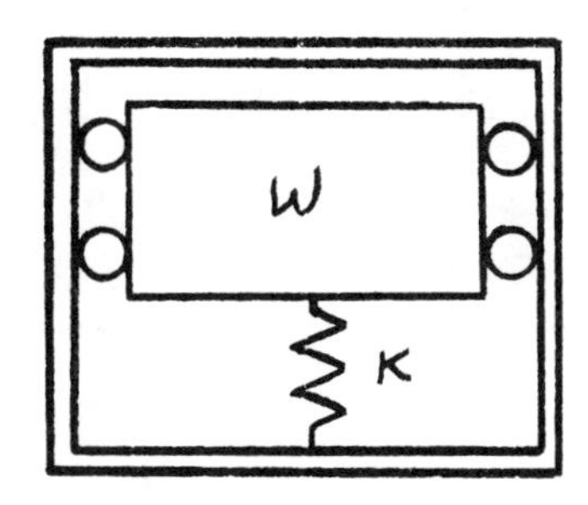

a) Assuming elastic behavior and no friction between Instrument and box.

$$\Delta U = \Delta K.E. + \Delta V_g + \Delta V_e$$

Between position shown and position where spring experiences max. deformation.

$$0 = 0 - 15\left[60 + y\right] + \tfrac{1}{2} K y^2$$

y = max. deformation in spring = 2 in.

$$0 = 0 - 15(62) + \tfrac{1}{2} K(2)^2$$

$$K = 930/2 = \underline{\underline{465 \text{ lb/in}}} \longleftarrow$$

b) F_s = spring force $= Ky = 465(2) = 930$ lb.

$$\Sigma F_y = ma_y = 930 - 15 = \frac{15}{g} a_y$$

$$a_y = \frac{915 g}{15} = \underline{\underline{61 g}} \longleftarrow$$

DYNAMICS AND VIBRATIONS 9

A large hangar door is required to be stopped at the end of its travel by a constant-force snubber. The door weighs 10,000 lbs and its speed is 1 ft per second when it strikes the snubber.

What constant force is required by the snubber to stop the door within a distance of 1 ft?

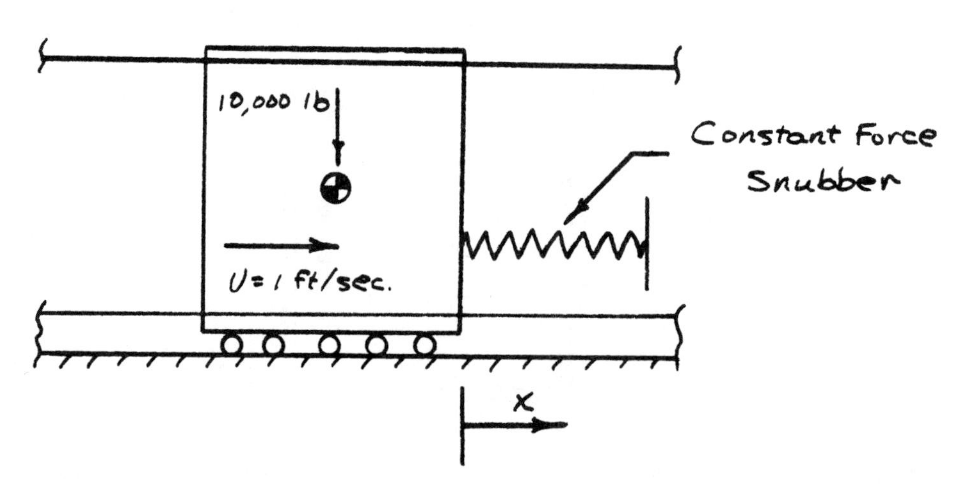

Assuming no friction and using Work Energy Principle

$$\Delta U = \Delta KE.$$

U - work

KE. - Kinetic Energy

$$-(\text{Snubber Force})(1') = -\frac{1}{2}\frac{10{,}000}{g}(1)^2 = -\frac{5000}{g}$$

$$\text{Snubber force} = \underline{\underline{155\ \text{lb}}}$$

DYNAMICS AND VIBRATIONS 10

A centrifuge in the form of an axially symmetric dish (cross-section illustrated) used for dynamic testing of transistors, etc., is to be accelerated to a speed of 12,000 rpm in 30 seconds. Assuming a constant acceleration over the entire speed range, determine the horsepower requirement for the drive motor.

What is the transistor acceleration?

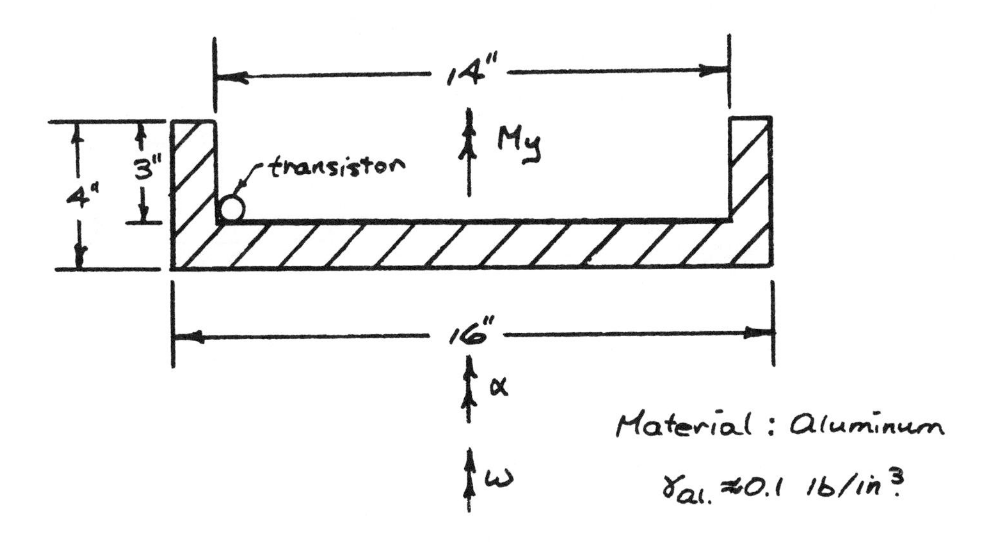

Horsepower Output of drive motor $= \dfrac{(M_y)(\omega_{max})}{550}$

ω_{max} – max. angular velocity $= \dfrac{12000}{60}(2\pi)$

$\omega_{max} = 1256$ rad/sec.

Neglecting friction, mass of transistor

M_y = motor torque required to acc. centrifuge

$= (I_y)_{centrifuge}(\alpha)$

$\alpha = \omega/t = 1256/30 = \underline{\underline{41.9 \text{ rad/sec}^2}}$

Mass Moment of Inertia of Centrifuge

$$I_y = \frac{1}{2}\frac{W_1 R_o^2}{g} - \frac{1}{2}\frac{W_2}{g} R_i^2$$

W_1 = wt. of 16" Dia. alum. disk, 4" thick

$= \pi (8)^2 (4)(0.1) = 81$ lb.

W_2 = wt. of 14" Dia. alum. disk 3" thick

$= \pi (7)^2 (3)(0.1) = 46.2$ lb.

$$I_y = \frac{1}{2g}\left[81\left(\frac{8}{12}\right)^2 - 46.2\left(\frac{7}{12}\right)^2\right] = \underline{\underline{0.314 \text{ slug ft}^2}}$$

Horsepower output of drive motor $= \dfrac{(0.314)(41.9)(1256)}{550}$

$= \underline{\underline{30 \text{ hp}}}$ ←

Transistor acceleration when $\omega = 1256$ rad/sec.

$\alpha = 41.9$ rad/sec^2

$$a_n \approx R_i \omega^2 \approx \frac{7(1256)^2}{12} = \underline{\underline{920{,}000 \text{ ft/sec}^2}}$$

$$a_t \approx R_i \alpha \approx \frac{7(41.9)}{12} = \underline{\underline{24.4 \text{ ft/sec}^2}}$$

Transistor Acceleration $\approx 920{,}000$ ft/sec^2

$= \underline{\underline{28{,}600 \text{ g}}}$ ←

DYNAMICS AND VIBRATIONS 11

The uniform bar ABC weighs 5 lbs, and a concentrated mass of 3 lbs is attached at C.

Find the natural frequency in cycles per second for angular motion about A. The static deflection of the spring under the conditions shown is 0.55 inches.

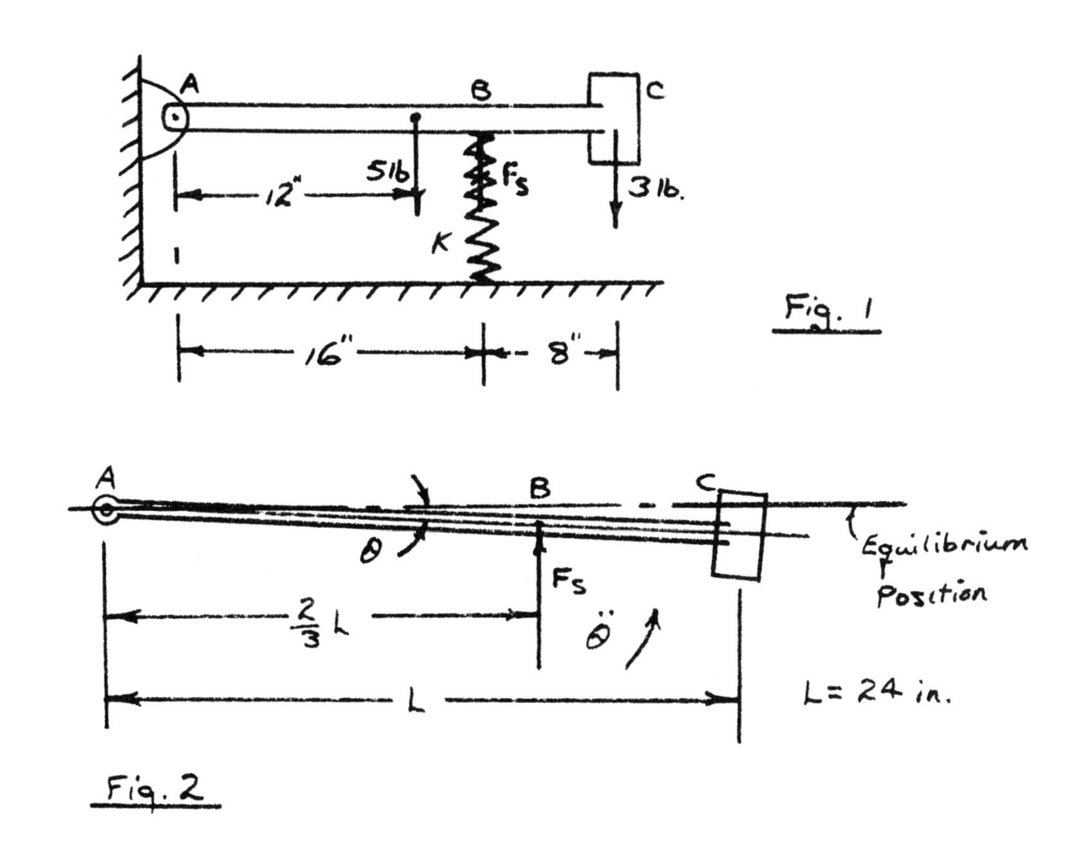

Determination of K

From Fig. 1 :

$$16\,F_s = 5(12) + 3(24) \qquad F_s = 8.25\ \text{lb}$$

$$F_s = Ky \qquad K = \frac{8.25}{0.55} = 15\ \text{lb/in.}$$

From Fig. 2 :

$$\Sigma M_A = I_A \ddot{\theta} = F_S(2/3 L) = -K(2/3 L\theta)(2/3 L)$$

$$I_A \ddot{\theta} + 4/9\, KL^2 \theta = 0$$

$$\ddot{\theta} + \left(4/9\, \frac{KL^2}{I_A}\right)\theta = 0$$

Let $p^2 = 4/9 \dfrac{KL^2}{I_A}$ p = circular freq. (rad/sec.)

natural freq. $f = \dfrac{p}{2\pi} = \dfrac{2}{3}\dfrac{L}{2\pi}\sqrt{K/I_A}$

$$f = \frac{L}{3\pi}\sqrt{K/I_A}$$

$$I_A = \frac{W_{bar} L^2}{3g} + \frac{W_c L^2}{g}$$

$$= \underbrace{\frac{5(2)^2}{3(32.2)}}_{0.207} + \underbrace{\frac{3(2)^2}{32.2}}_{0.372} = \underline{\underline{0.58 \text{ slug ft}^2}}$$

$$f = \frac{2}{3\pi}\sqrt{\frac{15(12)}{0.58}} = \underline{\underline{3.7 \text{ cps}}}$$

(for small values of θ)

DYNAMICS AND VIBRATIONS 12

A machine weighing 25 lbs is placed on rubber mounts which deflect 0.08 inches under its weight. Assume that the mounts have 10% of critical damping.

(a) Find the damped natural frequency for vertical motion, in cycles per second.

(b) When the machine speed is run slowly through resonance, the maximum double amplitude of vibration is observed to be 0.3 inches. Find the amplitude of the disturbing force in pounds.

25 lb

a) $K = \frac{25}{0.08} = 312$ lb/in

$$\zeta = 0.1 = \frac{\text{coeff. of damping}}{\text{coeff. of critical damping}} = \frac{C}{C_c}$$

Damped Circular freq. $q = p\sqrt{1 - \left(\frac{C}{C_c}\right)^2}$

where p = undamped circular freq. $= \sqrt{\frac{K}{m}}$

$$= \sqrt{\frac{312(386)}{25}} = 69.6 \text{ rad/sec.}$$

$$q = 69.6\sqrt{1 - 0.01} = 69.5 \text{ rad/sec}$$

Damped natural freq. $= q/2\pi = \underline{\underline{11.1 \text{ cps}}}$ ←

b) Amplitude of vibration for $q/p = 1$ is

$$y_0 = \frac{F_0/K}{\sqrt{\left[1-(q/p)^2\right]^2 + \left[2\zeta\, q/p\right]^2}}$$

where F_0 is the amplitude of disturbing force.

$$0.15 = \frac{F_0/K}{\sqrt{(1-1)^2 + (0.2)^2}}$$

$$F_0 = K(0.15)(0.2) = 0.03K = 0.03(312) = \underline{9.36 \text{ lb}}$$ ←

DYNAMICS AND VIBRATIONS 13

A mass traveling at 15.0 miles/sec is in a circular orbit about a second mass.

If the radius of this orbit is 70.0 x 10^6 miles, what specific potential energy change is required to change this orbit to another circular orbit with a 10 x 10^7 mile radius?

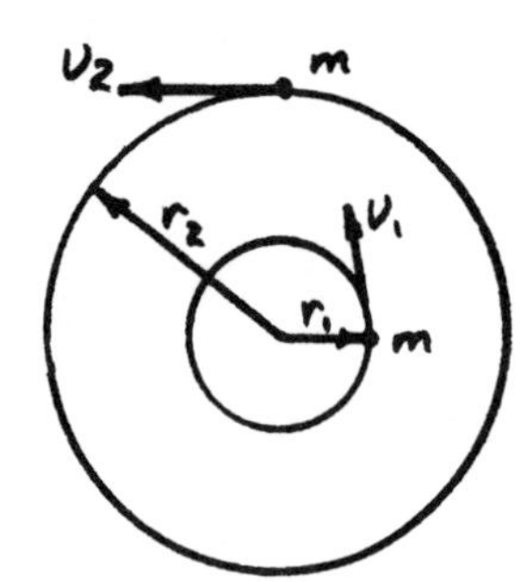

v_1 = 15 miles/sec.

r_1 = 70 x 10^6 miles

Potential Energy of mass m with respect to mass **M**

$$(U_g)_1 = -\frac{GMm}{r_1}$$

Force between m + M = F = ma = $\frac{GMm}{r_1^2}$

$$\frac{mv_1^2}{r_1} = \frac{GMm}{r_1^2} \quad ; \quad M = \frac{v_1^2 r_1}{G} = \frac{v_2^2 r_2}{G}$$

$$v_2^2 = v_1^2 \frac{r_1}{r_2}$$

$$(U_g)_1 = -\frac{Gm}{r_1}\left[\frac{v_1^2 r_1}{G}\right] = -mv_1^2$$

Specific Potential Energy Change $= -\frac{m}{m}\left[v_2^2 - v_1^2\right]$

$$= v_1^2 - v_1^2\frac{r_1}{r_2} = v_1^2\left[1 - \frac{r_1}{r_2}\right]$$

$$= (15)^2(5280)^2\left[1 - \frac{70(10)^6}{100(10)^6}\right]$$

$$= \underline{\underline{1.87 \times 10^9 \text{ ft}\cdot\text{lb/slug}}}$$

DYNAMICS AND VIBRATIONS 14

A spacecraft is spin-stabilized at a speed of 150 rpm. It is desired to reduce this spin speed, after boost, to 10 rpm. This is to be accomplished by the radial displacement of two masses as indicated in the accompanying diagram. The moment of inertia of the spacecraft about its spin axis is 10 slug-ft^2, excluding the de-spin masses. The mass of each of the de-spin masses is 0.5 slug.

Determine the required radial displacement R.

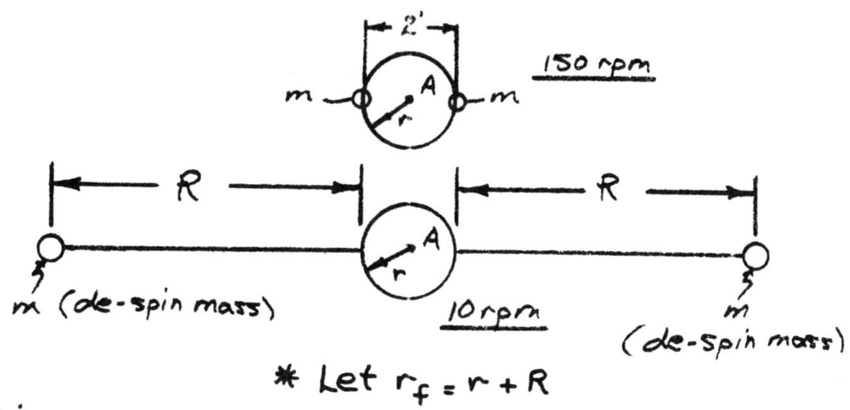

Solution

Initial Angular Momentum $(H_A)_i = I_A \omega_i + 2(m r \omega_i) r$

$$\omega_i = \frac{150}{60}(2\pi) = 5\pi \text{ rad/sec.}$$

$$H_A = 10(5\pi) + 2(0.5)(1)(5\pi)(1)$$
$$= 55\pi \ \frac{\text{slug ft}^2}{\text{sec.}}$$

$(H_A)_f$ = Final angular momentum = $(H_A)_i$ since moment about A i.e. $\Sigma M_A = 0$

$$\omega_f = (10/60)(2\pi) = \pi/3 \text{ rad/sec}$$

$$(H_A)_f = I_A \omega_f + 2(m r_f \omega_f)(r_f) = 10(\pi/3) + \pi/3\, r_f^2 = 55\pi$$

$$1/3\, r_f^2 = 55 - 3.33 = 51.67 \ : \ r_f = 12.45 \text{ ft.}$$

$$R = 12.45 - 1 = \underline{\underline{11.45 \text{ ft}}} \longleftarrow$$

DYNAMICS AND VIBRATIONS 15

A piece of putty weighing 25 lbs and traveling in a straight line at 15 mph is struck by another piece of putty weighing 35 lbs and traveling at 20 mph in a straight line in the opposite direction. How much kinetic energy is lost in the impact?

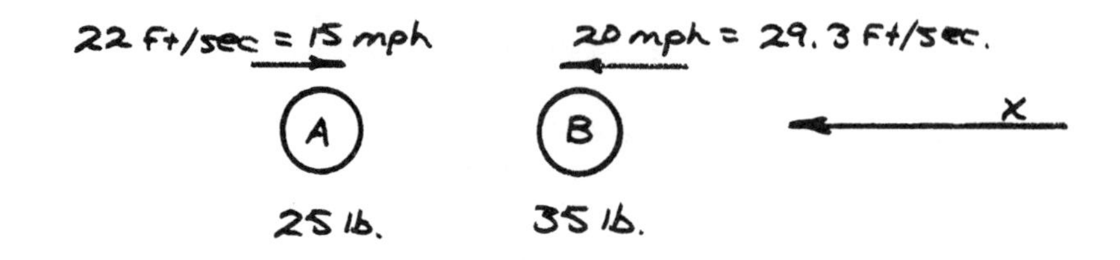

Solution

Assuming Plastic Impact

$$(v_A)_{final} = (v_B)_{final} = v$$

Initial linear Momentum of system

$$G_i = \frac{35(29.3)}{g} - \frac{25(22)}{g} = \frac{475}{g} \text{ lb. sec.}$$

$G_{final} = G_i$ since $\Sigma F_x = 0$

$$G_f = \frac{(35+25)}{g} v = \frac{475}{g} \quad ; \quad v = \underline{\underline{7.92 \text{ ft/sec}}}$$

$$(K.E.)_i = \tfrac{1}{2} m_A v_A^2 + \tfrac{1}{2} m_B v_B^2$$

$$= \tfrac{1}{2} \tfrac{25}{g} (22)^2 + \tfrac{1}{2} \tfrac{35}{g} (29.3)^2 = 188 + 466 = \underline{\underline{654 \text{ ft} \cdot \text{lb}}}$$

$$(K.E.)_f = \tfrac{1}{2} (m_A + m_B) v^2 = \frac{60}{2g} (7.92)^2 = \underline{\underline{58.5 \text{ ft} \cdot \text{lb}}}$$

$$\Delta K.E. = 654 - 58.5 \approx \underline{\underline{596 \text{ ft} \cdot \text{lb}}} \longleftarrow$$

DYNAMICS AND VIBRATIONS 16

Ship A of 4,000 tons weight is towing Ship B of 2,000 tons weight by means of a cable. At the instant the cable becomes taut, the velocity of Ship A is 12 ft/sec and that of Ship B is 8 ft/sec. The cable is constructed so that a pull of 2,000 lbs will elongate it 0.0024 in per foot of length and its safe load is 200,000 lbs. What is the least permissible length of the cable?

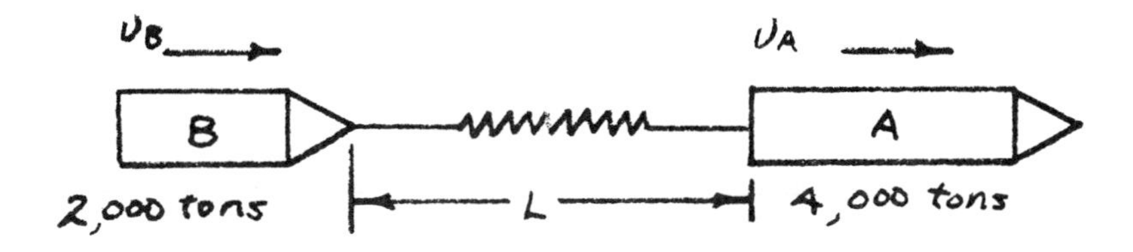

Solution

G_i = Initial Momentum of System

$$= m_A v_A + m_B v_B = \frac{4000(2000)(12)}{g} + \frac{2000(2000)8}{g}$$

$$= \frac{128 \times 10^6}{g} \text{ lb. sec.}$$

From Law of conservation of Linear Momentum

$$G_F = m_A v + m_B v = \frac{(4000+2000)(2000)}{g} v = \frac{128(10)^6}{g}$$

v = 10.7 ft/sec = vel. of each boat

From Law of conservation of Mechanical Energy

$$\tfrac{1}{2} m_A v_A^2 + \tfrac{1}{2} m_B v_B^2 = \tfrac{1}{2}(m_A + m_B) v^2 + \text{Strain Energy in cable}$$

- continued -

$$W_A v_A^2 + W_B v_B^2 = (W_A + W_B)v^2 + 2g(U_e)$$

$$8(10^6)(12)^2 + 4(10^6)(8)^2 = 12(10^6)(10.7)^2 + 2g\,U_e$$

$$1150(10^6) + 256(10^6) = 1380(10^6) + 2g\,U_e$$

$$U_e = 4.05(10)^5 \text{ ft}\cdot\text{lb} = \frac{P}{2}\,\Delta$$

Where : P = safe cable tension = 200,000 lb.

Δ = total elongation

$$U_e = 100{,}000\,\Delta = 100{,}000(\epsilon)(L) = 4.05 \times 10^5 \text{ ft}\cdot\text{lb.}$$

For cable tension = 2000 lb $\quad \epsilon = \dfrac{0.0024}{12}$

For cable tension = 200,000 lb $\quad \epsilon = \dfrac{0.24}{12} = \underline{\underline{0.02}}$

$$L = \frac{4.05 \times 10^5}{100{,}000(0.02)} = \underline{\underline{202.5 \text{ ft}}} \longleftarrow$$

DYNAMICS AND VIBRATIONS 17

A weight W and a coil spring with modulus K are attached to the end of a wooden cantilever beam of length L as shown.

W = 100 lbs
E = 2,000,000 psi for beam
I = 0.1 in^4 for beam
K = 100 lb/in

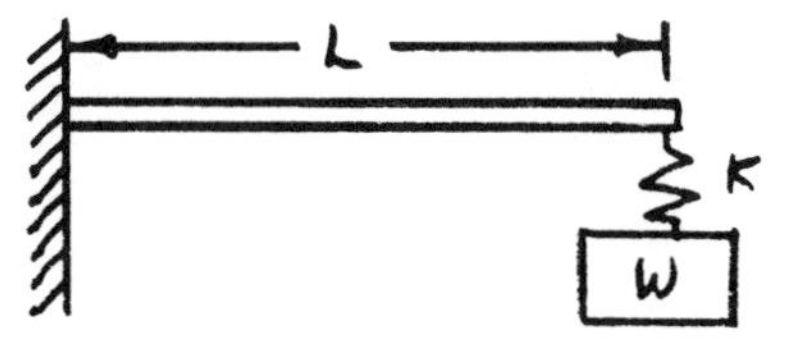

Determine the length L of the beam so that the natural frequency of the system will be 2 cycles per second. Neglect the mass of the spring and beam.

Solution

Weight W is supported by two springs in series, namely, the beam and the coil spring

$$K_{c.s.} = 100 \text{ lb/in} \; ; \; K_{beam} = \frac{3EI}{L^3} = \frac{3(2)(10^6)(\frac{1}{10})}{L^3}$$

$$= \frac{6(10^5)}{L^3} \text{ lb/in.}$$

Resultant Spring Constant

$$K = \frac{K_{c.s.}\, K_{Beam}}{K_{c.s.} + K_{Beam}} = \frac{6(10^7)}{L^3}\left[\frac{L^3}{100L^3 + 6(10^5)}\right]$$

$$= \frac{6(10^7)}{100L^3 + 6(10)^5}$$

$$\text{Natural Freq. } f = \frac{1}{2\pi}\left(\frac{gK}{W}\right)^{1/2}$$

$$f^2 = \frac{gK}{4\pi^2 W} = \frac{g}{4\pi^2}\,\frac{6(10^7)}{100}\left[\frac{1}{100L^3 + 6(10)^5}\right]$$

$$(2)^2 = 58.8(10^5)\left[\frac{1}{100L^3 + 6(10)^5}\right]$$

$$400L^3 + 24(10)^5 = 58.8(10)^5$$

$$400L^3 = 34.8(10)^5 = 3480(10^3)$$

$$L^3 = 8.7(10^3)$$

$$L = 2.06(10) = \underline{\underline{20.6 \text{ in}}}$$

Length of beam $\approx \underline{\underline{20\ 5/8 \text{ in}}}$ ←

4

Machine Design

EUGENE J. FISHER

This chapter is arranged to present typical Professional Engineering examination problems. Some will be preceded with an explanation, while in other cases only the problem and solution will appear. If there are solutions that are not clear, an excellent text on this subject is Mechanical Engineering Design by Shigley (published by McGraw-Hill). The references in this chapter are to the Fourth Edition.

The order of the chapter will be by problem type: Bolted Joints, Bearing and Shaft Loads, Interference Fits, Size of Shafts and Critical Speeds, Mechanisms, Gear Trains, Miscellaneous Problems.

Bolted joints that are loaded eccentrically may be analyzed by dividing the load per bolt into two components, one the direct shear and the second the moment load or secondary shear. The assumptions for bolted joints are that (a) none of the load is taken in friction between the two members, and (b) that the moment load on a bolt is directly proportional to its distance from the centroid of the bolt pattern.

To solve an eccentrically loaded bolted problem one may divide the load by the number of bolts to obtain the direct shear load per bolt, then determine the proportionality factor for obtaining the secondary shear force per bolt. This proportionality factor may be determined by solving the equations relating to the force per bolt and the distance to the centroid of the bolt pattern. Thus, $F_b = rk$, where F_b is the force on the bolt, r is the distance from the bolt centroid to this particular bolt and k is the unknown proportionality constant. Another equation may relate the total secondary shear load to each bolt:

$$Pe = F_1r_1 + F_2r_2 + F_3r_3 + F_4r_4 \;\cdots\cdots\; F_ur_u$$

In this equation the total eccentric load is P and the eccentricity from the center line of the bolt pattern is the distance e.

Substituting the first equation into the second equation for the force per bolt:

$$Pe = k(r_1^2 + r_2^2 + r_3^2 + r_4^2 \ldots\ldots r_u^2)$$

Thus, the only unknown is k.

MACHINE DESIGN 1

A removable bracket must be attached to the frame of a machine tool as a support. The need for interchangeability dictated bolting. Clearances and accessibility to the bolts required a pattern as shown. Mild steel bolts (shear strength 30,000 psi, bearing strength 120,000 psi) are to be used.

For a safety factor of three, what diameter standard bolt should be used?

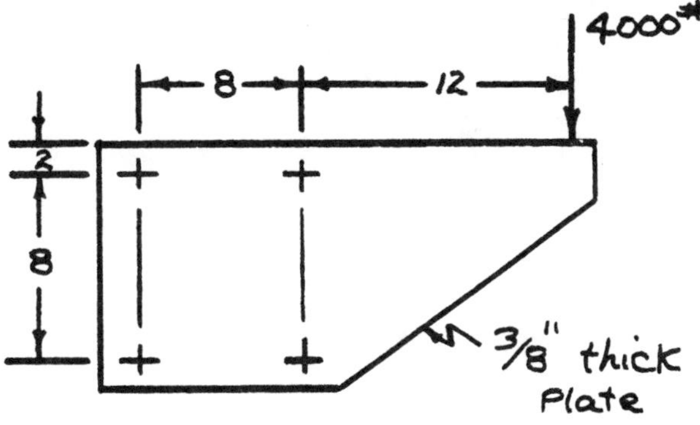

Solution

Determine the centroid of the bolt pattern to be 6" below the top edge and 4" to the right of the top left bolt.

Primary Shear

$$\frac{\text{Load}}{\#\text{ of bolts}} = \frac{4000}{4} = 1000\#/\text{bolt}$$

Secondary Shear

$$P_e = K(r_1^2 + r_2^2 + r_3^2 + r_4^2) = 4000(12+4)$$

$$= K(4)(4^2+4^2)$$

↑ # of bolts (the 4)

$$K = 500\#/\text{in.}$$

$$F_{secondary} = 500\sqrt{32} = \underline{\underline{2840\#/\text{bolt}}}$$

– continued –

These two shear loads will add vectorially, and by sketching these reactions one should see that the most highly loaded bolt is at position c or d.

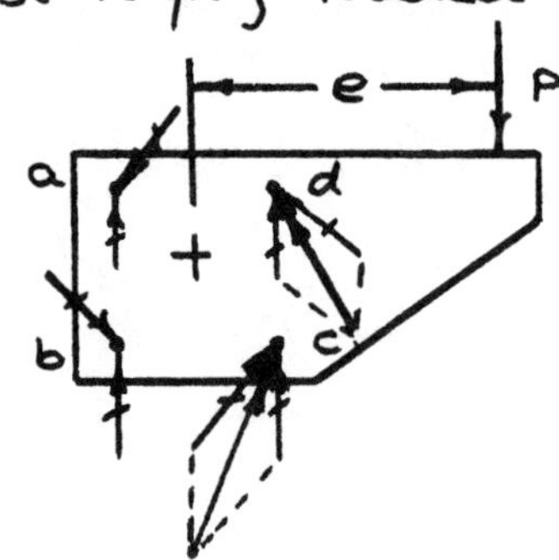

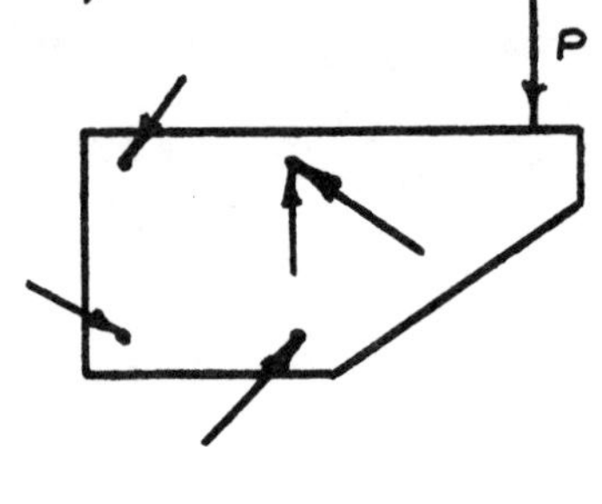

Breaking the secondary shear into its two components

$$F_{Horiz.} = F_{vert.} = 0.707(2840) = 2{,}010^{\#}$$

Thus,

$$F_{Total} = \sqrt{(2{,}010)^2 + (2010 + 1000)^2} = 3620^{\#}$$

Shearing Stress $\tau = P/A$

$$A = P/\tau = \frac{3620}{(30{,}000/3)}$$ ← factor of Safety (3)

$$\underline{\underline{A = 0.362 \text{ in}^2/\text{Bolt}}}$$

<u>Bolt Size (net)</u> - (Ref. P 361 Shigley)

3/4 Dia 0.302 in^2 for coarse series UNC

7/8 Dia 0.419 in^2

Therefore select 7/8" bolts

MACHINE DESIGN 2

The combustion chamber of a small attitude control rocket engine is attached to the injector head with four 8-32 bolts as shown. During lift off the engine is subjected to peak accelerations of 200 g's combined with high transverse air loads.

What is the minimum preload for the bolts to prevent fatigue stresses in the bolts and separation of the chamber from injector head?

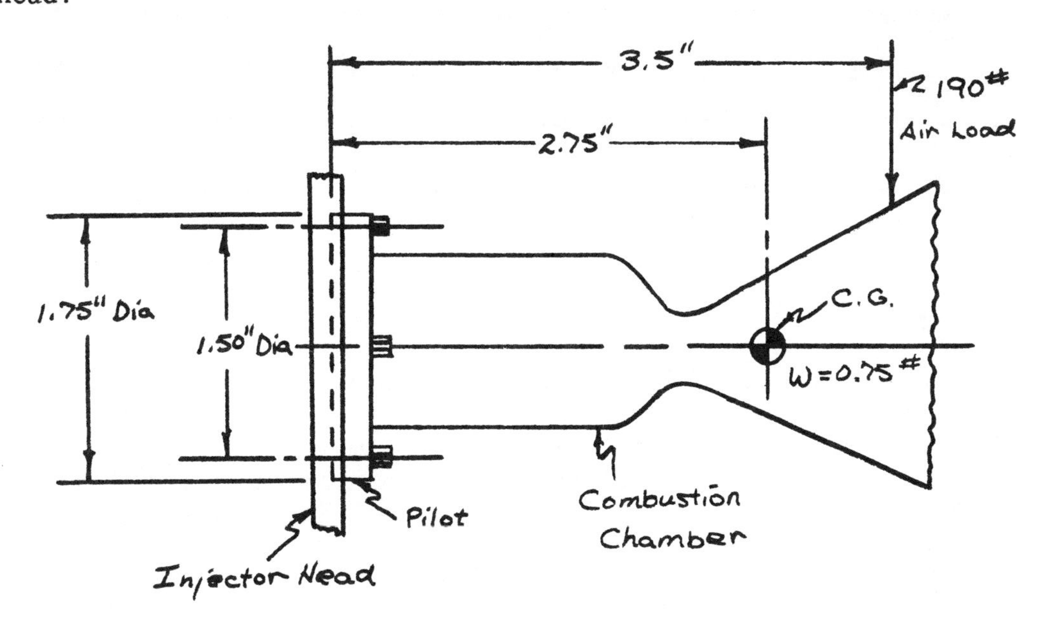

Solution

Assume that the pilot will support the primary shear, so that tension is the only load that the bolts will support. Assume also that the injector head is rigid so that the attitude control rocket will tend to rotate about the lower portion of the pilot. The load on the upper bolt would be:

$$P_1 e_1 + P_2 e_2 = K(r_1^2 + 2r_2^2 + r_3^2)$$

$$(190)(3.5) + [(0.75^{\#})(200)]\,2.75 = K(0.125^2 + 2(0.75 + .125)^2 + 1.625^2)$$

$$K = 257 \text{ \#/in}$$

$$F_{\text{Top bolt}} = K r_{top} = 257(1.625) = \underline{\underline{418 \text{ lb}}}$$

This value becomes equal to the minimum preload in order to prevent separation of the chamber from the injector head. This value would be lower if the materials and the geometry of the rocket base and the injector head were given. (see next problem)

MACHINE DESIGN 3

The cylinder head on a diesel engine is mounted with six equally-spaced studs as shown. These studs have upset threads with an effective shank diameter of 3/4 inch. The maximum gas pressure is 1500 psi. Metal-to-metal contact is maintained as the joint between cylinder and head. The initial tightening of the studs is to a torque of 200 lb-in.

(a) What is the maximum stress in the studs during operation?

(b) What is the effect of a gasket between the metal surfaces?

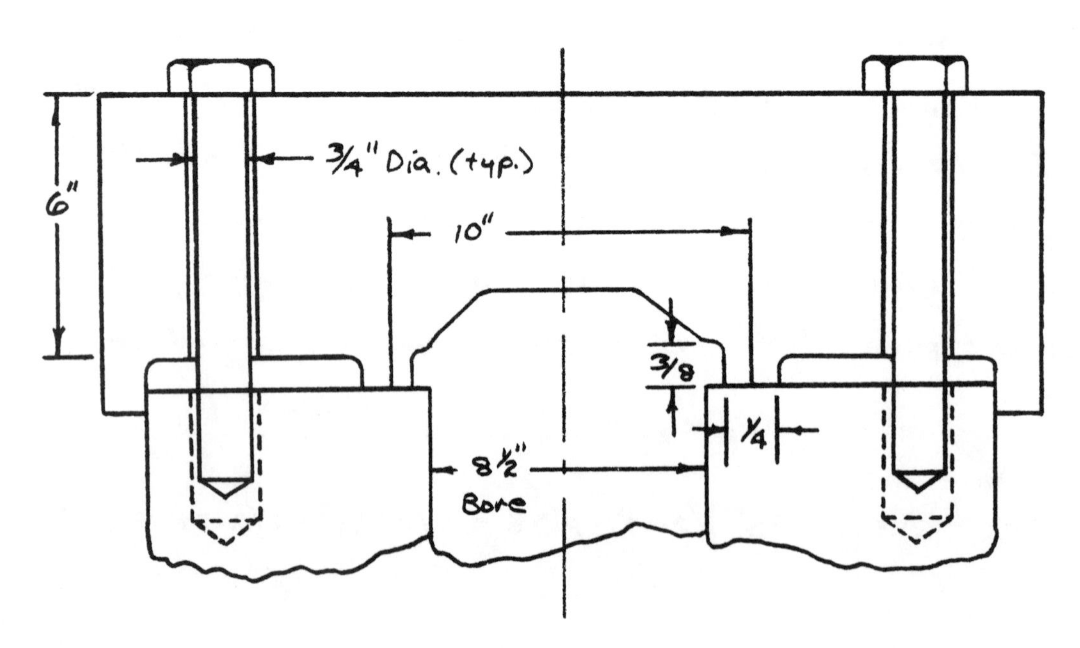

Solution

* Initial Force on bolts

$$\text{Torque} = .2\,F_{\text{initial}}\,d_m \quad (\text{Ref.- P. 378 Shigley})$$

$$F_i = \frac{T}{0.2(3/4)} = \frac{200 \text{ in}\cdot\text{lb}}{0.2(.75)} = 1{,}333 \text{ lb.}$$

(bolt preload should be equal to 90% of bolt proof load)

* Deflection of Parts (Ref. - P. 370 Shigley)

$$\delta = F\ell/AE \qquad \text{If } \delta = 1'', \quad F = k$$

$$\text{Thus } k = AE/\ell$$

$$k_b = AE/\ell$$

$$k_b = \frac{\left(\frac{\pi d^2}{4}\right)(30 \times 10^6 \text{ psi})(6)}{6\ 3/8} = 12.5 \times 10^6 \text{ lb/in.}$$

$$k_m = AE/\ell = \frac{\pi/4\left[(10\tfrac{1}{4})^2 - (9.75)^2\right] 12 \times 10^6 \text{ psi.}}{3/8}$$

E_{ASTM} Grade 25 cast iron (Ref. - P. 828 Shigley)

$k_m = 252 \times 10^6$ lb/in (considering only 3/8 x 1/4 ring because the rest of the head would be much stiffer since there is more area.)

$$c = \frac{k_b}{k_b + k_m} = \frac{12.5 \times 10^6}{(12.5 + 252)10^6} = 0.047$$

c may vary from 0 to 0.5.

-continued-

$$F_{\substack{Total\ per \\ bolt}} = CF_t + F_i = 0.047\left[\frac{1500\ psi\left(\frac{\pi}{4}(10)^2\right)}{6\ Bolts}\right] + 1333\ lb$$

$$= 930 + 1333 = \underline{\underline{2263\ lb/bolt}}\ \text{(During Operation)}$$

* Max. Stress

$$\sigma = P/A = 2263/0.334 = \underline{\underline{6,775\ psi}}$$

(max. stress on bolt during operation)

<u>* Part B</u>

Gasket is used to seal surfaces.

MACHINE DESIGN 4

The force on a pivot bracket casting bolted to the frame of a back hoe can be applied at any point on the 180° arc as shown. The design is such that the bolts "B" are not subjected to shear.

What is the maximum force on each of those bolts?

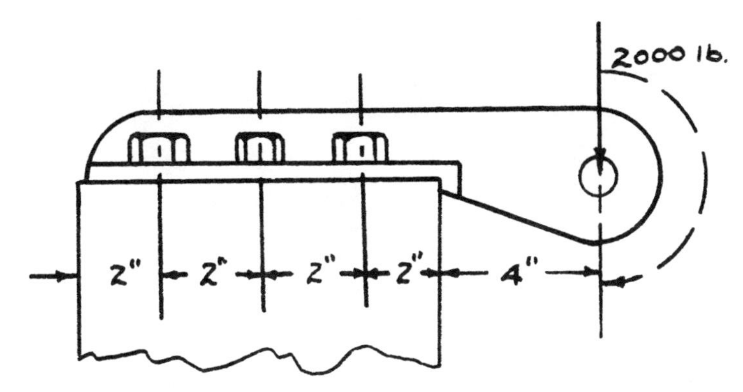

Solution

Assume that the structure is rigid and that it pivots about the point at the right or left side of the bracket. The longer lever arm will occur when the force is in the vertically upward direction, the shear will be absorbed by the pilot.

$$Pe = K(r_1^2 + r_2^2 + r_3^2) = 2000(12)$$

$$= K(2^2 + 4^2 + 6^2)$$

$$K = \frac{24000}{56} = 430 \text{ lb/in}$$

Thus, the max. force of each bolt is :

$F_1 = 430(2) = 860$ lb.

$F_2 = 430(4) = 1720$ lb.

$F_3 = 430(6) = 2580$ lb.

MACHINE DESIGN 5

A hydraulic actuator on a skip loader is mounted to the frame by means of a rigid cast steel bracket. The actuator under one condition exerts a force of 1250 lb as shown.

What are the loads on the cap screws?

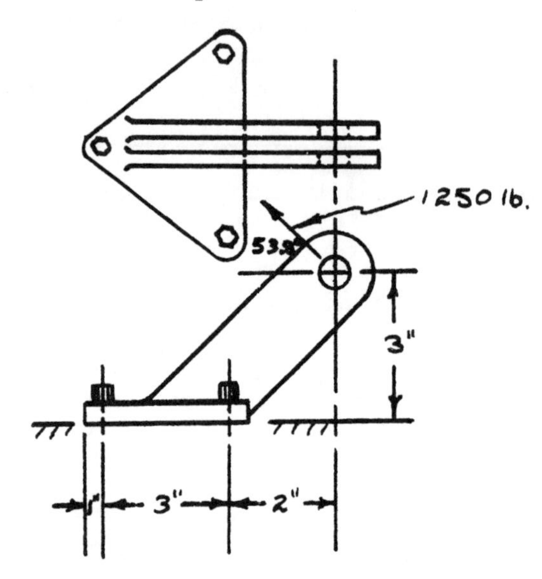

Solution

1) Determine the Horizontal and Vertical components of the force.

$$F_{Horiz.} = 1250(\cos 53.8°) = 750^{\#}$$

$$F_{vert.} = 1250(\sin 53.8°) = 1000^{\#}$$

2) take moment about left hand edge, assuming both bracket and mounting pad to be rigid.

$$F_{\text{right bolt}} = \frac{[1000(6) + 750(3)]}{4} = \frac{8250}{4}$$

$$= 2{,}060 \text{ lb/right two bolts}$$

$$F/\text{bolt} = 2{,}060/2 = 1030 \text{ lb/bolt}$$

(tensile load)

$$F_{Shear} = \frac{750}{3} = 250 \text{ lb/bolt}$$

The loads on these bolts would have been larger if one assumed the bracket to yield across the single left bolt hole. These are only the External loads, The other loads may be initial loads, residual and torsional.

MACHINE DESIGN 6

Under certain conditions a wheel and axle is subjected to the loading shown.

REQUIRED: (a) What are the loads acting on the axle at A-A?

(b) What maximum direct stresses are developed at that section?

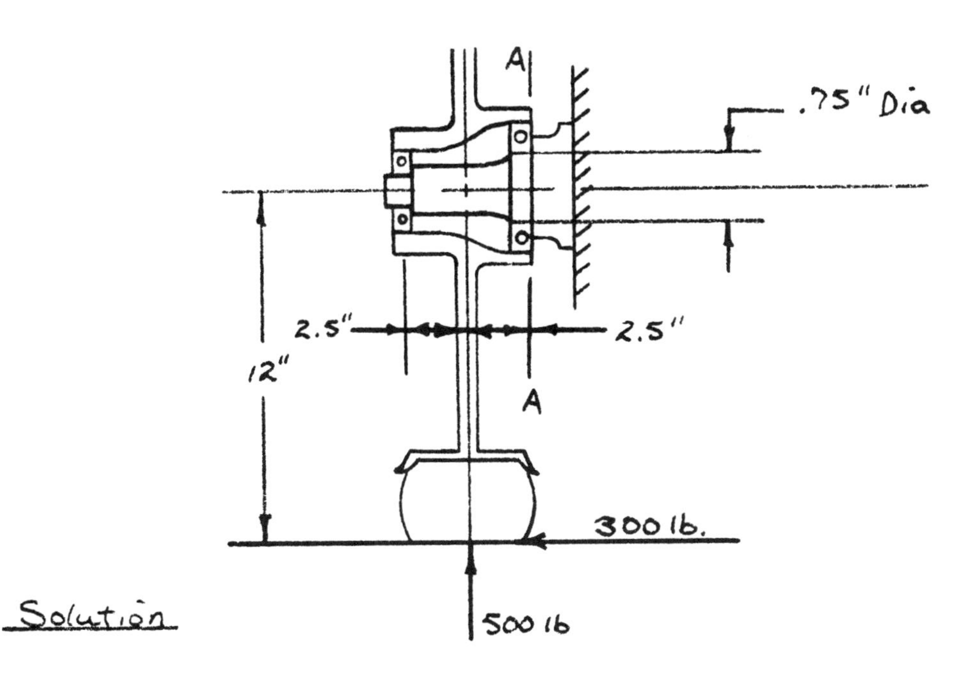

Solution

a) Axial tensile load - 300 lb

Direct Shear load - 500 lb

b) Bending Moment $= 500(2.5) + 300(12)$

$= 1250 + 3600 = 4850$ IN-lb

$$\sigma_{bending} = \frac{Mc}{I} = \frac{32M}{\pi d^3} = \frac{32(4850)}{\pi(.75)^3} = 117{,}000 \text{ psi}$$

$$\sigma = P/A = \frac{300}{\frac{\pi(.75)^2}{4}} = \frac{1200}{\pi(.75)^2} = 679 \text{ psi}$$

<u>Max. Stress at Section A-A</u>

$$\sigma = 117{,}000 + 679 = \underline{\underline{117{,}679 \text{ psi}}}$$

MACHINE DESIGN 7

A spiral bevel gear is mounted as shown. The mean working diameter Dm, tangential force TF, thrust force T_{tf}, and separating force S_{tf} are given. The angular contact bearings are a light press fit on the gear shaft and just slip into the housing. A spacer is matched to the bearings and the housing shoulder to provide a desired preload.

(a) If the preload was zero what are the bearing loads?

(b) What effect does warm-up during operation have on the preload?

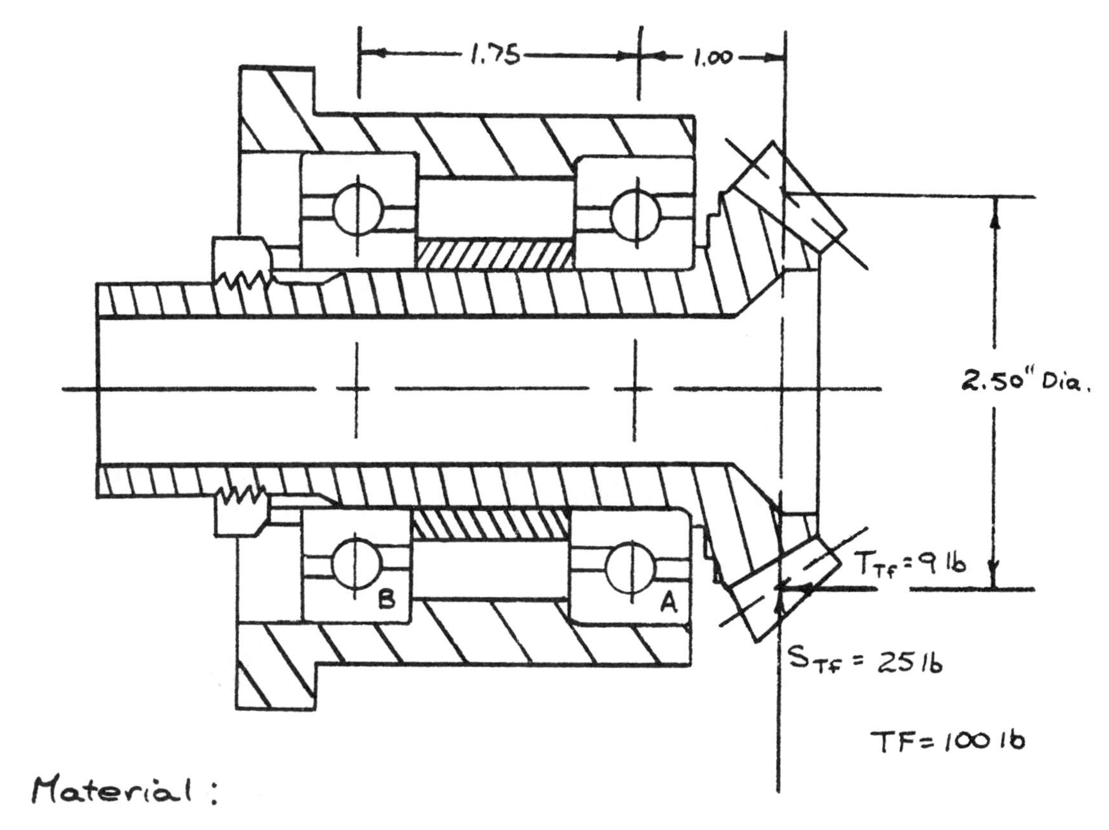

Material:

Gear + Spacer - Steel

Housing - Aluminum

Solution

Tangential tooth load

$$\text{Moment } \circlearrowright \text{ bearing B} = 0 = 100(2.75) - R_A\, 1.75$$

$\therefore \; R_{Z_A} = 157\text{ lb} \qquad R_{Z_B} = 57\text{ lb}$

Separating tooth load

$$R_{y_A} = \frac{25(2.75)}{1.75} = 39.3\text{ lb.}$$

$$R_{y_B} = 14.3\text{ lb.}$$

Thrust tooth load

$$R_{y_A} = R_{y_B} = \frac{9(1.25)}{1.75} = 6.4\text{ lb.}$$

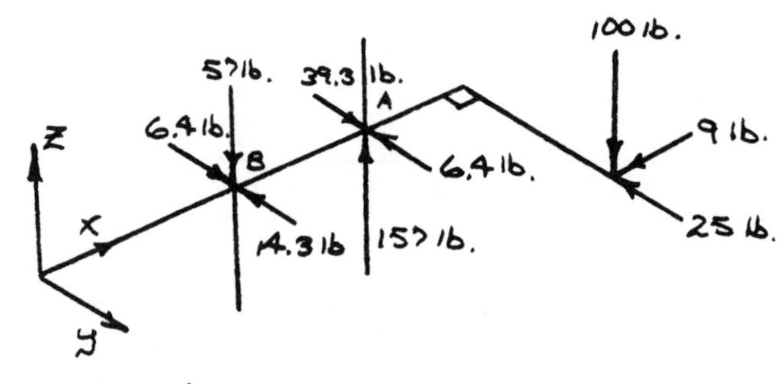

Resultant Radial loads

$$R_A = \left[157^2 + (39.3 - 6.4)^2\right]^{1/2} = 160\text{ lb}$$

$$R_B = \left[57^2 + (14.3 - 6.4)^2\right]^{1/2} = 58\text{ lb}$$

* In addition, bearing A would assume the thrust load of 9 lb.

Part b

Warmup increases preload since the aluminum housing will grow more rapidly than the steel gear and spacer.

MACHINE DESIGN 8

The vertical thrust bearing in a large turbine is designed with steel balls operating between thick horizontal steel plates. The sketch shows one of these 0.75 inch diameter balls being loaded by forces P.

Assuming elastic action, what value of P (the force on one ball) corresponds to a maximum contact pressure of 100,000 psi?

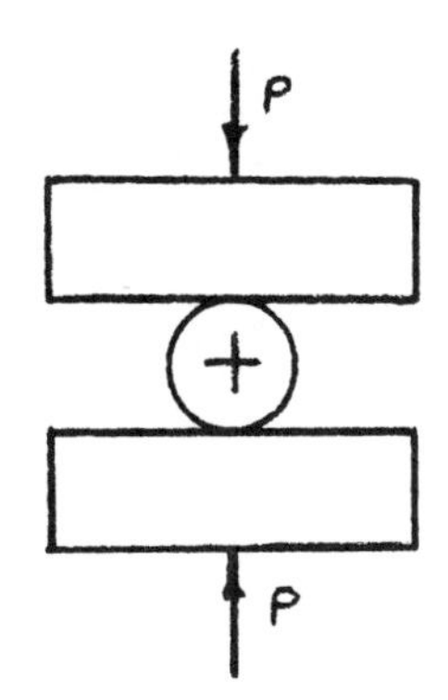

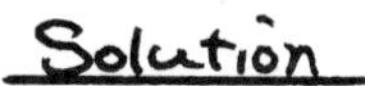

(Ref. - P 516 #1 Roark)

$$S_c = 0.616 \sqrt[3]{\frac{P E^2}{D^2}}$$

Solving for P

$$P = \left(\frac{S_c}{0.616}\right)^3 \frac{D^2}{E^2} = \left(\frac{10 \times 10^4}{0.616}\right) \frac{(0.75)^2}{(30 \times 10^6)^2}$$

$$= \underline{\underline{2.69 \text{ lb}}}$$

MACHINE DESIGN 9

A hardened steel seat is held in an aluminum valve body with an interference fit. The seat has an outside diameter of 0.6250 inches and the valve body has a wall thickness of 0.125 inches at this location.

REQUIRED: (a) What should be the nominal bore diameter of the valve body, when measured at 70°F, to be given an interference fit that will produce a stress of 15,000 psi in the aluminum? Consider all compressive stresses negligible.

(b) If the steel insert is cooled to -270°F with liquid nitrogen and the valve body heated to 180°F, what will be the diametral clearance during assembly?

Solution

a) $\sigma_{ot} = 15{,}000$ psi

* Employ Eq. # 2-66 P.76 Shigley - Ref.

$$\sigma_{ot} = p \frac{c^2 + b^2}{c^2 - b^2}$$ Solve for contact Pressure

$$p = \sigma_{ot} \frac{c^2 - b^2}{c^2 + b^2} = 15{,}000 \frac{(0.437)^2 - (0.312)^2}{(.191) + (.0974)}$$

$$= 4900 \text{ psi}$$

* Employ Eq. b P.77 Shigley

$$\delta_o = \frac{Pb}{E_o}\left(\frac{c^2 + b^2}{c^2 - b^2} + \mu\right)$$

$$= \frac{4900(.312)}{10 \times 10^6}\left[\frac{(.437)^2 + (.312)^2}{(.191) - (0.0974)} + 0.334\right]$$

$$= 1.53 \times 10^{-4}(3.3415) = 0.000523 \text{ in. (radial dimension)}$$

Thus the diameter of the valve body must be,

$$0.625 - 0.000523\,(2) = \underline{\underline{0.624 \text{ in}}} \longleftarrow$$

b) Ref. - p.79 Shigley,

$$\alpha_{STL} = 6\times10^{-6} \text{ in/in/°F}$$

$$\alpha_{alum.} = 13.3\times10^{-6} \text{ in./in./°F}$$

$$\delta_{STL} = 0.625\,(6\times10^{-6})(270+70) = 0.0013 \text{ in (Diametrical)}$$

$$\delta_{aLum} = (13.3\times10^{-6})(180-70)(0.624) = 0.0009 \text{ in (Diametrical)}$$

∴ Diametrical Clearance

$$\Sigma\delta = 0.0013 + 0.0009 - 0.001 = \underline{\underline{0.0012 \text{ in}}} \text{ (Dia.)}$$

↑ Initial Interference

MACHINE DESIGN 10

The bearings of a steel shaft are to be interference fitted with bronze sleeves as shown so the sleeves will not loosen during operation. Find the minimum and maximum diameters of the bushing.

Maximum operating temperature of bearings will be 160°F, and physical properties of the materials are as follows:

	Steel	Bronze
α	7×10^{-6} in/in°F	0.1×10^{-4} in/in°F
μ	0.30	0.35
E	30×10^{6} psi	16×10^{6} psi
Tensile Strength	80,000 psi	66,000 psi
Yield Strength	36,000 psi	28,000 psi

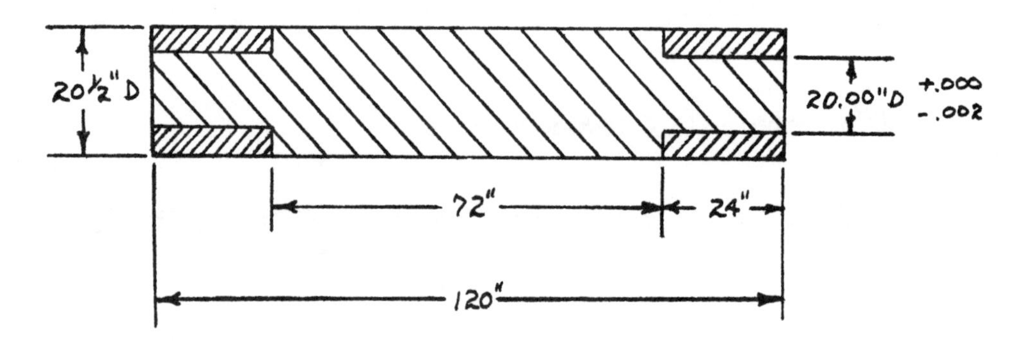

Solution

Ref. p. 76 Shigley

Check growth at 160°F

$$\Delta_{STEEL} = 20 \text{ in } (160-70)(7 \times 10^{-6}) = 0.0126 \text{ in.}$$

$$\Delta_{Bronze} = 20(90)(0.1 \times 10^{-4}) = 0.018 \text{ in}$$

$$\Delta_{Growth} = \Delta_{Bronze} - \Delta_{STEEL} = 0.018 - 0.0126$$

$$= \underline{\underline{0.0054 \text{ in}}}$$

-continued-

Try 0.005 in Interference fit at 160°F. At 70°F the minimum interference fit must be 0.005 + 0.0054 = 0.010 in.

Determine the allowable deflection so that the material will be within the yield stress.

$$\sigma_{ot} = P\left(\frac{c^2+b^2}{c^2-b^2}\right) \quad \& \quad P = \sigma_{ot}\left(\frac{c^2-b^2}{c^2+b^2}\right)$$

$$P = 28{,}000\left(\frac{(10.25)^2-(10)^2}{(10.25)^2+(10)^2}\right) = 28{,}000\left(\frac{5.06}{205}\right)$$

$$= 691 \text{ psi}$$

Now Consider Deformation (Eq. 2-69 - P.77 Shigley)

$$\delta = \delta_o - \delta_i = \frac{bP}{E_o}\left(\frac{c^2+b^2}{c^2-b^2} + \mu_o\right) + \frac{bP}{E_i}\left(\frac{b^2+a^2}{b^2-a^2} - \mu_i\right)$$

$$\delta = \frac{10(691)}{16\times10^6}\left(\frac{205}{5.06} + 0.35\right) + \frac{10(691)}{30\times10^6}(1-0.3)$$

$= 0.018$ in. Radial Allowable Bronze Deflection & still be within the yield stress

Thus, the min. interference at 70°F must be 0.010 in. and the max. interference must not be greater than 0.036 in. Diametrical

Size of Bushings

	Max.	Min.
Shaft Diameter	20.000	19.998
Bushing Diameter	19.964	19.988
	0.036	0.010

MACHINE DESIGN 11

The drive shaft on an outboard motor which rotates at speeds up to 5000 rpm is supported by a spline in the crankshaft at the upper end and a bearing at the lower end as shown.

(a) If the engine develops 20 hp at this speed, what diameter solid steel shaft should be used? The allowable tensional stress of 12,000 psi includes the safety factor, endurance limit and service condition considerations.

(b) Is the critical frequency of this shaft below 5,000 rpm?

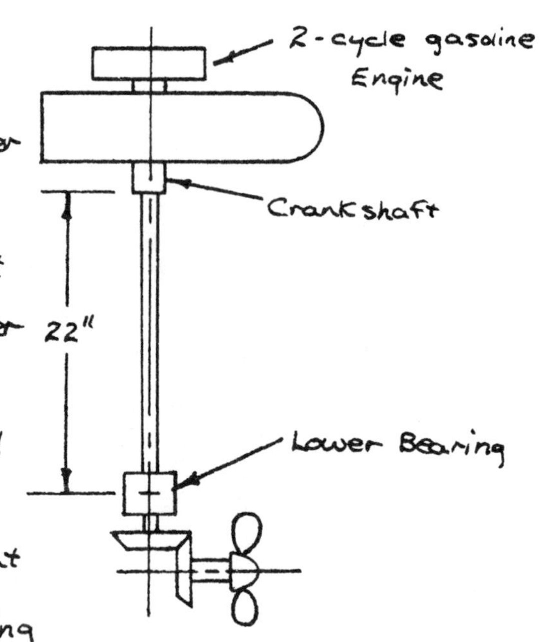

Solution

Assume that the upper end is a fixed support, since it is a splined shaft and assume that the lower end is free to rotate, this will yield a lower critical speed than if considered fixed. Neglect the fact that there will be an overturning moment due to the bevel gear load (can't determine it since no dimensions were given). Also neglect the axial load for the same reason, this load will also decrease the critical speed.

a)

$$T = \frac{63{,}025\,HP}{n} = \frac{63{,}025\,(20)}{5{,}000} = 252 \text{ in}\cdot\text{lb.}$$

$$\tau = \frac{16\,M_T}{\pi d^3} = \frac{16\,(252)}{\pi d^3} = 12{,}000 \text{ psi.}$$

Solving for Shaft Diameter:

$$d = \sqrt[3]{\frac{4030}{\pi(12\times10^3)}} = 0.475 \text{ in}$$

Employ a ½" diameter shaft (Not considering any residual stress or stress concentrations)

b) Employing the Energy method

$$\text{Area} = \frac{\pi}{4}\left(\frac{1}{2}\right)^2 = 0.197 \text{ in}^2$$

$$\rho = 0.28 \text{ lb/in}^3 \text{ for steel}$$

$$w = A\rho = (0.197)(0.28) = 0.055 \text{ lb/in}$$

$$\delta_{max} = -0.0054\left[\frac{0.055\,(22)^4}{30\times10^6\ \frac{\pi(0.5)^4}{64}}\right] = \underline{\underline{0.000756 \text{ in.}}}$$

Critical Speed

$$\omega_n = \sqrt{g/\delta} = \sqrt{\frac{386 \text{ in/sec}^2}{756\times10^{-6}}}$$

$$= 715 \text{ rad/sec}$$

$$\omega_n = \underline{\underline{6840 \text{ rpm}}}$$

MACHINE DESIGN 12

The drive shaft on an automobile composed of a mild steel tube (3.5" O.D. x 0.80" wall) welded to universal joint, yokes and a splined shaft as shown. If the engine develops 550 hp at 4,000 rpm in high gear, what is the stress in the tube? If the shaft is considered to have uniform properties, end to end, what is the critical speed of the shaft?

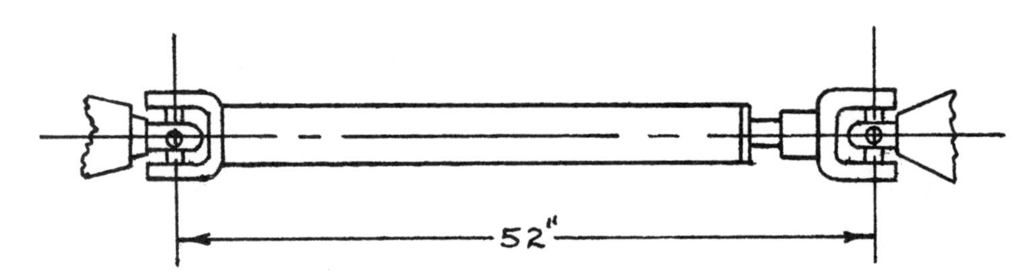

Solution

1) Stress in Shaft due to torsional shearing stress.

$$\tau = \frac{Tc}{J} = \frac{16\,T d_o}{\pi(d_o^4 - d_i^4)} \qquad T = \frac{63{,}000\,(HP)}{n}$$

$$= \frac{63{,}000\,(550)}{4000} = 8650 \text{ in·lb}$$

$$\tau = \frac{16\,(8650)\,3.5}{\pi(3.5^4 - 1.9^4)} = \underline{\underline{1127 \text{ psi}}} \longleftarrow$$

2) $\text{Weight/in} = \rho \times \text{Area} = 0.28 \text{ lb/in}^3 \, \frac{\pi}{4}\left[(3.5)^2 - (1.9)^2\right]$

$$= 1.9 \text{ lb/in}$$

consider both ends simply supported

$$\delta_{max} = \frac{5 w \ell^4}{384 EI} = \frac{5\,(1.9)(52)^4}{384(30\times10^6)\,\frac{\pi}{64}\left[3.5^4 - 1.9^4\right]} = 0.00089 \text{ in.}$$

Critical Speed

$$\omega_n = \sqrt{g/\delta_{static}}$$

$$\omega_n = \sqrt{\frac{386 \text{ in/sec}^2}{8.9\times10^{-4}}} = 630 \text{ rad/sec} = \underline{6020 \text{ rpm}}$$

MACHINE DESIGN 13

A scotch yoke mechanism is used to reciprocate a sorting tray in a food processing machine.

If the tray and yoke weigh 50 lbs, what are the peak loads on the slider when running at 100 rpm?

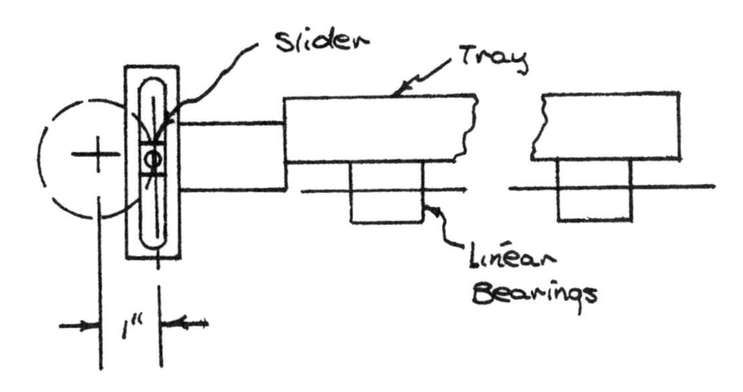

Solution

Assume that the linear bearings are frictionless since no value of μ was given.

$$\text{acceleration} = \omega^2 R \qquad \omega = \left(\frac{100}{60}\right) 2\pi = 10.5 \text{ rad/sec.}$$

$$A_{max} = (1)(10.5)^2 = 110 \text{ in/sec}^2$$

$$F = MA = \left(\frac{50 \text{ lb}}{386 \text{ in/sec}^2}\right) 110$$

$$= \underline{\underline{14.2 \text{ lb}}} \text{ max. force}$$

MACHINE DESIGN 14

The Geneva mechanism shown below is part of a binary digital counter. The wheel A rotates at a constant counterclockwise angular velocity of 100 revolutions per minute. Pin P drives the slotted wheel B so that wheel A turns 4 times before wheel B completes a revolution. The pin is located at a distance R = 1.5 inches from the center of wheel A. At the instant shown, wheel B has no angular velocity and the line OP makes a 45° angle with the vertical.

Find the angular velocity of wheel B .075 seconds later.

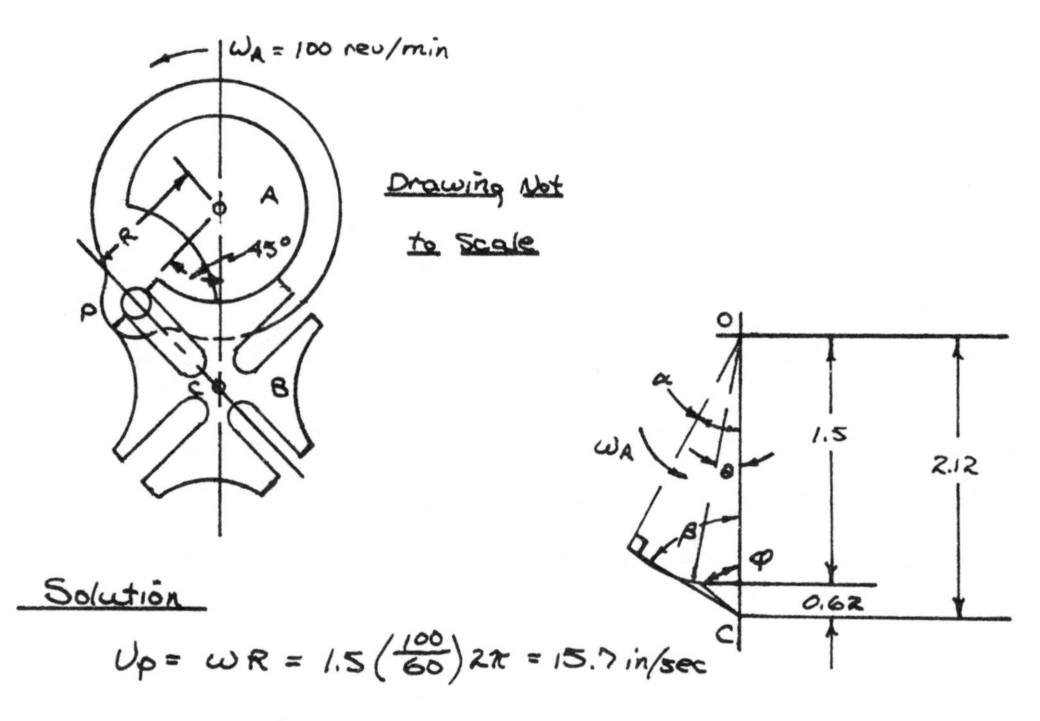

Solution

$$V_P = \omega R = 1.5\left(\frac{100}{60}\right)2\pi = 15.7 \text{ in/sec}$$

One could substitute into the equation of ω_B in terms of ω_A, α & θ where:

$$\omega_B = \omega_a\left(\frac{\cos\alpha\,(\cos\theta - \cos\alpha)}{1 - 2\cos\alpha\cos\theta + \cos^2\alpha}\right)$$

But lets try another approach:

$$\omega_A = \frac{100 \text{ rpm}}{60 \text{ sec/min}}(360°) = 600°/\text{sec} \qquad \text{Time} = \frac{45°}{600°} = .075 \text{ sec.}$$

Thus the point P is on the line of centers

$$\omega_B = \frac{V_P}{0.62} = \frac{15.7 \text{ in/sec}}{0.62} = 25.4 \text{ rad/sec} = \underline{\underline{242 \text{ rpm}}}$$

MACHINE DESIGN 15

The lift of a cycloidal cam is given by the equation:

$$Y = L\left(\frac{\theta}{\beta} - \frac{1}{2\pi}\sin\frac{2\pi\theta}{\beta}\right)$$

Where L is the maximum lift, θ is the cam angle, and β is the angle over which the lift occurs.

REQUIRED: If L = 0.75 in., β = 60° and the cam is turning at a constant speed of 1500 rpm, find the maximum acceleration of the follower in ft/sec².

Solution

$$y = l\left(\frac{\theta}{\beta} - \frac{1}{2\pi}\sin\frac{2\pi\theta}{\beta}\right)$$

$$\dot{y} = l\left(\frac{\omega}{\beta}\right)\left[1 - \cos\frac{2\pi\theta}{\beta}\right] \quad * \text{ Note}: \frac{d\theta}{dt} = \omega$$

$$\ddot{y} = 2\pi l\left(\frac{\omega}{\beta}\right)^2 \sin\frac{2\pi\theta}{\beta}$$

$$\ddot{y}_{max} = 2\pi l\left(\frac{\omega}{\beta}\right)^2 \quad \text{since } \sin\frac{2\pi\theta}{\beta} = 1 \text{ at } \ddot{y}_{max}.$$

For $l = 0.75$ in $\quad \beta = 60°$ $\quad \omega = \frac{1500\text{ rpm}}{60\text{ sec/min}}(2\pi)$

$$= 157 \text{ rad/sec}$$

$$\ddot{y} = 2\pi(0.75)\left(\frac{157 \text{ rad/sec}}{60/57.3 \text{ rad.}}\right)^2 = 106{,}000 \text{ in/sec}^2$$

$$= \underline{\underline{8{,}840 \text{ ft/sec}^2}}$$

MACHINE DESIGN 16

For the planetary gear system shown, arm A is the driver. Gears 2 and 3 are attached together, and gear 4 is held stationary. The numbers of teeth of the various gears are shown in parentheses.

If arm A is turning at 900 rpm clockwise as seen from the left, determine the rpm and direction of rotation of gear 1.

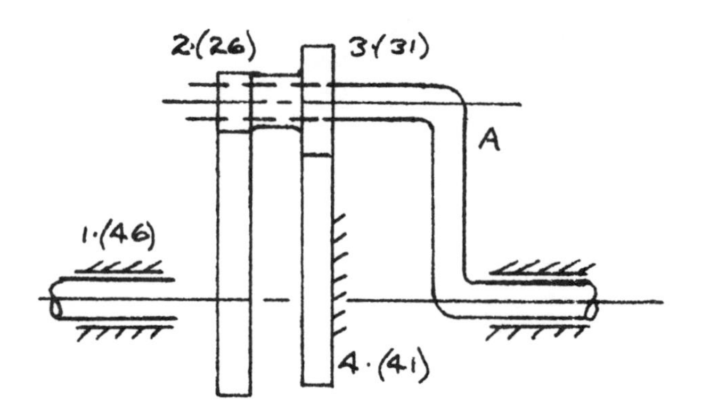

The solution for this type of planetary gear train may be obtained employing the relative velocity equation, the tabular method or looking up this particular arrangement in a handbook like Marks.

The first two methods will be illustrated here.

Reverted Gear Train - Tabular Method

Steps		1	2	3	4	ARM
1	Arm Held Stationary & Fixed Gear Rotate +1 Rev.	$+\frac{26 \times 41}{46 \times 31}$	$-41/31$	$-41/31$	+1	0
2	Lock all gears & rotate Everything back 1 rev.	−1	−1	−1	−1	−1
3	Total Each Column	$\frac{-1+.75}{-0.25}$			0	−1

Speed of gear 1 = + 0.25 (900) = **225 rpm** - same direction as arm - both were negative.

Relative Velocity Equation

$$e = \frac{n_l - n_a}{n_f - n_a}$$

where :

e - gear train value

n_a - angular vel. of arm

n_f = angular vel. of 1st gear

n_l = angular vel. of last gear

Gear Train Value

$$e = \left(\frac{46}{26}\right)\left(\frac{31}{41}\right) = \frac{n_l - n_a}{n_f - n_a} = \frac{0 - 900}{n_f - 900}$$

$$1.34\, n_f - 1200 = -900$$

$$n_f = \frac{300}{1.4} = \underline{\underline{225 \text{ rpm}}}$$

(same direction as the arm)

MACHINE DESIGN 17

A reverted epicyclic gear train is required to have an output speed of approximately 2400 rpm when driven from an electric motor at 1725 rpm. What are the number of teeth required on gears 2, 3 and 4? The motor pinion has 22 teeth, the total number of teeth on each reverted train is 60. Gear 1 has 40 teeth.

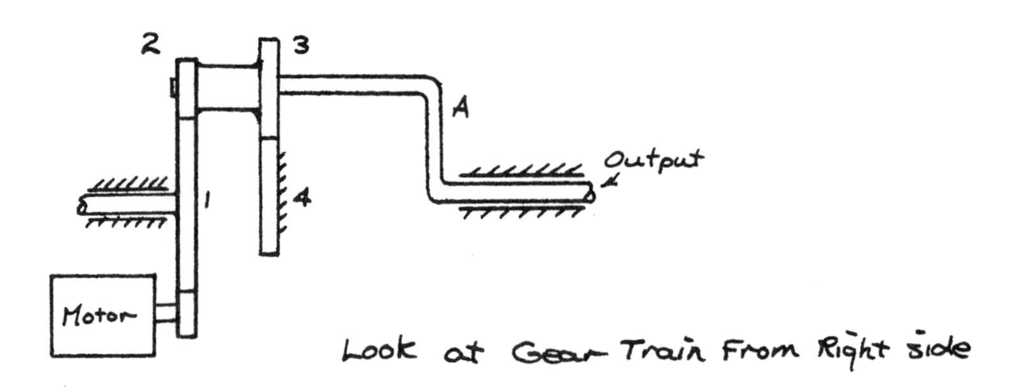

Look at Gear Train From Right side

Assume arm to rotate in a counterclockwise direction. Thus from sketch, the pinion must rotate in the same direction as the arm since gear no. 4 is fixed.

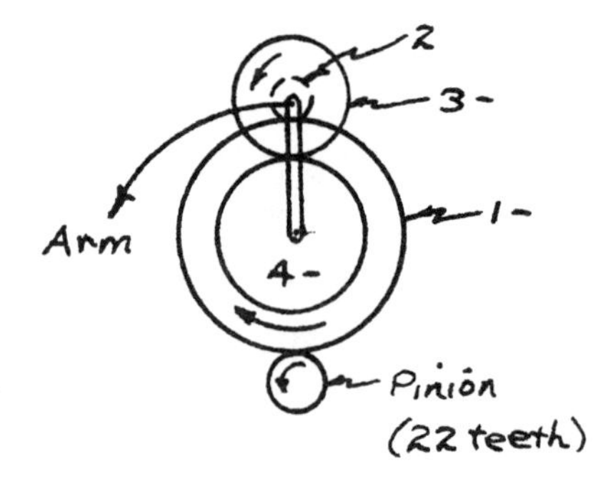

If gear no. 1 has 40 teeth & the total in the gear set is to be 60, then gear no. 2 must have 20 teeth.

Gear 1 rotates @ $1725 \times \frac{22}{40} = 950$ rpm ↻

$$e = \frac{n_\ell - n_a}{n_f - n_a}$$

assume:

n_f - gear 1

n_ℓ - gear 4

-continued-

$$e = \left(\frac{40}{20} \times {n_3}/{n_4}\right) = 2n_3/n_4 = \frac{0-(-2400)}{+950-(-2400)}$$

given output speed

Solving for n_3 in terms of n_4

$$n_3 = +\frac{2400}{6700}n_4 = +0.358\,n_4$$

Also, the sum of the teeth of gears 3 + 4 must be 60 or $n_3 + n_4 = 60$

$$n_4 = 60 - n_3$$

$$n_3 = 0.358\,(60 - n_3)$$

Solving, n_3 = 16 teeth - OK - no undercut if an involute profile

n_4 = 60 − 16 = 44 teeth

Check This Solution

$$e = \left(40/20\right)\left(16/44\right) = \frac{0-n_a}{+950-n_a}$$

Solve for $n_a = -\frac{690}{0.27} = -2550$ rpm opposite in direction to the input member - gear 1

* Note: If n_a were assumed to operate in the opposite direction, the gear 3 has 27 teeth & gear 4 has 33 teeth.

MACHINE DESIGN 18

A hydraulic actuator is needed to provide these forces: maximum force in compression - 4,000 lb, maximum force in tension - 8,000 lb. The rod is made of steel with a tension or compression yield strength of 40,000 psi.

What nominal (nearest 1/16 inch) diameter rod is required for a safety factor of 5 and what nominal bore?

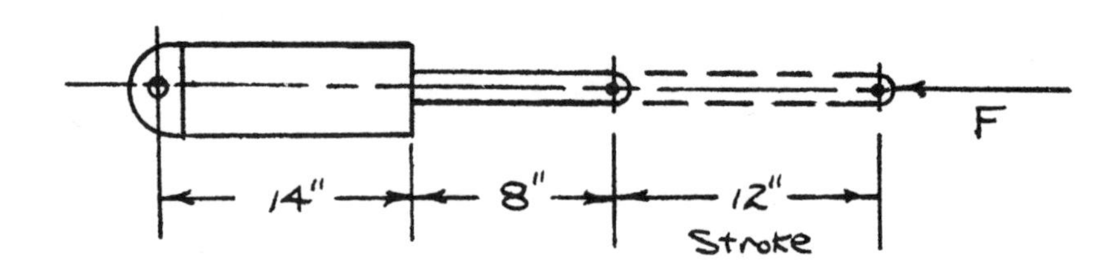

Assume a column - one end fixed and one end free since we know nothing of the linkage design.

$n = 1/4$ (Ref. - p. 145 Shigley)

$$P_{CR} = \frac{n\pi^2 EI}{\ell^2} = 4000(5) = \frac{\pi^2 (30 \times 10^6 \text{ psi}) \pi D^4}{4(20)^2 \, 64}$$

$D^4 = 2.2 \text{ in.}^4$ $\qquad D = 1.218$ in Use 1.25"

Design Stress $\qquad \sigma_{yield} = P/A = 40{,}000/5 = 8{,}000$ psi

$$A = \frac{\pi}{4}(1.25)^2 = 1.228 \text{ in.}^2$$

$$\sigma_{actual} = 8000/1.228 = 6{,}520 \text{ psi}$$

Design Stress is above this value

$$\text{Factor of Safety} = \frac{8000}{6520} = 1.23$$

—continued—

Net area required to provide the required tension,

$$A = \frac{8000\ lb}{2000\ psi} = 4\ in^2$$

← assumed hydraulic pressure.

∴ Area of piston must be $4 + 1.228 = 5.228\ in^2$

$$D_{piston} = \left[\frac{4(5.228)}{\pi}\right]^{1/2} = 2.6 \cong 2\tfrac{5}{8}\ in\ Dia.$$

Extra area helps to overcome seal friction.

MACHINE DESIGN 19

A pulley is keyed to a 2-1/2 inch diameter shaft by a 7/16 x 5/8 x 3 inch flat key. The shaft rotates at 50 rpm. The allowable shear stress for the key is 22 ksi. The allowable compressive stresses for the key, hub and shaft are 66 ksi, 59 ksi, and 72 ksi, respectively.

What is the maximum torque the pulley can safely deliver?

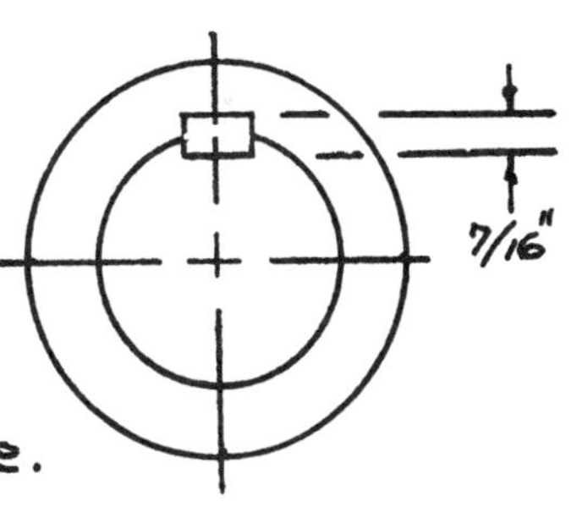

Solution

There are many ways that this key may fail - Calculate the various design stresses allowable.

Since we know nothing about the design loads, employ a safety factor of 2.

$$(\tau_{Design})_{key} = \left(\frac{S_{yield}}{2}\right)\left(\frac{1}{2}\right)$$

← safety factor

↑ Max. shear Theory of failure

$$(\tau_{Design})_{Key} = \frac{66,000}{4} = 16,500 \text{ psi}$$

$$(\tau_{Design})_{Hub} = \frac{59,000}{4} = 14,800 \text{ psi}$$

$$(\tau_{Design})_{Shaft} = \frac{72,000}{4} = 18,000 \text{ psi}$$

Shear of Key

$A_{Key} = 5/8 \times 3 = 1.87 \text{ in}^2$

Load = $1.87 \times 16,500 = 31,000$ lb (Permissible Shear load on Key)

Bearing Stress - Key

$$(S_{Design})_{Key} = \frac{66,000}{2} = 33,000 \text{ psi}$$

(2 ← Safety factor)

Area = $(7/16)(1/2)(3) = 0.658 \text{ in}^2$

Permissible Load - Crushing Key = $0.658(33,000)$ = 21,600 lb.

Bearing Stress - Hub

$$(S_{Design})_{Hub} = \frac{59,000}{2} = 29,500 \text{ psi}$$

Load = $0.658(29,500) = 19,400$ lb

Thus the maximum torque that may be transmitted is:

$$\text{Torque} = \left(\frac{2\frac{1}{2} \text{ in. Dia}}{2}\right)19,400 = \underline{\underline{24,100 \text{ in} \cdot \text{lb}}}$$

* Note: It is not possible to determine the permissible shear load on the Hub, since the dimensions of the hub were not given.

MACHINE DESIGN 20

An annular flange which is bonded uniformly along the periphery of a round bar is clamped to a fixed base as shown.

REQUIRED: Determine the maximum load P, which can be applied to the end of the bar without exceeding the maximum allowable shearing stress in the bond of 10,000 lbs/in^2. Neglect the weight of the bar.

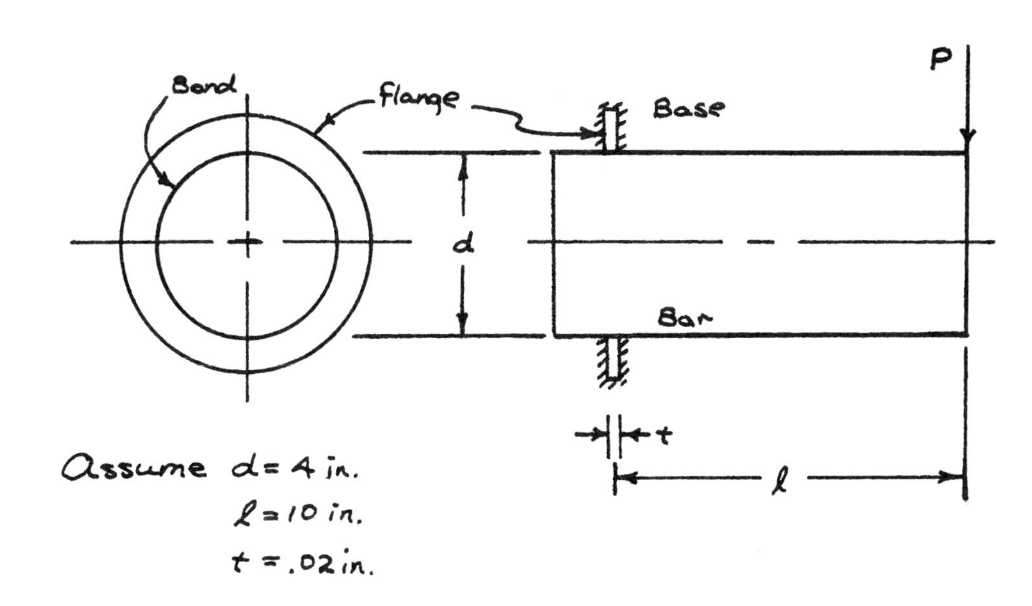

Assume $d = 4$ in.
$\ell = 10$ in.
$t = .02$ in.

Solution

$$\text{Moment} = P\ell = (\tau\, dA)(\rho) \quad \text{(See sketch below)}$$

(force: $\tau\, dA$; radius: ρ)

End view of Bar

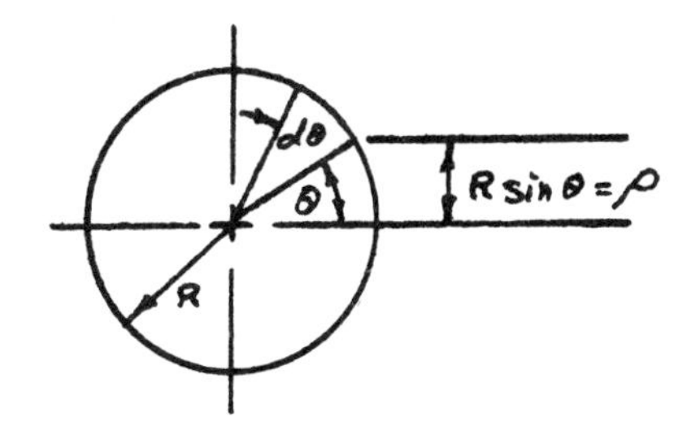

$$\frac{\tau}{\tau_{max}} = \frac{\rho}{R}$$

Assume that the shear varies directly as the distance from the neutral axis.

$$\tau = \tau_{max} \frac{\rho}{R}$$

Permissible Moment (Employing $M = \tau dA(\rho)$)

$$M = 2\int_0^{180^\circ} \tau \rho \, dA = 2\tau_{max}\int_0^{180^\circ} \frac{\rho^2 t R d\theta}{R}$$

Substituting $\tau = \tau_{max} \frac{\rho}{R}$

$$M = 2\tau_{max}\int_0^{180^\circ} t R^2 \sin^2\theta \, d\theta$$

$$M = 2\tau_{max} t R^2 \int_0^{180^\circ} \sin^2\theta \, d\theta = 2\tau_{max} t R^2 \left[\frac{\theta}{2} - \frac{\sin 2\theta}{4}\right]_0^{180^\circ}$$

$$M = 2\tau_{max} t R^2 (\pi/2)$$

Substituting $t = 0.02$ in, $\ell = 10$ in + $d = 4$ in

$$M = P\ell = P(10) = 2(10{,}000)(0.02)(2)^2(\pi/2)$$

$$P = \frac{200(4)(\pi)}{10} = 80\pi = \underline{\underline{251 \text{ lb}}}$$

MACHINE DESIGN 21

A snap ring is made of 0.177 inch diameter steel wire, bent into an arch of 270° with a mean radius of R = 1.5 in.

Find the tensile forces P required to spread the opening by 3/8 inch.

Use E = 30 x 10^6 psi.

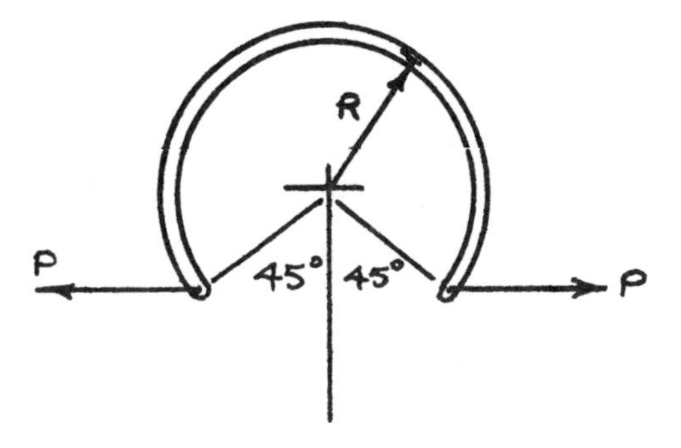

Solution

Employ half of the snap ring since it must be in equilibrium.

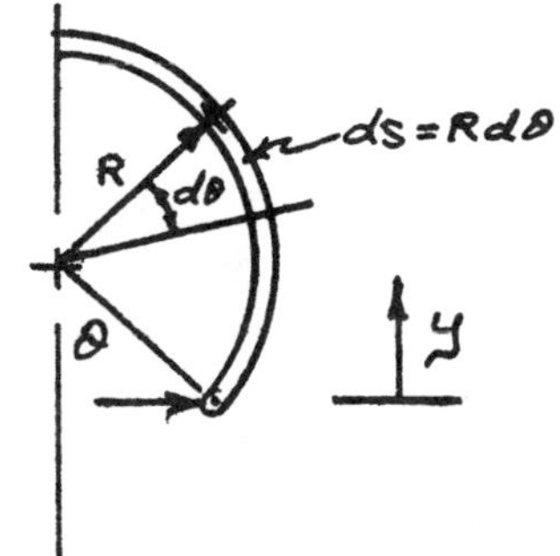

The moment will vary directly as y, so an expression of y as a function of θ & R is:

$$y = 0.707R - R\cos\theta$$

This is only valid between θ = 45° and θ = 180°

Employing Castigliano's Theorem

(Ref. - p.136 Shigley)

Strain Energy in Bending $U = \int \frac{M^2 ds}{2EI}$ & $ds = Rd\theta$

$$\delta_P = \int \frac{2M\left(\frac{\partial M}{\partial P}\right)ds}{2EI} = \int_{45^\circ}^{180^\circ} \frac{M\left(\frac{\partial M}{\partial P}\right)Rd\theta}{EI}$$

Bending Moment

$M = P(\text{Distance})$

$$M = P(0.707R - R\cos\theta)$$

$$\frac{\partial M}{\partial P} = (0.707R - R\cos\theta)$$

Substitute into δ_P formula:

$$\delta_P = \int_{45^\circ}^{180^\circ} \frac{P(0.707R - R\cos\theta)^2 R\,d\theta}{EI}$$

where: $I = \frac{\pi d^4}{64}$

$$I = \frac{\pi (0.177)^4}{64}$$

$$I = \underline{4.93 \times 10^{-5}\ in^4}$$

$$\delta_P = \frac{PR}{EI}\int_{45^\circ}^{180^\circ}(1.06 - 1.5\cos\theta)^2\,d\theta$$

$$\delta_P = \frac{PR}{EI}\left[1.123\theta - 3.18\sin\theta + 2.25\left(\tfrac{1}{2}\theta + \tfrac{1}{4}\sin 2\theta\right)\right]_{45^\circ}^{180^\circ}$$

$$\delta_P = (3/8)(1/2) = 0.188 = \frac{7RP}{EI}$$

considering only half of the snap ring

$$P = \frac{0.188(30\times10^6\ psi)(4.93\times10^{-5})}{(7.0)(1.5)} = \underline{\underline{26.5\ lb}} \leftarrow$$

MACHINE DESIGN 22

A horizontal circular bar, rigidly built in at one end, is to be loaded by a vertical force P = 9.42 kips. The force lies parallel to the X-Z plane shown. The bar's radius is 1.5 inches, its length 9 inches. The horizontal distance "A" is 5.75 inches. Assume a strong bond between lever arm and bar.

REQUIRED: (a) Find an element on the bar that is in pure shear and determine the maximum shear stress present.

(b) Using the maximum-shear stress theory of failure as a basis of design decision (the allowable shear stress is 15 ksi) will the bar fail with this size radius?

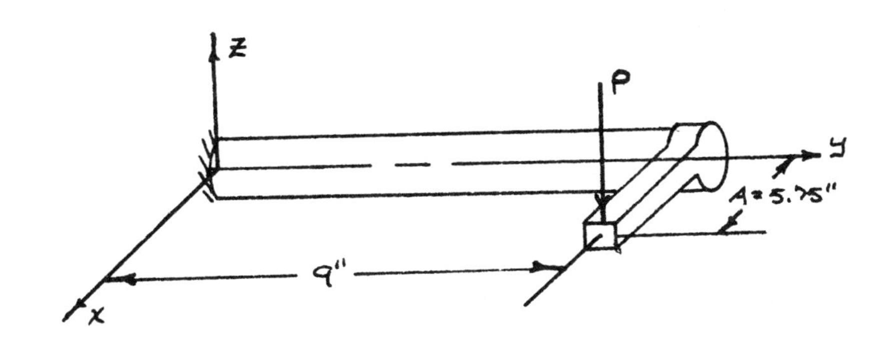

Solution

a) This is a combined stress problem.

$$\sigma_b = \frac{Mc}{I} \;+\!\!\rightarrow\; \frac{Tc}{J} = \left(\frac{32M}{\pi d^3} \;+\!\!\rightarrow\; \frac{16T}{\pi d^3}\right)_{\text{circular shaft}}$$

where: $M = (9420\text{ lb})(9\text{ in}) = 84{,}780\text{ in}\cdot\text{lb.}$

$T = (9420)(5.75) = 54{,}165\text{ in}\cdot\text{lb.}$

$$\sigma_{\text{Bending-Max at X-Z plane}} = \frac{32(84{,}780)}{\pi(3)^3} \;+\!\!\rightarrow\; \frac{16(54{,}165)}{\pi(27)}$$

$$= 31900 \;+\!\!\rightarrow\; 10{,}217$$

–continued–

$$\tau_{max} = \left[\left(\frac{31,900}{2}\right)^2 + (10,217)^2\right]^{1/2} = \underline{\underline{18900}} \text{ psi}$$

$$\sigma_1 \text{ principal stress} = \frac{31,900}{2} + 18,900 = \underline{\underline{34,840}} \text{ psi}$$

One could plot Mohr's Circle or simply employ a sketch & determine the location of the plane of pure shear.

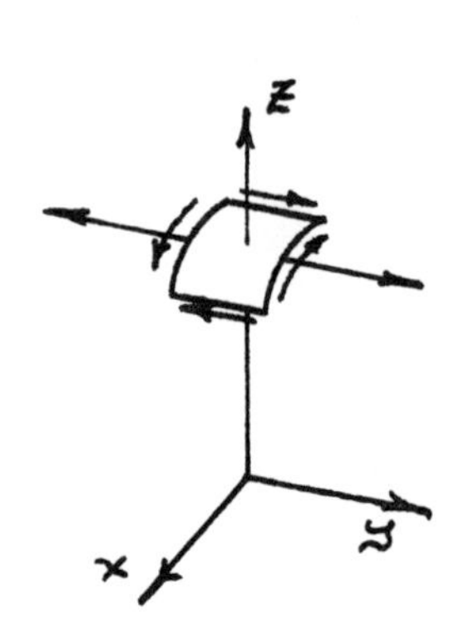

Removing an element from the top of the bar - Notice that the shear on the right hand surface tends to rotate the element counter-clockwise ∴ is a negative shear (Timoshenko Notation).

Therefore, the origin of the circle appears in the 4th quadrant and is $\varphi°$ below the Horizontal axis.

$$\tan 2\varphi = \frac{10,217 \text{ psi}}{15,950 \text{ psi}} = 0.6405 \qquad \varphi = 16° - 19'$$

Pure shear will occur where the bending stress is zero.

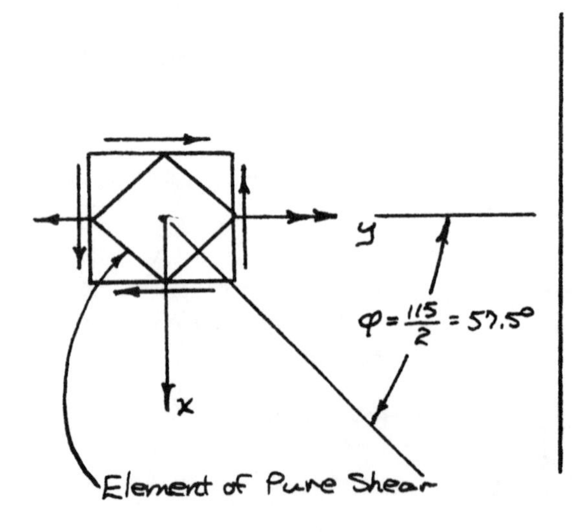

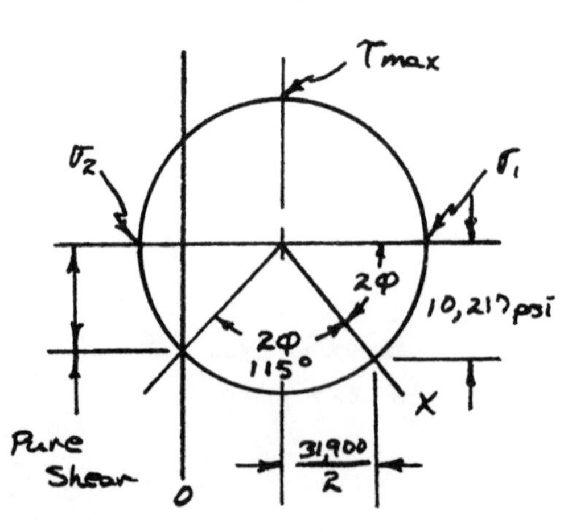

b) This theory asserts that the breakdown of the material depends only on the maximum shearing stresses that are attained in an element.

$$\tau_{max} = \tfrac{1}{2}(\sigma_1 - \sigma_2) = \underline{\underline{18{,}900 \text{ psi}}}$$

∴ this part will fail since this stress is larger than the allowable stress of $\underline{\underline{15{,}000 \text{ psi}}}$.

MACHINE DESIGN 23

A punch press is to have a capacity of 8000 lb through a 1/4 inch stroke. For efficiency and cost a small motor with a flywheel was used.

What flywheel effect is required to maintain the flywheel within 20% of its no-load speed of 600 rpm?

Solution

Kinetic Energy of a rotating body

$$K.E. = \tfrac{1}{2} I_o \omega^2 = \tfrac{1}{2} I_o (\omega_1^2 - \omega_2^2)$$

assume that all the energy must be stored in the flywheel or that the frequency of this operation is low.

Energy to be Stored

$$8000 \text{ lb} (\tfrac{1}{4} \text{ in}) = 2000 \text{ in}\cdot\text{lb}$$

$$\omega_1 = \frac{2\pi N}{60} = \frac{2\pi(600)}{60} = 62.8 \text{ rad/sec.}$$

* a 20% drop in speed is permissible

$$\Delta\omega = 20\%(62.8) = 12.56 \text{ rad/sec} \qquad \omega_2 = 62.8 - 12.56 = 50.24 \text{ sec}^{-1}$$

$$K.E. = 2000 \text{ in}\cdot\text{lb} = \tfrac{1}{2} I_o (62.8^2 - 50.24^2)$$

Solving for I_o = $\underline{\underline{2.82 \text{ lb}\cdot\text{in}\cdot\text{sec}^2}}$

MACHINE DESIGN 24

A jack screw was designed as shown. It was well lubricated, and the nut had a sufficient number of threads so that the calculated bearing pressures were well within the allowable limits. The materials were compatible for this usage, and all other calculated stresses were reasonable. Nevertheless, this screw galled on the first usage. Why?

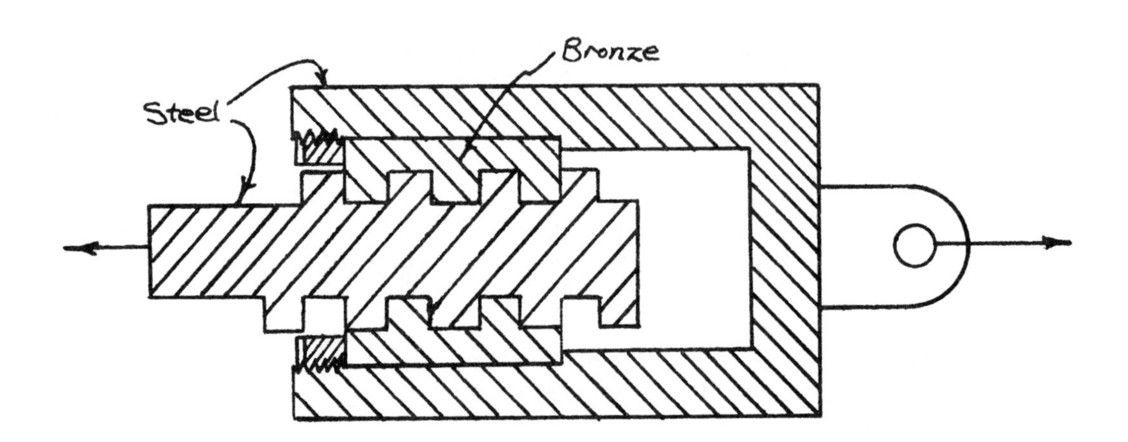

The statement, "That the calculated bearing pressures were well within the allowable limits", bears analysis.

Since we don't have the calculations before us, we can only surmise the assumptions. If the designer neglected to consider the deformation of the threads and considered them all equally sharing the load, this will lead to a high load on the first thread in the nut and a low load on the last thread (Ref.- p382 Shigley). Thus the load on the first thread may be in the range to cause galling.

Another more remote possibility may be operating at a higher temperature environment such that the bronze will "grow" faster than the steel and thus cause galling.

Perhaps the nut was excessively tightened on the bronze threaded insert and this caused the spaces between the bronze threads to deform and increase the bearing pressure above that anticipated.

One assumes that the parts were made to the drawing and that clean lubricant was employed.

MACHINE DESIGN 25

An instrumentation torque limiting device is made as shown.

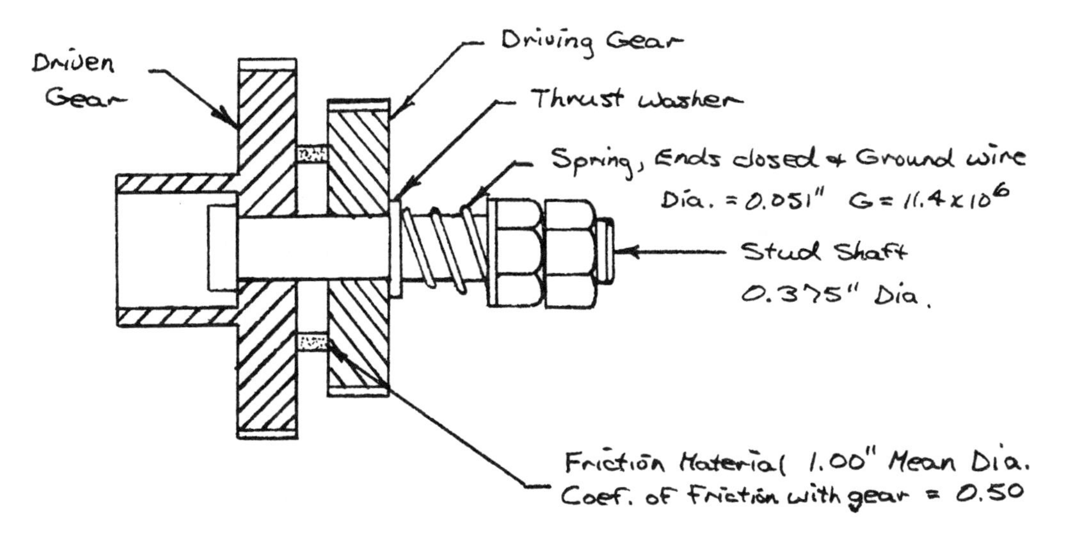

REQUIRED: If the torque is to be adjustable from 0 to 6 in. oz. and the maximum diameter of the thrust washer is 0.50, what are the dimensions of the spring?

Comment: There are numerous springs that will meet the requirements varying in free length, mean diameter and number of active coils. Since this requires a knowledge of spring design and the judgment to make some reasonable assumptions, it is not a problem in which the solution can be obtained by substituting numbers in a handbook equation.

Solution

Equations and Relationships to be used:

$$\delta_{spring} = \frac{8FD^3N}{d^4G} \quad \text{(Ref.- p 446 shigley)}$$

Torque - Integrate the Frictional force times radius (Ref.- p.734 Shigley) uniform wear

$$T = \int_{d/2}^{D/2} 2\pi f p r^2 dr = \frac{\pi f p_a d}{8}(D^2 - d^2) = \frac{Ff}{4}(D+d)$$

Normal Force (Ref-p.734 Shigley)

$$F = \int_{d/2}^{D/2} 2\pi p r dr = \frac{\pi p_a d}{2}(D-d)$$

Combining the two above equations gives,

$$T = \frac{Ff}{4}(D+d)$$

Assume an O.D. of the spring equal to that of the thrust washer, also employ a 16 awg gage wire (0.051 in Dia)

$d = 0.051$ $\quad D = OD - d = .500 - .051 = .449$ in.

I.D. = .500 - .102 = .398 in - ok -

ID Less than 3/8 in. Dia.

N = 4 active coils - Thus must add 2 coils for solid height since spring ends are closed and ground.

Solid Height = (4+2) 0.051 = 0.306 in

$$\text{Torque} = 6 \text{ in}\cdot\text{oz.} = 0.375 \text{ in}\cdot\text{lb}$$

Solve for F in the torque Eq. above.

$$F = \frac{4T}{f(D+d)} = \frac{4(0.375)}{0.5\ (2)} \qquad D_{mean} = \frac{D+d}{2}$$

$$D+d = 2D_{mean}$$

$$F = 1.5 \text{ lb.}$$

$$F = 1.5\text{lb} = \frac{\delta d^4 G}{8 D^3 N} = \frac{\delta (0.051)^4 (11.4\times10^6 \text{psi})}{8(.449)^3(4)}$$

$$\delta = \frac{1.5\ (2.88)}{0.0666\ (11.4\times10^6)} = \underline{\underline{0.0568 \text{ in.}}}$$

Thus for zero torque, the spring is free from the thrust washer. For max. torque, the spring must be deflected 0.057 in.

The deflection of the parts has been ignored in the analysis.

5

Fluid Mechanics

R. IAN MURRAY

FLUID STATICS

A fluid continuously deforms when subjected to shear forces. Therefore, if a fluid is at rest or if it moves as a rigid body, there can be no shear stresses within or on the boundaries of the fluid. The differential equation of the pressure field is then given by

$$dp = -\gamma\left[\frac{a_x}{g}\,dx + \frac{a_y}{g}\,dy + \left(1 + \frac{a_z}{g}\right)dz\right]$$

p = pressure

$\gamma = \rho g$ = specific weight (i.e., weight density; ρ is mass density)

a_x, a_y, a_z = components of acceleration

g = gravitational force per unit mass

z-axis is parallel to $\vec{g}$ but directed upward

If the density and acceleration are uniform throughout the body of fluid under consideration, the above equation is easily integrated to give the pressure distribution. For the important special case in which the density is uniform and the acceleration is zero, the pressure distribution is given by

$$\frac{p}{\gamma} + z = C$$

C is a constant that can be evaluated if the density is known and the pressure is known at some reference elevation. From the pressure distribution the resultant force on a submerged surface can readily be determined by applying the principles of statics for distributed loads. The difference in pressure between discrete points in the fluid can also be determined and is important in instrumentation problems.

$$p_1 - p_2 = \gamma\,(z_2 - z_1)$$

IDEAL FLUID FLOW

When a fluid flows, shear stresses are produced within the fluid and at the boundaries. However, there are many important problems in which the effect of the shear stresses can be neglected. This tends to be true for low viscosity fluids when the distance along the flow path is short and the streamlines under consideration are not close to a fixed boundary. If, in addition, the flow is steady and the variation of density is negligible*, Bernoulli's equation is obtained:

$$H_1 = H_2 \quad \text{where} \quad H = z + \frac{p}{\gamma} + \frac{V^2}{2g} = \text{total head}$$

1 and 2 refer to two points on the same streamline.

$z + \frac{p}{\gamma}$ is the piezometric head; it decreases toward the center of curvature when streamlines are curved.

z is the potential head.

$\frac{p}{\gamma}$ is the pressure head.

$\frac{V^2}{2g}$ is the velocity head.

When changes of elevation are negligible Bernoulli's equation may be written:

$$p_1 + \rho \frac{V_1^2}{2} = p_2 + \rho \frac{V_2^2}{2} = p_o$$

Here p is the static pressure, $\rho \frac{V^2}{2}$ is the dynamic (or velocity) pressure, and p_o is the stagnation (or total) pressure.

REAL FLUID FLOW

When the effect of shear stress cannot be neglected, several alternative procedures are available:

1. Solution of the general equations of motion. This is beyond the scope of this review.

2. Use of empirical coefficients to modify ideal fluid flow solutions. Appropriate coefficients will be defined as they are encountered in the review problems. These coefficients may be used to cope with difficult geometric conditions as well as to correct for the effect of assuming ideal fluid flow conditions.

---

*For problems in which the variation of density is not negligible, see Chapter 8.

3. Augmentation of Bernoulli's equation with a head loss term for flow in pipes and ducts,

$$H_1 = H_2 + H_L$$

$$H_L = f\,\frac{\ell}{D}\,\frac{V^2}{2g} + \Sigma\,K_L\,\frac{V^2}{2g} = f\left(\frac{\ell}{D} + \Sigma\,\frac{L}{D}\right)\frac{V^2}{2g} = \text{head loss}$$

f is the Darcy friction factor and is a function of Reynolds' number and relative roughness. $f = f(N_R, e/D)$ can be obtained from a Moody diagram.

ℓ is the total length of pipe of diameter D along a single flow path connecting points 1 and 2.

K_L is a loos coefficient for a valve or fitting.

L is an equivalent length of pipe for a valve or fitting

$$N_R = \frac{VD\rho}{\mu} = \frac{VD}{\nu}$$

ρ = mass density

μ = dynamic viscosity

ν = kinematic viscosity

e = equivalent sand grain roughness

Good numerical data can be obtained from "Technical Paper 410", Crane Co., from "Standard Handbook for Mechanical Engineers" by Baumeister & Marks, or from most textbooks on fluid mechanics.

For noncircular ducts, D can be interpreted as the hydraulic diameter defined as 4A/P where A is the flow cross-sectional area and P is the wetted perimeter.

4. Use of experimental coefficients based on a dimensional analysis of the problem. In particular, for the problem of drag,

$C_D = f(N_R, N_F, N_M)$ for geometrically similar systems.

$$C_D = \frac{F_D}{A\rho\,\frac{V^2}{2}}$$ = drag coefficient; A is a characteristic area.

F_D is drag force.

Letting L represent a characteristic dimension and a represent the velocity of sound,

$$N_R = \frac{VL}{\nu} = \text{Reynolds' number.}$$

$$N_F = \frac{V}{\sqrt{Lg}} = \text{Froude number}$$

$$N_M = \frac{V}{a} = \text{Mach number}$$

Note: Froude number enters into problems involving free surfaces; Mach number enters into problems of compressible flow. It is difficult to imagine a problem in which both would be involved. Frequently, the drag coefficient may be considered constant (limited range of independent variables).

CONSERVATION PRINCIPLES

A control volume is an open system whose boundaries are fixed with respect to an inertial coordinate system. An open system is a system across whose boundaries mass, energy, and momentum can flow. When the principles of mass, energy and momentum conservation are employed to obtain equations applicable to a control volume, the resulting equations relate changes within the control volume to conditions at the boundaries. For steady state the rate of change of mass, energy, and momentum <u>within</u> the control volume is zero.

In the following equations summations (Σ) are to be interpreted as addition for inflow quantities, subtraction for outflow quantities. In expressions for evaluating terms, one-dimensional flow is assumed. t is time.

1. Mass (continuity)

$$\Sigma\dot{m} = \frac{dm}{dt} \qquad \dot{m} = \rho AV = \rho Q = \text{mass flow rate}$$

A is area normal to V, Q is volumetric flow rate, $\frac{dm}{dt}$ is the rate of change of mass <u>within</u> the control volume.

2. Energy (1st law of thermodynamics)

$$\dot{Q} + P + \Sigma(\dot{m}e) = \frac{dE}{dt}$$

Q is net heat transfer rate (positive for net input). Avoid confusion with volumetric flow rate.

P is power (mechanical or electrical) (positive for net input).

Note: For a pump $P = \frac{Q\gamma H}{\eta}$ where η = pump efficiency and H is the <u>increase</u> in total head produced by the pump.

$\frac{dE}{dt}$ is rate of change of energy <u>within</u> the control volume.

e = specific energy associated with a flow

$= u + \frac{p}{\rho} + gz + \frac{V^2}{2} = u + gH$ (H = total head at section of boundary under consideration)

u = specific internal energy: $u + \frac{p}{\rho} = h$ = specific enthalpy

3. Momentum (Newton's 2nd law of motion)

a.) $\Sigma\vec{F} + \Sigma(\dot{m}\vec{V}) = \frac{d\vec{M}}{dt}$ (Linear Momentum Equation)

$\vec{F}$ = external force acting on the control volume.

$\vec{M}$ = momentum within the control volume.

Note: This is a vector equation. Summations must be made vectorially or else the equation must be resolved into three algebraic component equations.

b.) $\Sigma T + \Sigma\dot{m}\vec{r} \times \vec{V} = \frac{d\vec{\mathcal{H}}}{dt}$ (Angular Momentum Equation)

$\vec{T}$ = external torque or moment of force with respect to a chosen moment center.

$\vec{r}$ = position vector (relative to the moment center) of the centroid of the area over which the velocity is V.

$\vec{\mathcal{H}}$ = angular momentum within the control volume.

FLUID MECHANICS 1

A cylindrical tank in a rocket contains 10 gallons of water. The internal tank dimensions are 8" in diameter and 100" long. The tank is 3000 feet above sea level and is being accelerated parallel to its length at 20 feet per second at an angle 20 degrees above horizontal. The space not occupied by water is air at 100 psia.

Determine the angle of the liquid surface exposed to the air.

Solution

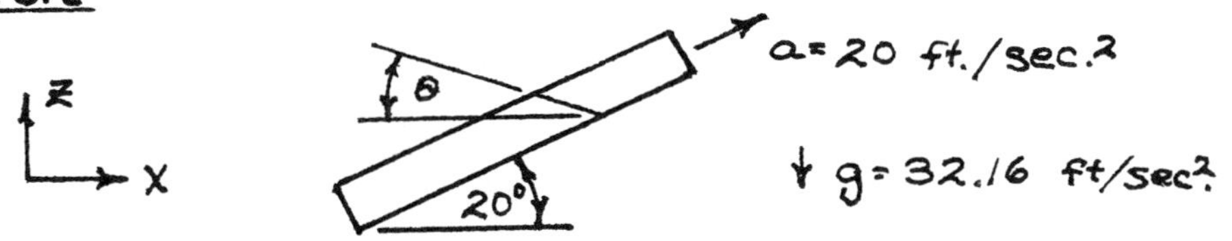

z, x

θ, 20°

$a = 20$ ft./sec.^{2}

$g = 32.16$ ft/sec.^{2}

$$dp = -\gamma\left[\frac{a_x}{g}\,dx + \left(1 + \frac{a_z}{g}\right)dz\right] = 0 \quad \text{(on interface)}$$

$$\theta = \tan^{-1}\left(-\frac{dz}{dx}\right) = \tan^{-1}\frac{a_x}{g + a_z}$$

$$a_x = a\cos 20^\circ = 18.80 \text{ ft./sec.}^2$$

$$a_z = a\sin 20^\circ = 6.85 \text{ ft./sec.}^2$$

$$\theta = \tan^{-1}\frac{18.80}{39.01} = \underline{25.7^\circ}$$

FLUID MECHANICS 2

Water flows through a perfect nozzle in the side of a water tank. The water level in the tank is held constant 20 feet above the ground level.

What height y should the nozzle be to make the stream from the orifice travel a maximum horizontal distance before it strikes the ground?

Ignore any friction.

Show all calculating and reasoning.

Solution

at nozzle exit: $u^2/2g = H = 20\text{ ft} - y$

$$\therefore\ u = \sqrt{2g(20\text{ ft.} - y)}$$

$x = ut$ and $y = \frac{1}{2}gt^2 \rightarrow t = \sqrt{2y/g}$

$$x = \sqrt{2g(20\text{ ft.} - y)(2y/g)} = 2\sqrt{20\text{ ft.}\ y - y^2}$$

$\frac{dx}{dy} = 0$ when $20\text{ ft.} - 2y = 0$

$$y = \underline{10\text{ ft.}}$$

FLUID MECHANICS 3

Six measurements are made on a ventilating duct system as shown in the sketch below. The measurements are all made within two feet of the fan. The air velocity and duct resistance on the suction side of the fan are 2,000 feet per minute and 0.80 inches of water respectively. The air velocity and duct resistance on the discharge side of the fan are 4,005 feet per minute and 2.5 inches of water respectively. The inlet of the suction duct is in the weather and the discharged air is in spaces maintained at atmospheric pressure.

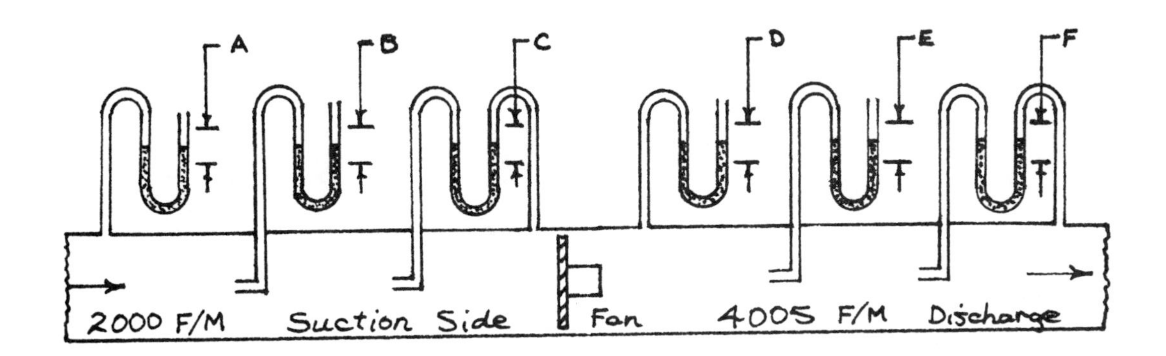

To find:

(a) With the fan in operation, what will be the manometer readings in inches of water at A, B, C, D, E, and F?

(b) Which of the manometers in the sketch indicates velocity pressure, total pressure, and static pressure for both the suction and discharge side of the fan?

Solution

Let subscripts refer to conditions at the point to which the left leg of the manometer of the subscript is attached. Let the subscript zero refer to the atmosphere.

a. $$\frac{V_A^2}{2g} = \frac{\left(\frac{2000\ \text{ft.}}{60\ \text{sec}}\right)^2}{64.4\ \text{ft/sec}^2} = 17.25\ \text{ft. of air} \times \frac{.075}{62.4} \times \frac{12\ \text{in}}{\text{ft.}}$$

$$= 0.25 \text{ inches of water}$$

$$\frac{U_D^2}{2g} = 4\frac{V_A^2}{2g} = 1.00 \text{ inch. of water}$$

$$\frac{P_O}{\gamma} = \frac{P_A}{\gamma} + \frac{V_A^2}{2g} + H_{L_{O-B}} = \frac{P_B}{\gamma} + H_{L_{O-B}}$$

$$\frac{P_O - P_A}{\gamma} = \frac{U_A^2}{2g} + H_{L_{O-B}} = 0.25 + 0.80$$

$$= \underline{1.05 \text{ inches of water}}$$

$$\frac{P_O - P_B}{\gamma} = H_{L_{O-B}} = \underline{0.80 \text{ inches of water}}$$

$$\frac{P_B - P_A}{\gamma} = \frac{V_A^2}{2g} = \underline{0.25 \text{ inches of water}}$$

* Note : Assume that the size of the duct on the discharge side of the fan is the same as on the suction side. The higher velocity on the discharge side is due to the blockage effect of the fan's wake. FUTHER DOWNSTREAM THE VELOCITY WILL AGAIN BE EQUAL TO V_A

$$\frac{P_E}{\gamma} = \frac{P_D}{\gamma} + \frac{U_D^2}{2g} = \frac{P_O}{\gamma} + \frac{U_A^2}{2g} + H_{L_{E-O}}$$

AN ALTERNATE ASSUMPTION WOULD BE THAT THE DUCT AREA ON THE DISCHARGE SIDE IS HALF THAT ON THE SUCTION SIDE. THIS WOULD YIELD A DIFFRENT SOLUTION

$$\frac{P_D - P_O}{\gamma} = H_{L_{E-O}} - \frac{V_D^2 - V_A^2}{2g} = 2.50 - (1.00 - 0.25)$$

$$= \underline{1.75 \text{ inches of water}}$$

$$\frac{P_E - P_O}{\gamma} = H_{L_{E-O}} + \frac{V_A^2}{2g} = 2.50 + 0.25$$

$$= \underline{2.75 \text{ inches of water}}$$

$$\frac{P_E - P_D}{\gamma} = \frac{V_D^2}{2g} = \underline{1.00 \text{ inches of water}}$$

b. Velocity Pressure : C & F

Total Pressure : B & E

Static Pressure : A & D

FLUID MECHANICS 4

Your plant operates a sheet plastic extrusion press with a nozzle .005" thick x 96" wide. You wish to increase production by installing a larger motor, speed reducer and pump combination. At 10 ft per second extrusion velocity, the pressure at the nozzle will be 1250 psi.

What size electric motor and what pump capacity are required to obtain the 10 ft per second rate?

Solution

$$Q = AV = 0.005 \text{ in} \times 96 \text{ in.} \times 10 \text{ ft/sec} = 4.80 \text{ in}^2 \frac{\text{ft.}}{\text{sec.}}$$

$$= 4.80 \text{ in}^2 \frac{\text{ft}}{\text{sec}} \times \frac{\text{ft}^2}{144 \text{ in}^2} \times \frac{7.48 \text{ gal.}}{\text{ft.}^3} \times 60 \frac{\text{sec.}}{\text{min.}}$$

$$= \underline{15 \text{ gpm}}$$

$$P_{ideal} = \dot{m}\frac{\Delta P}{\rho} = Q\Delta p$$

$$= \frac{4.80\ in^2\ \frac{ft.}{sec.} \times 1250\ \frac{lb}{in.^2}}{550\ \frac{ft \cdot lb}{hp \cdot sec}} = \underline{10.9\ Hp}$$

* Assuming reasonable efficiencies for pump, and speed reducer, a 20 Hp motor will be required.

FLUID MECHANICS 5

A 16" x 16" x 24" steel piping tee is to carry 10,000 gpm water (5,000 gpm from each branch) at 50 psi pressure. The joints are to be made with Dresser-type couplings that will not transmit axial thrust.

Find the reaction at its anchor.

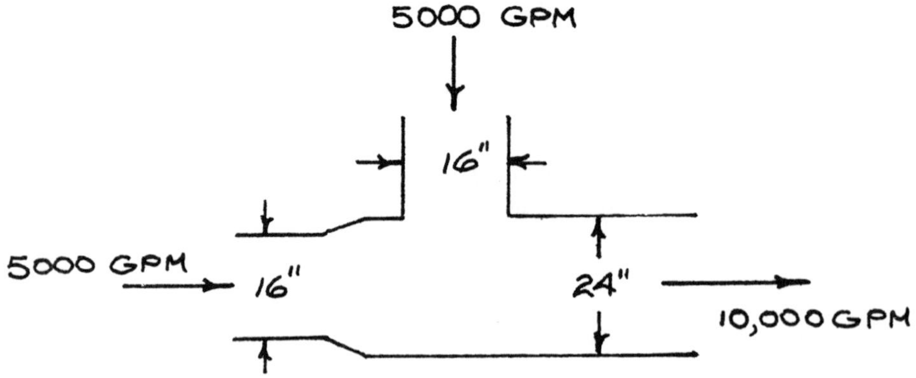

5000 GPM
16"
5000 GPM
16"
24"
10,000 GPM

Solution

$$V_1 = V_2 = \frac{Q_1}{A_1} = 7.97\ ft./sec.$$

$$V_3 = \frac{Q_3}{A_3} = 7.09\ ft./sec.$$

* Note: Since changes in velocity are small, assume uniform pressure.

$F_1 = F_2 = p A_1 = 50 \text{ lb/in}^2 \times \frac{\pi}{4} (16)^2 \text{ in}^2 = 10050 \text{ lb}$

$F_3 = p A_3 = 50 \times \pi/4 \, (24)^2 = 22650 \text{ lb}$

$$(\dot{m} V)_3 = \rho A_3 V_3^2 = \frac{\gamma A_3 V_3^2}{g} = \frac{62.4 \times \pi/4 \, (2)^2 (7.09)^2}{32.2}$$

$$= 307 \text{ lb.}$$

* Momentum changes are negligible

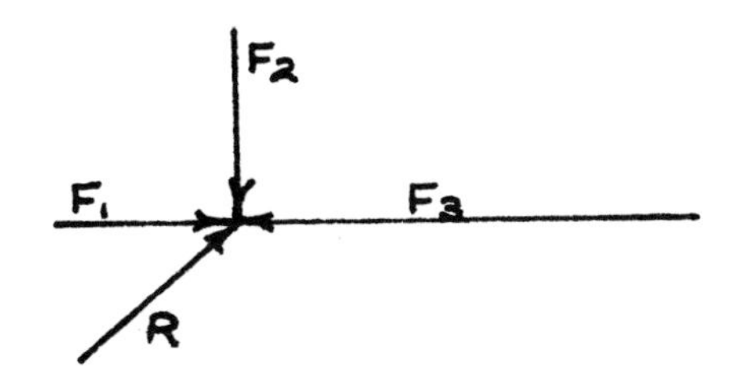

R

$F_2 = 10050$ lb

θ

$F_3 - F_1 = 12600$ lb

$$\theta = \tan^{-1} \frac{10050}{12600} = \underline{38.6°}$$

$$R = F_2 / \sin\theta = \underline{16100 \text{ lb}}$$

FLUID MECHANICS 6

A recent book describing World War II submarines is "Iron Coffins", 1st edition, Holt, Rinehart & Winston, New York, 1969, by H. A. Werner. On page 15 of the book it is stated that "Two electric motors, operating on gigantic storage batteries, ran the ship when she was submerged; they would propel the boat for one hour at the top speed of nine knots, or for three days at a cruising speed of one or two knots".

Give a technical explanation of the variation of underwater duration with speed noted by Captain Werner.

Solution

$$\text{Power output} = -\eta \frac{dE}{dt} \qquad \text{where } \eta = \text{efficiency}$$

$$= F_D V = \tfrac{1}{2} C_D A \rho V^3$$

$$-\frac{dE}{dt} = \frac{C_D A \rho}{2\eta} V^3$$

$$E_{max} - E_{min} = \frac{C_D A \rho}{2\eta} \int_0^t V^3 \, dt$$

* if $\frac{C_D A \rho}{2\eta}$ is constant (an assumption)

then $$\int_0^{t_1} V_1^3 \, dt = \int_0^{t_2} V_2^3 \, dt$$

If V_1 and V_2 are constant (given),

then $V_1^3 t_1 = V_2^3 t_2$

$$V_2 = \sqrt[3]{\frac{t_1}{t_2}} \, V_1 = \frac{9 \text{ knots}}{\sqrt[3]{72}} = 2.16 \text{ knots}$$

Therefore, 1 to 2 knots for 3 days is feasible.

FLUID MECHANICS 7

A high-rise 54-story office building in San Francisco has 2 water tanks located in the penthouse. The centrifugal pumps for supplying water to these tanks are located on the basement floor. Each pump discharges 400 gallons per minute through a 6-inch schedule 40 pipe. The pressure of the water entering the pump is 60 pounds per square inch.

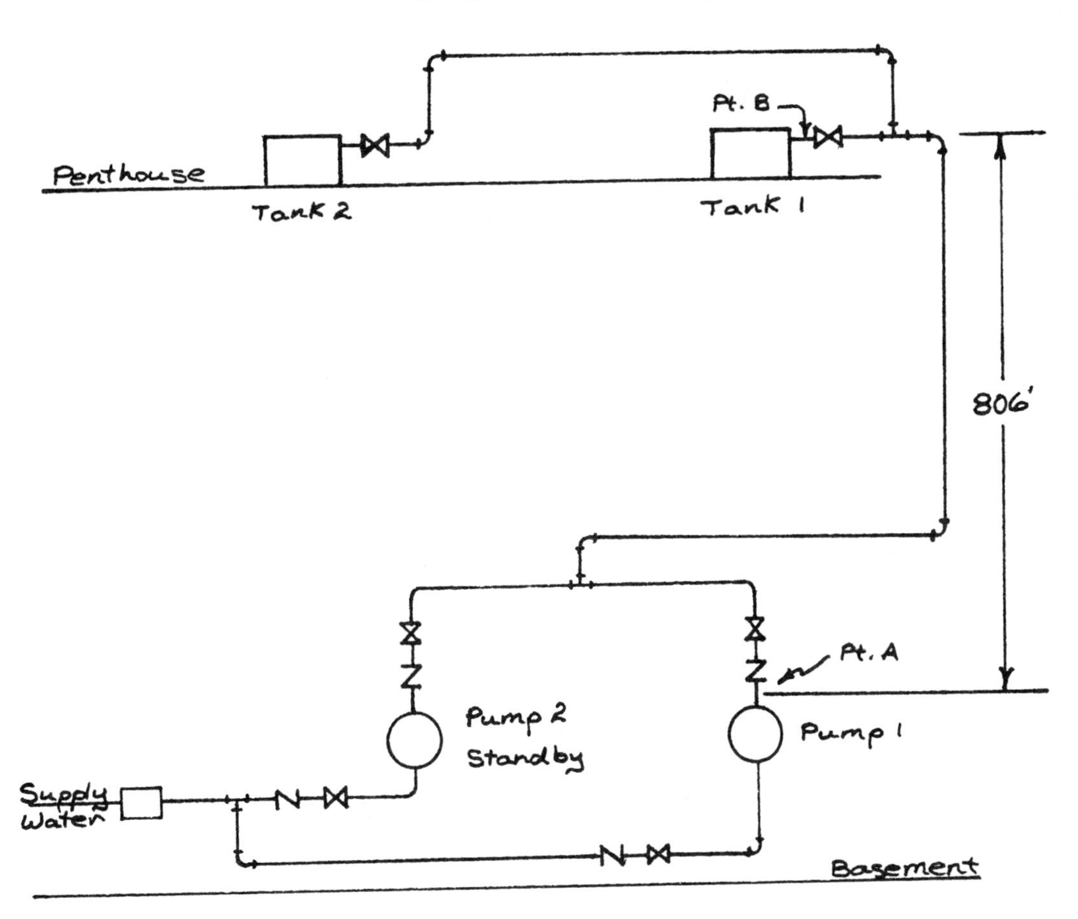

Point B is 806 ft above Point A.

Between Point A and Point B there is 910 ft of 6-inch (6.065" I.D.) piping and the head loss for the fittings and valves between these two points is 1.15 ft of water.

The water temperature is 60 degrees Fahrenheit.

The kinematic viscosity of the water is 1.13 centistokes.

The efficiency of the pump is 70 per cent.

The relative roughness of the pipe is $\frac{e}{D} = 0.001$.

Determine:
- a. Kinematic viscosity of the water in ft^2/sec.
- b. Velocity of the water in the pipe in feet per second.
- c. Reynolds number.
- d. Friction factor (f).
- e. Total head drop between Point A and Point B.
- f. The pump horsepower required.

Solution

a. $1.13 \text{ centistokes} \times \dfrac{cm^2}{100 \text{ centistoke sec}} \times \dfrac{in^2}{(2.54 cm)^2} \times \dfrac{ft^2}{144 in^2}$

$$= \underline{1.216 \times 10^{-5} \text{ ft}^2/\text{sec.}}$$

b. $V = Q/A = \dfrac{400 \text{ gal/min} \times ft^3/7.48 \text{ gal} \times \text{min}/60 \text{ sec}}{\pi/4 \left(\frac{6.065}{12}\right)^2 ft^2}$

$$= \frac{0.891 \text{ ft}^3/\text{sec}}{0.2006 \text{ ft}^2} = \underline{4.45 \text{ ft/sec}}$$

c. $N_R = \dfrac{VD}{\nu} = \dfrac{4.45 \times 6.065}{1.216 \times 10^{-5} \times 12} = \underline{1.85 \times 10^5}$

d. $f = \underline{0.0212}$ (from Moody Diagram)($\frac{e}{D} = 0.001$)

e. $H_L = f \dfrac{l}{D} \dfrac{V^2}{2g} + 1.15 \text{ ft}$

$$= 0.0212 \times \frac{910 \times 12}{6.065} \times \frac{(4.45)^2}{64.4} + 1.15$$

$$= 11.77 + 1.15 = \underline{12.9 \text{ ft}}$$

f. The pressure in the tank is not specified; it must be assumed. Assume 60 psig. Then the total head that the pump must supply is:

$$H = 806 \text{ ft.} + 13.3 \text{ ft.} \approx 820 \text{ ft}$$

$$P = \dot{m}gH + [\dot{m}(u_2 - u_1) - \dot{Q}]$$

$$P = \frac{\dot{m}gH}{\eta} = \frac{Q\gamma H}{\eta}$$

$$= \frac{0.891 \text{ ft}^3/\text{sec} \times 62.4 \text{ lb/ft}^3 \times 820 \text{ ft.}}{0.70 \times 550 \text{ ft} \cdot \text{lb/hp} \cdot \text{sec.}}$$

$$= \underline{118 \text{ hp}}$$

FLUID MECHANICS 8

A ranch complex requires a maximum water supply rate of 300 cubic feet per minute. The supply is a reservoir with its surface 250 ft above the point of delivery, and one-quarter mile distant. Minimum pressure required at point of delivery is 50 psig.

What minimum size of steel pipe would be needed for this pipe line from the reservoir to the delivery point?

Solution

$$H_1 = H_2 + H_L$$

$$250 \text{ ft} = \frac{50 \times 144}{62.4} + \frac{V^2}{2g} + \left(f\frac{\ell}{D} + \Sigma K_L\right)\frac{V^2}{2g}$$

$$\left(f\frac{\ell}{D} + 1.5\right)\frac{V^2}{2g} = 250 - 115 = 135 \text{ ft.}$$

(assuming a sharp-edged entrance)

$$V = \frac{Q}{A} = \frac{300 \text{ ft}^3/\text{min} \times \frac{1 \text{ min}}{60 \text{ sec}}}{\frac{\pi}{4} \times \frac{D^2}{144}} \qquad \begin{cases} V - \text{ft/sec} \\ D - \text{inches} \end{cases}$$

* Try D = 10 inches : $V = 9.16$ ft/sec , $\frac{V^2}{2g} = 1.30$ ft

$VD = 91.6$

$\frac{e}{D} = 0.00018 \qquad f = 0.0149$

$$f\frac{\ell}{D} = \frac{0.0149 \times 5280 \times 12}{4 \times 10} = 23.6$$

$$(23.6 + 1.5) \times 1.30 = 32.6 < 135$$

* Try D = <u>8 inches</u> : $100 < 135$

FLUID MECHANICS 9

Determine a standard size pipe for the drain line for the tank system shown on the sketch so that it will drain a full tank of water in as close to 9 hours as possible.

Show your calculations.

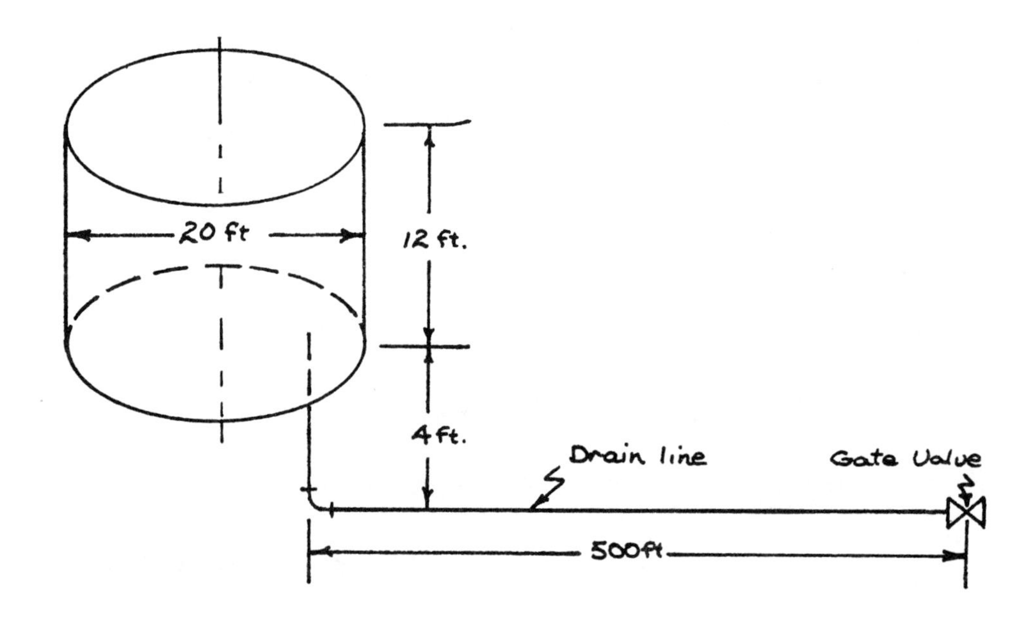

Solution

* Assume quasi-steady flow ; i.e., Bernoulli's augmented equation holds at any time t even though H changes with time.

$$H_1 = H \;,\; H_2 = \frac{V^2}{2g} = \frac{Q^2}{2gA^2} \;,\; H_L = f\left(\frac{\ell}{D} + \Sigma \frac{L}{D}\right)\frac{V^2}{2g}$$

Thus,

$$H = \left[1 + f\left(\frac{\ell}{D} + \Sigma \frac{L}{D}\right)\right]\frac{Q^2}{2gA^2} = KQ^2, \text{ assuming } f \text{ is constant}$$

By Continuity $Q = \frac{\pi(20\text{ ft})^2}{4}\left(-\frac{dH}{dt}\right) = \sqrt{\frac{H}{K}}$

$$\int_{16\text{ ft}}^{4\text{ ft}} H^{-1/2}\, dH = \int_{0}^{9\text{ hrs.}} \frac{-dt}{100\pi\sqrt{K}\ \text{ft}^2}$$

$$2H^{1/2}\Big]_{16\text{ ft.}}^{4\text{ ft.}} = \frac{-9\text{ hrs}}{100\,\pi\sqrt{K}\ \text{ft}^2} = 2(2-4)\text{ ft}^{1/2} = -4\text{ ft}^{1/2}$$

$$\sqrt{K} = \frac{9\text{ hrs.} \times 3600\text{ sec/hr}}{400\,\pi\ \text{ft}^{5/2}} = 25.8\text{ sec. ft}^{-5/2}$$

$$K = 665\text{ sec}^2\text{ ft}^{-5}$$

$$Q_{min} = \sqrt{\frac{H_{min}}{K}} = \frac{\sqrt{4\text{ ft}}}{25.8\text{ sec ft.}^{-5/2}} = 0.0775\text{ ft}^3/\text{sec}$$

$$Q_{max} = \sqrt{\frac{H_{max}}{K}} = \frac{4}{25.8} = 2Q_{min}$$

* Assume: wide open gate valve $L/D = 13$

Standard 90° Elbow $L/D = 30$

Sharp-Edged Entrance $K_L = 0.5$

* Crane : "Flow of Fluids" Tech. paper 410

Then,

$$K = \frac{1.5 + f\left(\frac{504\text{ ft}}{D} + 43\right)}{2g\left(\frac{\pi}{4}\right)^2 D^4}$$

$$\frac{1.5 + f\left(\frac{504\text{ ft}}{D} + 43\right)}{D^4} = 665 \times 64.4\left(\frac{\pi^2}{16}\right)$$

$$= 26{,}500\text{ ft}^{-4}$$

* Assume $D = 3'' = .25$ ft and Steel

$$V_{min} = Q/A = \frac{.0775}{\frac{\pi}{4}\left(\frac{1}{4}\right)^2} = 1.58\text{ ft/sec.}$$

$$VD_{min} = 4.74\ \frac{\text{ft}}{\text{sec}}\text{ in} \qquad VD_{max} = 9.48\ \frac{\text{ft}}{\text{sec}}\text{ in}$$

$$\frac{e}{D} = .0006 \qquad f_{mean} = .023$$

$$[1.5 + .023(2016 + 43)] \times 256\text{ ft}^{-4} = 12{,}500 < 26{,}500\text{ ft}^{-4}$$

*Assume $D = 2\frac{1}{2}''$

$$U_{min} = 1.58 \times \left(\frac{3}{2.5}\right)^2 = 2.28 \text{ ft/sec}$$

$UD_{min} = 5.7$ $\qquad UD_{max} = 11.4$

$e/D = .0007$ $\qquad f_{mean} = 0.023$

$$[1.5 + .023(2420 + 43)]\left(\frac{12}{2.5}\right)^4 = 30{,}900 \text{ ft}^{-4} > 26{,}500 \text{ ft}^{-4}$$

Therefore, use $2\frac{1}{2}''$ steel pipe

FLUID MECHANICS 10

A pipe line with diameter 36 in and length 2.5 miles, connects two fresh-water open reservoirs which have their water surfaces at elevations 240 ft and 165 ft, respectively. In order to increase the rate of flow between the reservoirs by exactly 30 per cent, it is decided to lay an additional 36 in diameter pipe line from the upper reservoir. The second pipe line is to lie parallel to the original pipe line and is to be connected to the latter at the point which produces this increase in flow.

Determine this point of connection, assuming that the friction factor, f, has the value 0.01 for each pipe line.

Minor losses may be considered to be negligible.

Solution

$$H_L = f \frac{\ell}{D} \frac{U^2}{2g} = \left(\frac{8f}{\pi^2 g D^5}\right) \ell Q^2$$

By assumption $\frac{8f}{\pi^2 g D^5}$ is the same throughout old and new systems.

Let x = Length of new pipe and note that total head loss is the same for both systems.

Thus, $\ell Q^2 = x\left(\frac{1.3}{2} Q\right)^2 + (\ell - x)(1.3Q)^2$ and $x = 0.544\,\ell$

$= 1.36$ mi.

FLUID MECHANICS 11

Fuel oil is to be removed from a barge by a portable pump at a rate of 200 gallons per minute. The pump is to discharge the oil into a 6-inch schedule 40 steel pipe oil transfer line that runs from the dock to a storage tank, as shown schematically. Point A and Point B are at the same elevation

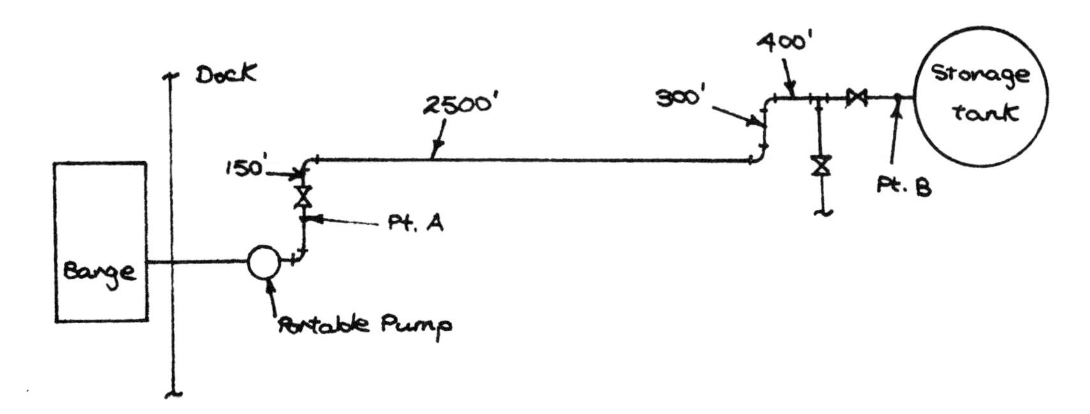

Required fittings and valves:

- 4 long-radius flanged 90-degree ells
- 1 tee, flanged
- 2 gate valves, flanged, wide open

Oil tempeature is 50°F, and its viscosity is 3400 Saybolt Seconds Universal. Specific gravity of the oil at 50°F is 0.934.

Determine the pump power and head required to perform this task.

Solution

* Assume entrance and exit losses are negligible and that there is no change in elevation.

$$H_L = f\left(\frac{\ell}{D} + \Sigma \frac{L}{D}\right)\frac{v^2}{2g} = H \text{ (required from pump)}$$

$$\ell/D = \frac{150 + 2500 + 300 + 400}{1/2} = 6700$$

$L/D = 4 \times 20 = 80$ 90° ells (long radius)

$1 \times 20 = 20$ Tee (flow through run)

$2 \times 13 = \underline{26}$ Gate values (wide open)

$\Sigma L/D = 126$ (turbulent flow)

$$U = Q/A = 200\ \text{gal/min.} \times \frac{\text{ft}^3}{7.48\ \text{gal.}} \times \frac{\text{min}}{60\ \text{sec}} \times \frac{4 \times 144}{\pi (6.06)^2\ \text{ft}^2}$$

$$= 2.23\ \text{ft/sec}$$

$$\nu = \frac{3400\ \text{sec. S.U.}}{4.635\ \frac{\text{sec. S.U.}}{\text{centistoke}}} \times 1.076 \times 10^{-5}\ \frac{\text{ft}^2}{\text{sec centistoke}}$$

$$= 7.88 \times 10^{-3}\ \text{ft}^2/\text{sec}$$

$$N_R = \frac{UD}{\nu} = \frac{2.23 \times 6.06}{7.88 \times 10^{-3} \times 12} = 143 \quad (\therefore \text{Laminar flow})$$

$$f = \frac{64}{N_R} = \frac{64}{143} = 0.448$$

$$(L/D)_{\text{Laminar}} = \frac{N_R}{1000} \left(\frac{L}{D}\right)_{\text{Turbulent}} \quad \text{for } N_R < 1000$$

(from Crane, technical paper 410)

$$\Sigma L/D = 0.143 \times 126 = 18$$

$$H_L = 0.448 \times 6720 \times \frac{(2.23)^2}{64.4} = 233\ \text{ft.} = H$$

$$P = \frac{Q \gamma H}{\eta} \quad \text{assume } \eta = 60\%$$

$$P = \frac{200\ \text{gal/min}}{7.48\ \text{gal/ft}^3} \times \frac{.934 \times 62.4\ \text{lb/ft}^3 \times 233\ \text{ft.}}{.6 \times 33000\ \frac{\text{ft.lb}}{\text{hp}\cdot\text{min}}}$$

$$= \underline{18.3\ \text{hp}}$$

FLUID MECHANICS 12

A hydro-pneumatic tank is to be used as an accumulator in a five-story building water system, as shown.

The pressures are a maximum of 90 psig on the first floor and a minimum of 50 psig on the fifth floor. First floor elevation is 120 feet, fifth floor, 160 feet, and tank, 170 feet. The pump is started and stopped by a pressure switch on the tank. The pump has a constant flow of 100 gpm when it runs. The system will use water at anywhere from zero to 100 gpm.

What capacity tank would you specify to allow a maximum of 6 pump starts per hour? Neglect friction loss.

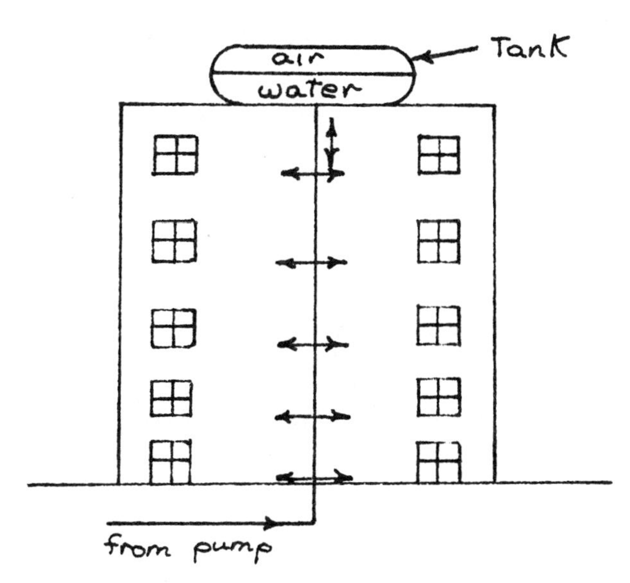

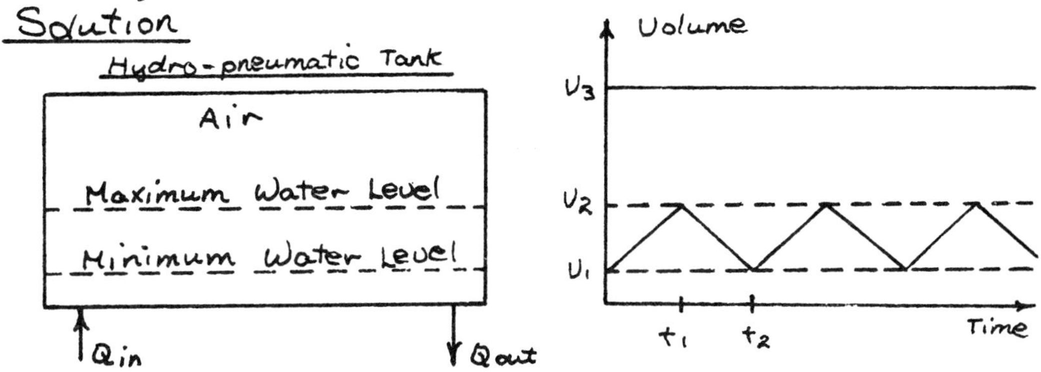

$$\Sigma \dot{m} = \frac{dm}{dt} \; ; \quad \text{for } \rho = \text{constant} \; \Sigma Q = \frac{dV}{dt} \; ; \quad V = \text{volume}$$

$$(Q_{in} - Q_{out})t_1 = Q_{out}(t_2 - t_1) = V_2 - V_1$$

Let $x = \frac{Q_{out}}{Q_{in}}$ Then $(1-x)t_1 = x(t_2 - t_1)$ and $t_1 = x t_2$

$$V_2 - V_1 = Q_{in}(1-x)x t_2 \; ; \quad \frac{d(V_2 - V_1)}{dx} = Q_{in} t_2 (1 - 2x)$$

$(1 - 2x) = 0 \rightarrow x = \frac{1}{2}$ for maximum value of $V_2 - V_1$

$$V_2 - V_1 = 100 \text{ gpm} \times \tfrac{1}{2} \times \tfrac{1}{2} \times 10 \text{ min} = 250 \text{ gal.}$$

---

* Assume the change of pressure directly due to variation of water surface elevation is negligible.

$$90 \text{ psig} = (P_{air})_{max} + \frac{62.4 \text{ lb/ft}^3 \times 50 \text{ ft}}{144 \text{ in}^2/\text{ft}^2} \rightarrow (P_{air})_{max} = 68.3 \text{ psig}$$

$$50 \text{ psig} = (P_{air})_{min} + \frac{62.4 \times 10}{144} \rightarrow (P_{air})_{min} = 45.7 \text{ psig}$$

---

* Assume perfect gas and constant temperature

Then $p_1(V_3 - V_1) = p_2(V_3 - V_2)$

$$= P_2[(V_3 - V_1) - (V_2 - V_1)]$$

$$V_3 - V_1 = \frac{P_2}{P_2 - P_1}(V_2 - V_1)$$

$$= \frac{68.3 + 14.7}{22.6} \times 250 \text{ gal} = 918 \text{ gal.}$$

Let $V_3 = \underline{1000 \text{ gal}}$

FLUID MECHANICS 13

Two identical pumps, each pump 800 gallons per minute of water against a total delivery head of 475 feet, as shown.

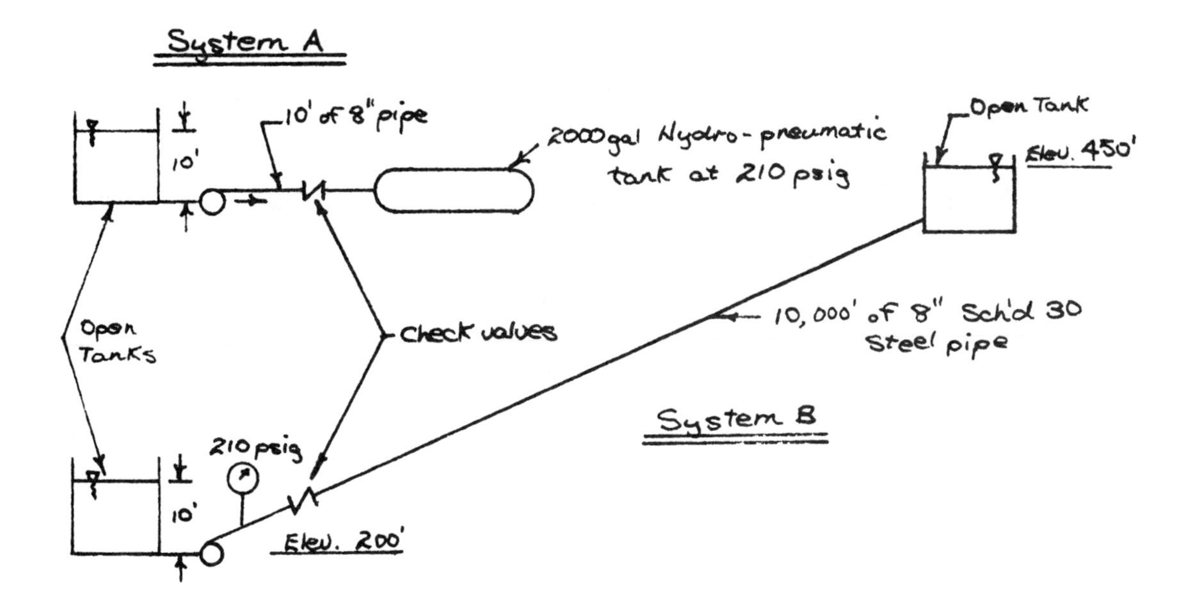

An observer is puzzled that the swing check valve on A slams noisily when the pump stops, while an identical valve on B closes quite acceptably. He feels that this is contrary to his understanding that water hammer is a problem only on long pipe lines.

(a) Explain the phenomenon.

(b) The valve flanges are rated for 400 psi. Is this adequate?

(c) Is the interior of the pipe line of system B in good, fair, or poor condition?

Solution

a. The noisy closing of a valve does not necessarily imply water hammer, nor does the quiet closing of a valve imply that there is no water hammer problem. Water hammer is associated with large pressure fluctuations due to the inertial effects in a transcient flow. In system A there is very little inertia, the flow quickly reverses and the valve slams shut; but the pressure is maintained at approximately 210 psi. In system B there is a great deal of inertia; the flow reverses more slowly and the valve closes quietly. Nevertheless, there may be large pressure fluctuations.

b. There is not sufficient data given to determine what actually happens in system B, but if the valve closes within

$$\frac{2L}{a}\text{ sec} \simeq \frac{20{,}000\text{ ft}}{4\,000\text{ ft/sec}} = 5\text{ sec}$$

then variations of head given by the following equation could occur.

$$\Delta H = \frac{aU}{g} \approx \frac{4000 \times 5\ \text{ft}^2/\text{sec}^2}{32.2\ \text{ft/sec}^2} = 620\text{ ft.}$$

* See below for calculation of U.

This corresponds to a pressure variation of 269 psi. This much pressure reduction would not occur, for the pressure can not fall below the vapor pressure. Column separation occurs and the pipe would probably collapse. In any event, on rejoining a large pressure rise occurs. The flanges will be subject to pressures greater than their rating for short periods of time. They probably will not fail, however, because of the safety factor in the rating.

c. $$Q = 800\text{ gpm} \times \frac{\text{ft}^3}{7.48\text{ gal}} \times \frac{\text{min}}{60\text{ sec.}} = 1.78\ \text{ft}^3/\text{sec}$$

$$U = Q/A = \frac{1.78}{\frac{\pi}{4}\left(\frac{8.071}{12}\right)^2} = 5.02\ \text{ft/sec}$$

$$\frac{V^2}{2g} = 0.391 \text{ ft.}$$

$$H_L = 475 - (450 - 210) = 235 \text{ ft.}$$

* assume negligible minor losses

$$H_L = f \frac{\ell}{D} \frac{V^2}{2g} \qquad f = \frac{235}{15000 \times 0.391} = 0.04$$

* For new steel pipe :

$$e/D = \frac{0.00015}{2/3} = 0.000225$$

$$VD = 5 \text{ ft/sec} \times 8 = 40$$

$f = 0.017$ (Moody Diagram)

$$0.04 >> 0.017$$

pipe is in poor condition

FLUID MECHANICS 14

You are selecting the pipe diameter for an 800-foot pipe transmission line with one outlet to irrigate a field of yet undetermined size. The required flow is 8 gallons per minute per acre and the consumption is 4.1 acre-feet per acre, per year. The water is pumped with a pump of 75 per cent overall efficiency at a cost of 0.7¢ per kilowatt-hour. At 325 gallons per minute, friction is 101 feet per 1000 feet in 4 in pipe, and 12 feet per 1000 feet in 6 in pipe. Economic life is 40 years and money is worth 7%. The 4 in pipe costs $1.25 and the 6 in pipe costs $1.50 per foot installed.

Find the economic break-even point between the two pipe sizes.

Solution

Let A be the number of acres to be irrigated

Let T be the pumping time per year

Then $Q = 8A$ gpm ; $QT = 4.1A$ acre-ft./year

$4.1/8$ acre·ft/gpm $= 2783$ hrs./year $= T$

Note : 1 acre·ft = 43560 ft^3 ; 7.48 gal. = 1 ft^3

$$\frac{H_4}{101\text{ ft}} = \frac{H_6}{12\text{ ft.}} = \frac{800}{1000}\left(\frac{8A}{325}\right)^2 = 4.847\times10^{-4}\,A^2$$

Note: the above is from $H_L = f\,\frac{L}{D}\,\frac{U^2}{2g}$ and $U = \frac{4Q}{\pi D^2}$

assumes f is independent of N_R

$$P = \frac{Q\gamma H}{\eta} \quad \therefore \frac{P_4}{101\text{ ft}} = \frac{P_6}{12\text{ ft.}} = \frac{62.4\text{ lb/ft}^3 \times 8 \times 4.847\times10^{-4}A^3\text{ gpm}}{0.75}$$

$$\frac{62.4\frac{\text{lb}}{\text{ft}^3}\times 8 \times 4.847\times10^{-4}A^3\frac{\text{gal}}{\text{min}}}{0.75}\times\frac{\text{ft}^3}{7.48\text{ gal}}\times\frac{\text{hp}\cdot\text{min}}{33\times10^3\text{ ft}\cdot\text{lb}}\times\frac{0.746\text{ Kw}}{\text{hp}}$$

$$\frac{P_4}{101} = \frac{P_6}{12} = 9.75\times10^{-7}\,A^3\text{ Kw}$$

Annual pumping cost due to pipe friction $= PT \times \$.007 = \$19.48\,P$

4" pipe : $(\$19.48 \times 9.75\times10^{-7}A^3)\times 101 = \$1.918\times10^{-3}\,A^3$

6" pipe : $(\$1.899\times10^{-5}\,A^3)\times 12 = \$0.228\times10^{-3}\,A^3$

Annual investment cost : (capital recovery factor = .0750)

4" pipe : 800 ft × \$1.25/ft × .0750 = \$75

6" pipe : $\$75 \times \frac{1.50}{1.25}$ = \$90

For break-even point, increment of annual investment cost equals increment of savings in operating costs. Thus

$\$15 = \$1.690\times10^{-3}A^3 \rightarrow$ <u>A = 20.7 acres</u>

6

Thermofluid Mechanics

RICHARD K. PEFLEY

The more elaborate problems associated with this topic may require a complex mix of determinations involving system behavior, physical principles and property relationships. A brief summary of the major features relating to such problems follows:

A. SOLUTION PROCEDURE CHECK LIST

1. Establish whether the system is an open (control volume) or closed system--recall that open and closed relates to matter (working substance) crossing the system's boundaries.

2. Specify whether the event being examined constitutes a process or cycle.

3. Determine the phase region in which the working substance is situated.

4. Identify and apply the physical principles (commonly Newton's second law of motion, the first and second laws of thermodynamics, and conservation matter) that give insight to the problem.

5. Introduce working substance property relationships and process or cycle specifications.

6. Achieve a solution and reflect to see if the answer makes sense.

B. POINTS TO REMEMBER RELATIVE TO PURE SUBSTANCE MOLECULAR STATE BEHAVIOR

1. In the single phase region the substance has two degrees of freedom, i.e., two properties such as p and T may be independently varied.

2. If two phases are in equilibrium, each phase component loses one degree of freedom, i.e., pressure and temperature become dependent and specifying one establishes all other properties such as density and viscosity.

3. If three phases coexist, the molecular state of each phase component is fixed, i.e., each phase has zero degrees of freedom and only the amount of each phase can be varied.

4. For most processes involving pure solids or liquids, the density and specific volume can usually be treated as constants. As a second approximation the coefficient of compressibility (κ) and cubical expansion (β) can be treated as constants which allows the density and volume change to be determined by

$$\ell n \frac{\nu_2}{\nu_1} = \ell n \frac{\rho_1}{\rho_2} = \kappa(p_2 - p_1) + \beta(T_2 - T_1)$$

5. For the perfect gas region the density is related to pressure and temperature by

$$\frac{p}{\rho} = RT$$

C. PROCESS DESCRIPTIONS — PROCESSES ARE DESCRIBED BY PROPERTIES THAT REMAIN CONSTANT

isothermal - constant temperature
isobaric - constant pressure
isometric - constant volume
isentropic - constant entropy
throttling - constant enthalpy

Note that processes produce different effects depending upon where they take place in the phase regions. For example, a throttling process results in a slight increase in temperature for liquids, a drop in temperature for a mixture of liquid and vapor and a constant temperature for a perfect gas.

D. PROCESS EFFICIENCIES ARE PROCESS RATINGS

For examples consider volumetric and isentropic efficiencies of a reciprocating compressor.

$$\eta_{vol} = \frac{\text{actual volume of fluid displaced}}{\text{ideal volume of fluid displaced}}$$

$$\eta_{isen} = \frac{\text{reversible adiabatic work of compression}}{\text{actual work of compression}}$$

One efficiency has to do with the breathing quality of the process while the other has to do with energy use efficiency.

E. CYCLE AND CYCLE RATINGS

The Carnot cycle is the basic cycle for considering conversion of heat to work--a heat engine, or for pumping heat--a heat pump.

Cycle rating definitions and their particular values for the Carnot cycle are as follows:

$$\eta_{t_{\text{heat eng.}}} = \frac{W_{net}}{Q_{added}} = 1 - \frac{T_c}{T_h}$$

$$\text{C.O.P.}_{\text{heat pump}} = \frac{Q_{delivered}}{W_{net}} = \frac{1}{\eta_t} = \frac{1}{1 - \frac{T_c}{T_h}}$$

$$\text{C.O.P.}_{\text{refrigerator}} = \frac{Q_{\text{picked up}}}{W_{net}} = \text{C.O.P.}_{\text{heat pump}} - 1$$

Note that all quantities here are cyclic quantities and are easily identified on the T-s plane since the Carnot cycle is a rectangle on this plane.

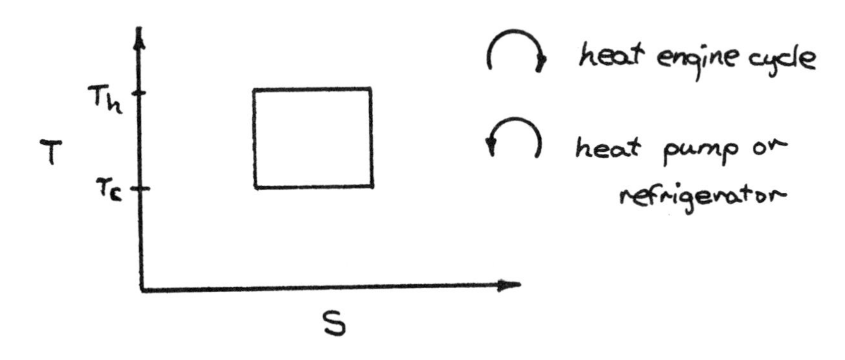

F. SUMMARY OF PHYSICAL PRINCIPLES AND PROPERTY RELATIONSHIPS

1. Property Relationships for a Pure Substance

   a. p-v-T surface of state

   b. Incompressible fluid - ρ is a constant:

   $$du = c\,dT \qquad ds = \frac{c\,dT}{T}$$

   $$dh = du + v\,dp$$

   c. Mixed phase:

   $$x = \frac{m_g}{m_{f+g}} \qquad v_x = v_f + x v_{fg} \qquad \text{same for } h_x,\ u_x,\ s_x$$

   d. Perfect gas:

   $$pv = \frac{R_o}{M} T \quad du = c_v dT,\ ds = \frac{c_v dT}{T} + R\frac{dv}{v},\ k = c_p/c_v$$

   $$c_p - c_v = R, \quad dh = c_p dT,\ ds = \frac{c_p dT}{T} - R\frac{dp}{p}$$

2. Continuity Principle

$$\Sigma\ \dot{m} = \frac{dm}{d\tau}$$

3. First Law of Thermodynamics

a. Closed system

$$Q + W = \Delta E \begin{matrix}\text{final}\\ \text{initial}\end{matrix}$$

$$\Delta E = \Delta U + \Delta KE + \Delta PE$$

b. Open system or control volume

$$\frac{dQ}{d\tau} + P + \Sigma\ \dot{m}e)_{\text{inflow}} = \frac{dE}{d\tau}$$

$$e = u + pv + z\ \frac{g}{g_c} + \frac{V^2}{2g_c} = h + z\ \frac{g}{g_c} + \frac{V^2}{2g_c}$$

4. Momentum Principle

a. Closed system

$$\vec{F} = \frac{1}{g_c}\frac{d}{d\tau}\ (m\vec{V}) \qquad \Sigma\vec{T} = \frac{1}{g_c}\frac{d}{d\tau}\ (I\vec{\omega})$$

b. Open system or control volume

$$\Sigma\vec{F} + \frac{1}{g_c}\ \Sigma\ \dot{m}\vec{V})_{\text{inflow}} = \frac{1}{g_c}\frac{d}{d\tau}\ (m\vec{V})_{\text{system}}$$

5. Second Law of Thermodynamics

a. Closed system

$$\frac{dQ}{T}_{\text{surroundings}} \le dS_{\text{system}}$$

b. Open system or control volume

$$\frac{1}{T_{sur}}\frac{dQ}{d\tau} + \Sigma\ \dot{m}s \le \frac{dS}{d\tau})_{\text{system}}$$

G. TERMINOLOGY FOR THERMOFLUID MECHANICS REVIEW

c, c_p, c_v	specific heats
E	total energy
e	specific energy
F	force
f	Darcy-Stanton friction factor
g	acceleration of gravity
g_c	proportionality in Newton's second law of motion
h	enthalpy

H	head
H_L	head loss
I	mass moment of inertia
k	specific heat ratio
m	mass
$\dot{m}$	mass flow rate
M	mass equivalent of molecular weight
p	pressure
P	power
Q	heat
R	gas constant
R_o	universal gas constant
r	radius
T	temperature
s	specific entropy
v	specific volume
V	velocity
W	work
x	quality
z	elevation
u	internal thermal energy
U	total internal thermal energy
ρ	density
γ	specific weight
ω	angular velocity
τ	time
η	efficiency

THERMOFLUID MECHANICS 1

Air is compressed reversibly and adiabatically to 3 atmospheres. It is then cooled back to room temperature and stored in a reservoir until used. If it expands reversibly and adiabatically through a positive displacement motor:

(a) What is the ratio of work input to work output?

(b) Explain whether or not it would be advantageous to prevent the air from cooling back to room temperature before entering the air motor.

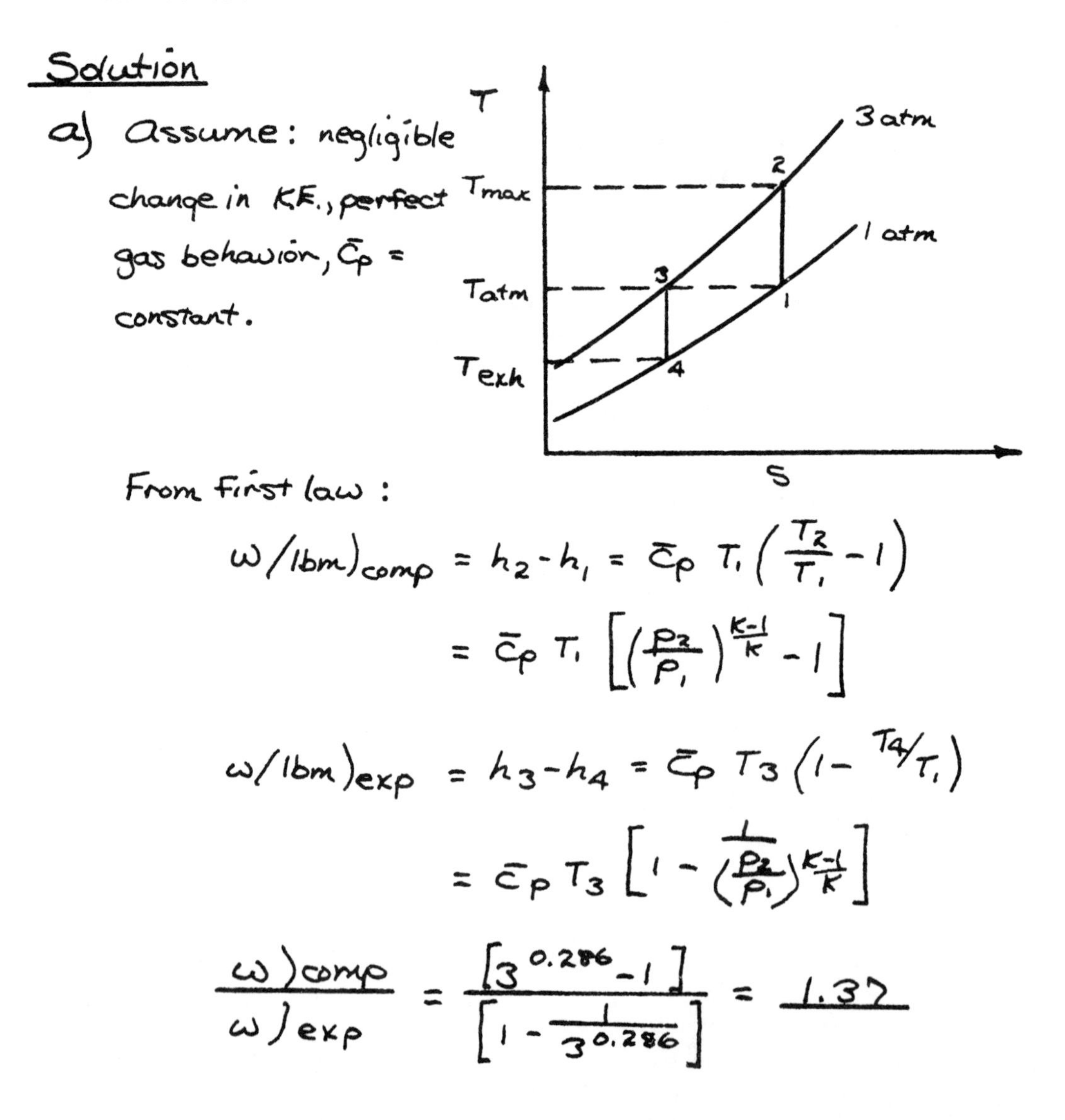

Solution

a) Assume: negligible change in K.E., perfect gas behavior, $\bar{C}_p$ = constant.

From First law:

$$w/lbm)_{comp} = h_2 - h_1 = \bar{C}_p T_1 \left(\frac{T_2}{T_1} - 1\right)$$

$$= \bar{C}_p T_1 \left[\left(\frac{P_2}{P_1}\right)^{\frac{k-1}{k}} - 1\right]$$

$$w/lbm)_{exp} = h_3 - h_4 = \bar{C}_p T_3 \left(1 - T_4/T_1\right)$$

$$= \bar{C}_p T_3 \left[1 - \frac{1}{\left(\frac{P_2}{P_1}\right)^{\frac{k-1}{k}}}\right]$$

$$\frac{w)_{comp}}{w)_{exp}} = \frac{\left[3^{0.286} - 1\right]}{\left[1 - \frac{1}{3^{0.286}}\right]} = \underline{1.37}$$

b) Process will yield more work per pound by

—continued—

not cooling. Condensation and possibly icing in lines and expander may be avoided. The expander will run hotter which if handled may be a disadvantage.

THERMOFLUID MECHANICS 2

A hilsch tube is a device whereby compressed air at room temperature is caused to spin by discharging through a nozzle into a cylinder. The core air is bled off through one exhaust line and is found to be cold and the peripheral air is bled off through another exhaust pipe and is found to be warm. During this process there is no heat transfer or work exchange with the surroundings. The steady state data taken from this system follows:

	Inlet Air	Hot Outlet Air	Cold Outlet Air
Velocity (ft/sec)	10.0	30.0	20.0
Pressure (psia)	40.0	21.8	14.7
Temperature (F)	70.0	112.0	- 60.0
Flow Area (in^2)	2.00	1.00	0.50

(a) What physical laws are available for checking the authenticity of this data?

(b) Use these laws to check the data to confirm its authenticity.

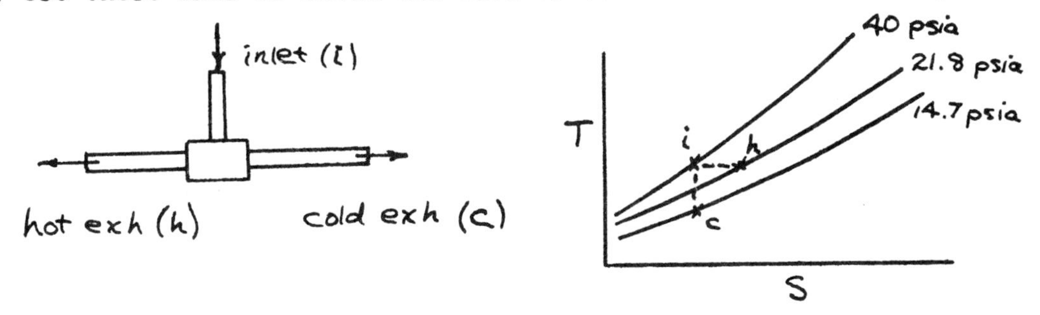

Solution

Principles: a) conservation of matter

b) 1st Law of thermodynamics

c) 2nd Law of thermodynamics

Assume : steady, 1-dimensional, adiabatic flow of a perfect gas.

Chk a) $\dot{m}_i = U\rho A = \frac{U p A}{RT} = \frac{10(40)2}{53.3(530)} = 0.0283 \frac{lbm}{sec}$

$\dot{m}_c = \frac{20(14.7)\,0.5}{53.3(400)} = 0.0069 \frac{lbm}{sec}$

$\dot{m}_h = \frac{30(21.8)\,1.0}{53.3(572)} = 0.0214 \frac{lbm}{sec}$

$\Sigma\, \dot{m}_c + \dot{m}_h = 0.0283 \frac{lbm}{sec}$

(0.0128 Kg/SEC)

a) OK ⟵

Chk b) $\dot{m}_h \left[c_p (T_h - T_i) + \frac{(U_h^2 - U_i^2)}{2g_c} \right]$ (neg.)

$= \dot{m}_c \left[c_p (T_i - T_c) + \frac{(U_i^2 - U_c^2)}{2g_c} \right]$ (neg.)

$0.0214\left[0.24\,(112-70)\right] \stackrel{?}{=} 0.0069\left[0.24\,(70+60)\right]$

$0.215 = .215$ BTU/SEC

(0.227 KJ/SEC) b) OK ⟵

Chk c) $\Sigma\, \dot{S}_{outflow} - \dot{S}_{inflow} \geq 0$

$\dot{m}_h \left[c_p \ln \frac{T_h}{T_i} - R \ln\left(\frac{p_h}{p_i}\right) \right] + \dot{m}_c \left[c_p \ln\left(\frac{T_c}{T_i}\right) - R \ln\left(\frac{p_c}{p_i}\right) \right] \geq 0$

$0.214\left[0.0185 + .0418\right] + 0.0069\left[-0.067 + 0.069\right] > 0$

Note: $\frac{BTU}{SEC\,°R} \times 1.90$ YIELDS $\frac{KJ}{SEC\,°K}$

c) OK ⟵

THERMOFLUID MECHANICS 3

A two stage compressor develops a 1.27:1 pressure ratio in the first stage with a total-to-total efficiency of 90% and a 1.31:1 pressure ratio in the second stage with a total-to-total efficiency of 88%. If the specific heat ratio was 1.400, what is the compressor efficiency?

Solution

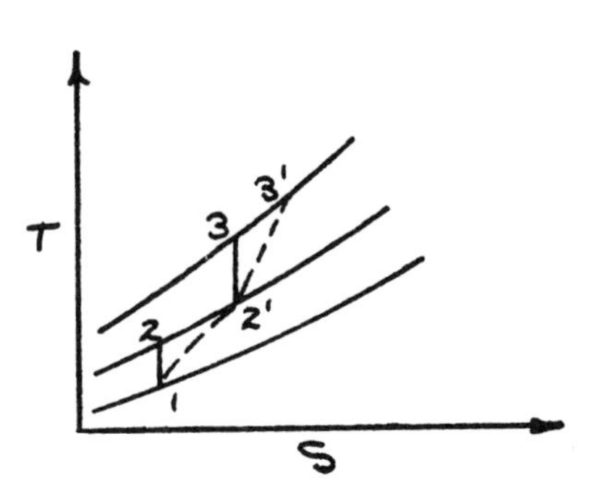

Assume: Total to total means stagnation enthalpy to stagnation enthalpy, perfect gas, constant c_p

η is isentropic efficiency

$$\eta_{overall} = \frac{\omega_{(s)}}{\omega_{actual}} = \frac{\omega_{(s)}}{\omega_{1^{st}stage} + \omega_{2^{nd}stage}}$$

Let $P_2/P_1 = r_{p_1}$, $P_3/P_2 = r_{p_2}$, $\frac{k-1}{k} = \alpha$

$$\omega_{1^{st}stage} = h_{2'} - h_1 = \frac{c_p T_1 (T_2/T_1 - 1)}{\eta_1} = \frac{c_p T_1 (r_{p_1}^{\alpha} - 1)}{\eta_1}$$

$$\omega_{2^{nd}stage} = \frac{c_p T_{2'} (r_{p_2}^{\alpha} - 1)}{\eta_2} \quad \text{but } T_{2'} - T_1 = \frac{T_2 - T_1}{\eta_1}$$

$$\therefore \omega_{2^{nd}stage} = \frac{c_p (r_{p_2}^{\alpha} - 1)}{\eta_2} (T_1)\left(1 + \frac{r_{p_1}^{\alpha} - 1}{\eta_1}\right)$$

$$= \frac{c_p T_1}{\eta_1 \eta_2} (r_{p_2}^{\alpha} - 1)(\eta_1 + r_{p_1}^{\alpha} - 1)$$

Then
$$\eta_{overall} = \frac{c_p T_1 \left[(r_{p_1} r_{p_2})^{\alpha} - 1\right]}{\frac{c_p T_1}{\eta_1}(r_{p_1}^{\alpha} - 1) + \frac{c_p T_1}{\eta_1 \eta_2}(r_{p_2}^{\alpha} - 1)(\eta_1 + r_{p_1}^{\alpha} - 1)}$$

$$\eta_{overall} = \frac{\left[(r_{p_1} r_{p_2})^{\alpha} - 1\right] \eta_1 \eta_2}{\eta_2 (r_{p_1}^{\alpha} - 1) + \eta_1 (r_{p_2}^{\alpha} - 1) + (r_{p_2}^{\alpha} - 1)(r_{p_1}^{\alpha} - 1)}$$

$$= \frac{\left[(1.27 \times 1.31)^{0.286} - 1\right] 0.90 \times 0.88}{0.88(1.27^{0.286} - 1) + 0.90(1.31^{0.286} - 1) + (1.27^{.286} - 1)(1.31^{.286} - 1)}$$

$$= \underline{\underline{88.5\,\%}}$$

THERMOFLUID MECHANICS 4

A single-state air compressor having 7.0% clearance is driven by a constant speed electric motor. It has been tested in San Francisco (at sea level) at an air temperature of 60°F, at atmospheric pressure, and found to deliver 100 cubic feet per minute of dry air when the discharge pressure is 100 psig (7.64 pounds/minute of free air).

What will be the capacity of this air compressor, in pounds of dry air per minute, if it is operated in a mountain location at an altitude of 10,000 feet, where the pressure is 10.10 psia and the temperature is 60°F?

Assume pressure drop at suction is negligible, polytropic exponent of compression is n = 1.3, and discharge pressure remains 100 psig.

Solution

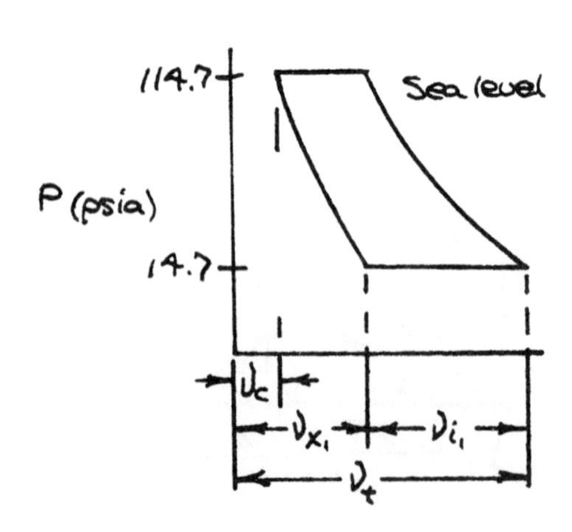

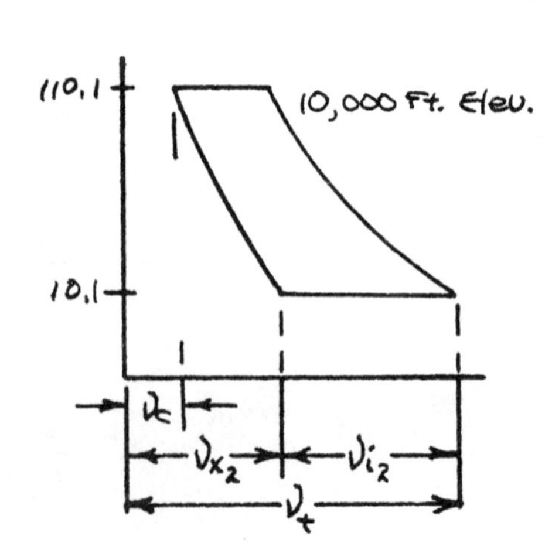

assume: 1. $\mathcal{V}_c = 0.07\,\mathcal{V}_t$

2. $p\mathcal{V}^{1.3} = C$ is ok for all comp. & exp. processes

$$\therefore \mathcal{V}_{x_1} = \mathcal{V}_c \left(\frac{114.7}{14.7}\right)^{\frac{1}{1.3}} = 0.336\,\mathcal{V}_t$$

$$\mathcal{V}_{x_2} = \mathcal{V}_c \left(\frac{110.1}{10.1}\right)^{\frac{1}{1.3}} = 0.442\,\mathcal{V}_t$$

$$\mathcal{V}_{i_1} = 0.664\,\mathcal{V}_t \qquad \mathcal{V}_{i_2} = 0.558\,\mathcal{V}_t$$

$$\text{new intake rate} = \frac{\mathcal{V}_{i_2}}{\mathcal{V}_{i_1}} \times 100 = 84.5 \text{ cfm}$$

$$\dot{m} = \text{cfm}\,\rho)_{intake} = \text{cfm}\,\frac{p}{RT}$$

$$= \underline{\underline{4.38 \frac{\text{lbm.}}{\text{min.}}}} \longleftarrow$$

$$(1.99 \text{ Kg/min})$$

THERMOFLUID MECHANICS 5

A large hotel has a boiler room where steam is generated for heating, laundry, and culinary purposes. The two boilers in this boiler room were formerly oil-fired but are being converted to natural gas. It is necessary to run a gas pipe line a distance of 265 feet from the meter to the gas burners in the boilers.

What size gas line would you specify, based on the recommendation of the Uniform Plumbing Code? Allow for a steaming rate of 200 per cent rated.

Specifications:

Boilers	-	two in number, each rated at 1500 lbs steam per hour
Steam	-	125 psia and 95 per cent quality
Feedwater	-	100°F
Natural Gas	-	1100 Btu per cu ft; 0.65 specific gravity
Gas pressure leaving the meter	-	5 psig
Gas pressure required at the burners	-	1.5 psig
Gas line length	-	265. feet

Solution

$$\dot{E}_{steam} = \dot{m}_s (h_{out} - h_{in})$$

$$= 6000 \ \frac{lbm}{hr} (h_f + 0.95 h_{fg} - h_{in})$$

$$= 6000(315.6 + 0.95 \times 875.4 - 68.0)$$

$$= 6.46 \times 10^6 \ Btu/hr.$$

Assume $\eta_{boiler} = 85\%$

$$\dot{E}_{gas} = \frac{6.46 \times 10^6}{0.85} = 7.59 \times 10^6 \ \frac{Btu}{hr.}$$

$$Q_{gas} = \frac{7.59 \times 10^6}{1100} = 6900 \ \frac{ft^3}{hr} \Big)_{stp}$$

$$= 5{,}150 \ ft^3/hr \Big)_{5\,psig}$$

$$= 6{,}250 \ ft^3/hr \Big)_{1.5\,psig}$$

Code is not specific for these conditions
(Suggests 4" - 6" d)

Estimate $V = 20$ ft/sec @ 1.5 psig & check Δp

$$VA = Q = \frac{6250}{3600} \text{ ft}^3/\text{sec}$$

$$\therefore d = 4.0''$$

$$H_L = f \frac{\ell}{d} \frac{V^2}{2g} \qquad f = .03 \text{ assuming fully turbulent}$$

$$= .03 \times \frac{265}{1/3} \times \frac{400}{64.4} = 148 \text{ ft.}$$

Since $\rho \cong .05 \frac{\text{lbm}}{\text{ft}^3}$ $\quad \Delta p = .05$ psi

4" d ok

3" d might be ok

THERMOFLUID MECHANICS 6

It is proposed to use energy radiated from the sun as an energy supply for a heat engine on the surface of the earth. The best engine cycle will approximate the Carnot cycle. It will receive energy from a solar energy collector and reject waste energy to the surrounding atmosphere. Assuming that the temperature of the sun is 10,000 R, the temperature of the surroundings is 60°F and the solar collector is perfectly insulated from the surroundings, determine the desired operating temperature of the collector for a maximum power output from the prime mover.

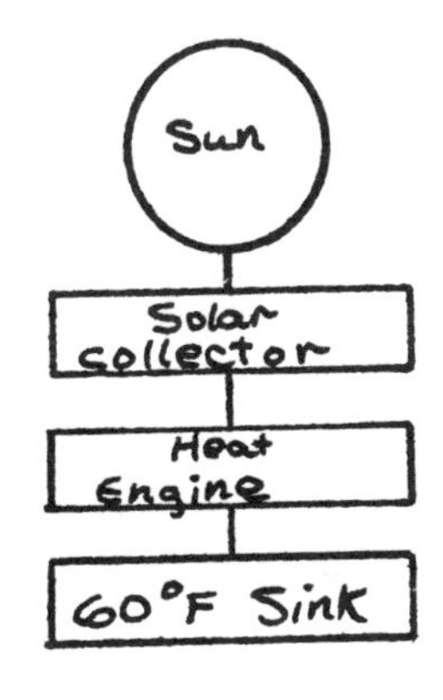

Solution

$$q_{input} = c_1 \left(T_S^4 - T_h^4 \right)$$

$$\text{Power} = \eta_t \, q_{input} = \left(1 - \frac{T_S}{T_h}\right) c_1 \left(T_S^4 - T_h^4 \right)$$

want P to be max. as T_h is varied

$$\text{Let } T' = \frac{T}{1000}$$

Then $P/c_1 = \left(1 - \frac{0.520}{T_h'}\right)\left(10^4 - T_h'^4\right)$ T_h' is $\frac{T_{collector}}{1000}$

Substitute values for T_h' and plot three or four points.

P/c is found to be a maximum when $T_h \simeq$ 4200°R (2333°K)

THERMOFLUID MECHANICS 7

An evaporator operating at a shell pressure of 38 psia is producing 25,000 lbs/hr of water vapor with a purity of 25 ppm solids. Make up water to the evaporator is softened city water at 60°F with 774 ppm solids. Saturated water from a boiler operating at 2100 psia is being added to the shell of the evaporator at a rate of 1000 lbs/hr and 30 ppm solids. Terminal difference at the blow down heat exchanger is 40°F. The heating steam to the evaporator is supplied at 103.5 psia and 653° and is discharged at a saturation pressure of 99.4 psia.

Below is a flow diagram of the evaporator.

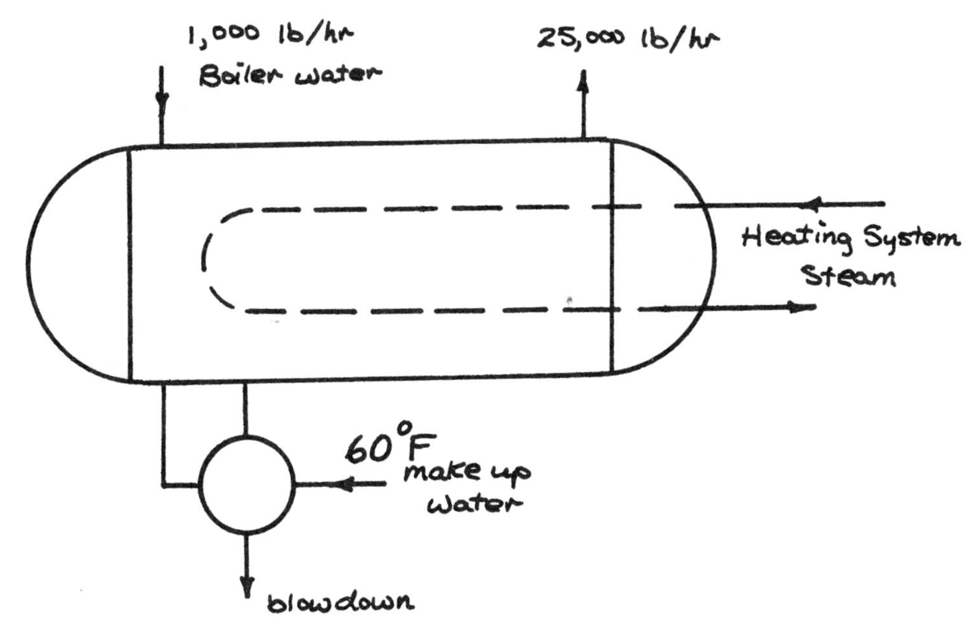

(a) What rate of evaporator blow down (in lbs/hrs) is required to maintain a 3000 ppm solids concentration in the evaporator shell?

(b) What rate of heating steam flow is necessary to produce the 25,000 lbs/hr of water vapor under the above conditions?

Solution

Assume steady state

$$C_w \dot{m}_w + C_s \dot{m}_s = C_b \dot{m}_b + C_v \dot{m}_v$$

$$\dot{m}_w = \dot{m}_v + \dot{m}_b - \dot{m}_s$$

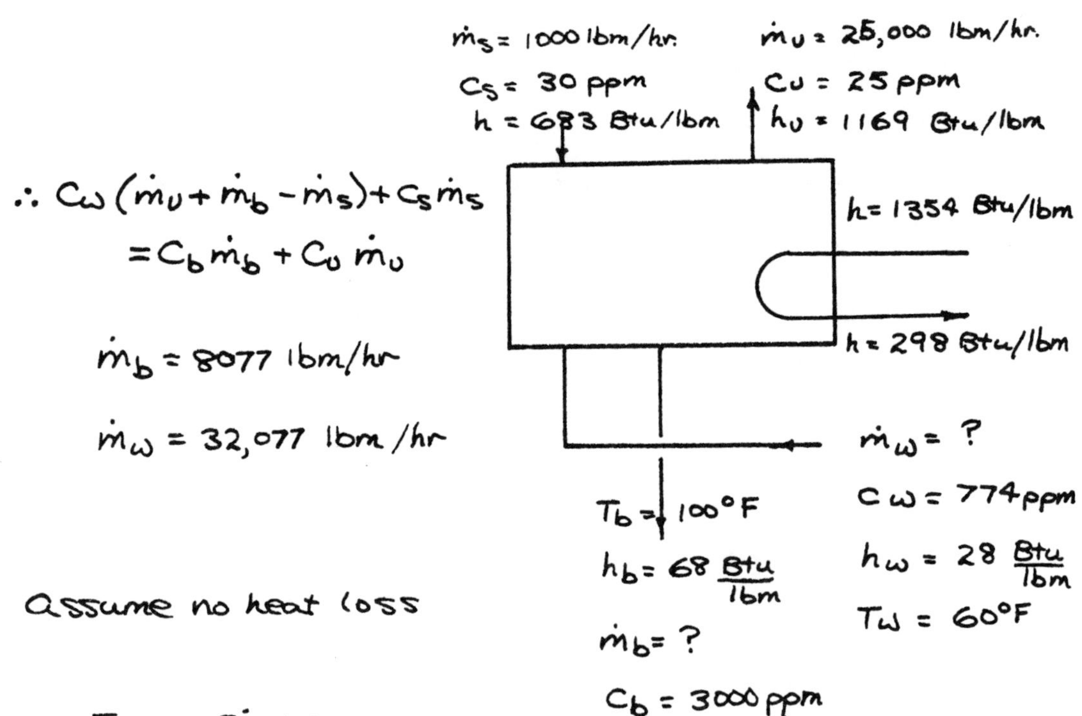

$$\therefore C_w(\dot{m}_v + \dot{m}_b - \dot{m}_s) + C_s \dot{m}_s = C_b \dot{m}_b + C_v \dot{m}_v$$

$\dot{m}_b = 8077$ lbm/hr

$\dot{m}_w = 32{,}077$ lbm/hr

Assume no heat loss

From First Law:

$$\dot{m}_v h_v + \dot{m}_b h_b - \dot{m}_s h_s - \dot{m}_w h_w = \dot{m}\,\Delta h)_{\text{steam heat source}}$$

$$\dot{m}_v h_v = 25{,}000 \times 1169 = 29.2 \times 10^6 \text{ Btu/hr}$$

$$\dot{m}_b h_b = 8{,}100 \times 68 = \underline{0.6 \times 10^6 \text{ Btu/hr}}$$

$$29.8 \times 10^6 \text{ Btu/hr}$$

$$\dot{m}_s h_s = 1{,}000 \times 683 = 0.7 \times 10^6 \text{ Btu/hr}$$

$$\dot{m}_w h_w = 32{,}100 \times 28 = \underline{0.9 \times 10^6 \text{ Btu/hr}}$$

$$1.6 \times 10^6 \text{ Btu/hr}$$

$$\Sigma = 28.2 \times 10^6 \text{ Btu/hr}$$

$$m_{\text{steam}} = \frac{2820 \times 10^4}{1354 - 298} = \underline{\underline{26{,}600 \text{ lbm/hr}}}$$

$$(12{,}065 \text{ Kg/hr})$$

THERMOFLUID MECHANICS 8

The figure illustrates the general arrangement of an experimental vapor compression still.

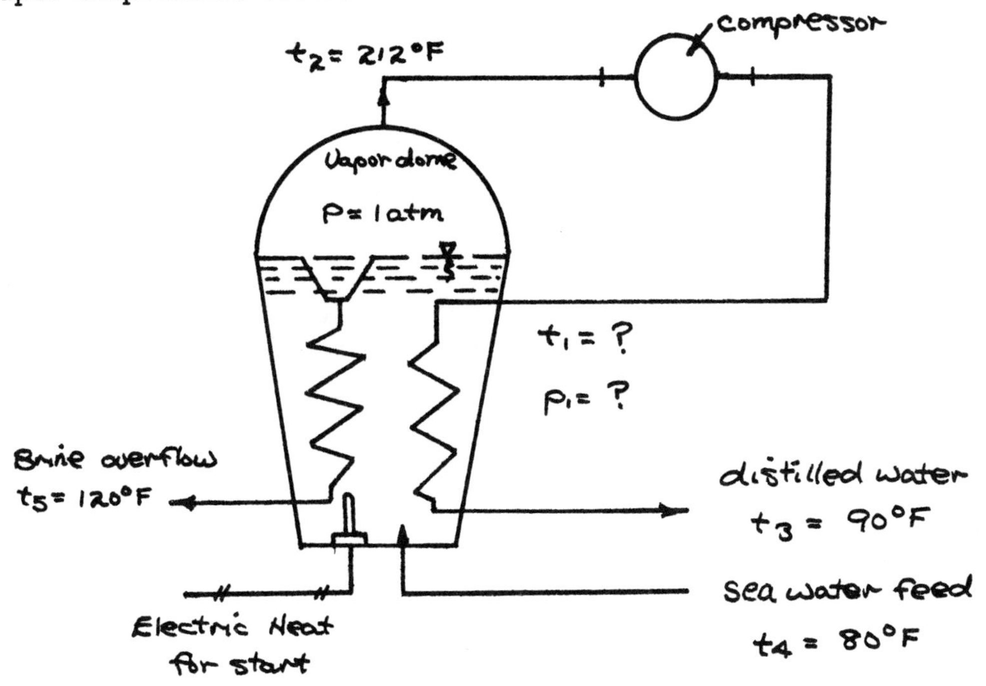

(a) Analyze the process to determine:

 (i) The work expended by the compressor per pound of water distilled.

 (ii) The pressure and the temperature of the compressed vapor.

(b) Enumerate the advantages and disadvantages of the process.

(c) If you were given the project for improving the system for commercial operation, what would be your recommendations?

Assume that the temperature of the inlet sea water feed is 80°F, the brine overflow is 120°F and the vapor chamber is 212°F. The brine overflow is 200% of the distillate (i.e., each 3 lbs of feed yields 1 lb of distillate). The specific heat of sea water is 0.97.

Solution

Assume: Steady State, adiabatic operation, isentropic compression, vapor is a perfect gas.

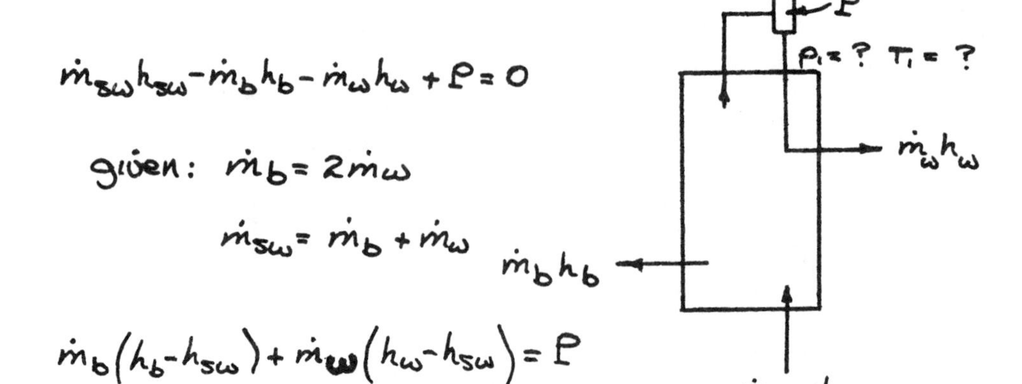

$$\dot{m}_{sw}h_{sw} - \dot{m}_b h_b - \dot{m}_w h_w + P = 0$$

given: $\dot{m}_b = 2\dot{m}_w$

$$\dot{m}_{sw} = \dot{m}_b + \dot{m}_w$$

$$\dot{m}_b(h_b - h_{sw}) + \dot{m}_w(h_w - h_{sw}) = P$$

* for liquids experiencing temp. change and small pressure change, $\Delta h \simeq c\Delta T$

$$\frac{\dot{m}_b}{\dot{m}_w} c(T_b - T_{sw}) + c(T_w - T_{sw}) = P/\dot{m}_w = W/\text{lbm}$$

From given values

$$2(0.97)(120-80) + (0.97)(90-80) = w = 87.2 \ \frac{\text{Btu}}{\text{lbm}}$$

$$(202.8 \ \text{kJ/kg})$$

for compressor

$$w = h_1 - h_{g,212} = c_p(T_1 - 212°F)$$

$$T_1 = \frac{87.2}{0.45} + 212°F = \underline{\underline{406°F}}$$

for isentropic compression

$$\frac{p_1}{p_{14.7}} = \left(\frac{T_2}{T_1}\right)^{k/k-1} = \left(\frac{866}{672}\right)^{\frac{1.33}{0.33}}$$

$$\underline{\underline{p_1 = 39.6 \ \text{psia}}}$$

$$(273 \ \text{kN/m}^2)$$

b) advantage – no high pressure, temperature or vacuum.

disadvantage – large mechanical work investment and compressor size

c) Water cool compressors, and recover some of waste energy in exhaust streams.

THERMOFLUID MECHANICS 9

Water is forced out of Tank No. 2 by gas under pressure in Tank No. 1. The desired water pressure at the outlet valve is 500 psig when Tank No. 2 is empty. Both tanks are insulated.

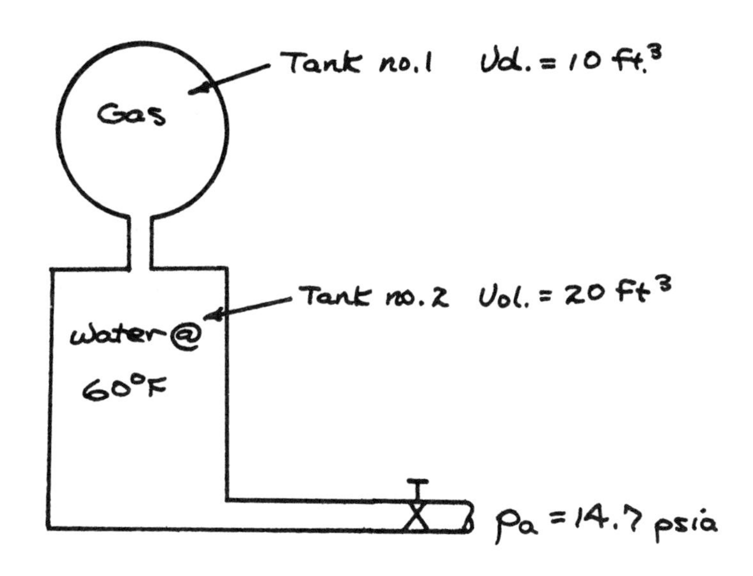

(a) If the pressurizing gas is N_2, starting at 60°F, what should the starting gas pressure be?

(b) If the gas is helium under the same conditions, what would be the starting pressure?

(c) What will be the weight of gas used in each case described above?

Solution

Assume isothermal expansion, as mass of gas is small compared to mass of tank and water, and blowdown will not be abrupt so there is time for heat transfer.

a) $\frac{V_2}{V_1} = \frac{p_1}{p_2}$ $\quad p_1 = \frac{515 \times 30}{10} = \underline{\underline{1545 \text{ psia}}}$

b) same as a)

c) $p_1 \upsilon_1 = m \frac{R_o}{M} T_1$

$$m_{N_2} = \frac{1545 \times 144 \times 10 \times 28}{1544 \times 520} = \frac{77 \text{ lbm}}{(34.9 \text{ Kg})}$$

$$m_{He} = \frac{1545 \times 144 \times 10 \times 4}{1544 \times 520} = \frac{11 \text{ lbm}}{(4.99 \text{ Kg})}$$

An alternate assumption would be adiabatic expansion which in general will not be as close to reality.

For this case, $\left(\frac{\upsilon_2}{\upsilon_1}\right)^k = \frac{p_1}{p_2}$.

THERMOFLUID MECHANICS 10

In a particular steam power plant air is believed to leak into the condenser. To check whether or not this is so, the plant is run until conditions are steady, whereupon the steam supply from the engine is shut off. Simultaneously, the air and condensate extraction pumps are closed down so that the condenser is isolated. At shutdown the temperature and vacuum in the condenser are observed to be 101.7°F and 27.3 in of Hg. After 5 minutes the values are 82.5°F and 19.25 in of Hg. The barometer reading is 29.7 in of Hg. The effective volume of the condenser is 12 cu ft.

Determine from this data:

(a) The weight of air leakage into the condenser during the observed period.

(b) The weight of water vapor condensed during this same time period.

Assume (the gas constant) R for air is 53.3 (Ft lbf/lbm °R)

Solution

Assume that at any instant T is constant throughout the condenser.

$\boxed{\mathcal{V} = 12\ ft^3}$

At Shutdown,

$T = 101.7°F \quad p = 29.7 - 27.3 = 2.4$ in Hg abs.

$= 1.18$ psia

For $T = 101.7°F$, $p_{sat} = 1$ psia

* Therefore, some air in condenser at shutdown.

$$m_{i,a} = \frac{p\mathcal{V}}{RT} = \frac{.18 \times 144 \times 12}{53.3 \times 561.7} = 0.0104 \text{ lbm}$$

After 5 minutes,

$T = 82.5°F \quad p = 29.7 - 19.3 = 10.4$ in Hg abs

$= 5.11$ psia

$p_{sat} = .55$ psia

$$m_{f,a} = \frac{4.56 \times 144 \times 12}{53.3 \times 542.5} = 0.0272 \text{ lbm}$$

a) $\Delta m_a = 0.0272 - 0.0104 = \underline{\underline{0.0168 \text{ lbm}}}$

b) $$\Delta m_w = \frac{\mathcal{V}}{R}\left(\frac{p_{i,s}}{T_i} - \frac{p_{f,s}}{T_f}\right) = \frac{12 \times 144}{85.6}\left(\frac{1}{461.7} - \frac{.55}{542.5}\right)$$

$$= 20.2\,(.00217 - .00101)$$

$$= \underline{\underline{0.0232 \text{ lbm}}}$$

(0.0105 Kg)

7

Heat Transfer

RICHARD K. PEFLEY

BASIC HEAT TRANSFER RELATIONSHIPS

1. Conductive Heat Transfer

   a. Defining equation

$$dq = -k\, dA \frac{\partial t}{\partial n}$$

   b. Temperature field equation in rectangular and cylindrical coordinates

$$\frac{\partial^2 t}{\partial x^2} + \frac{\partial^2 t}{\partial y^2} + \frac{\partial^2 t}{\partial z^2} = \frac{1}{\alpha}\frac{\partial}{\partial \tau} - \frac{p}{k}$$

$$\frac{\partial^2 t}{\partial r^2} + \frac{1}{r}\frac{\partial t}{\partial r} + \frac{1}{r^2}\frac{\partial^2 t}{\partial \theta^2} + \frac{\partial^2 t}{\partial z^2} = \frac{1}{\alpha}\frac{\partial t}{\partial \tau} - \frac{p}{k}$$

   c. Thermal conductive resistance for one-dimensional heat flow in a rectangular slab, cylinder and a sphere.

$$\frac{w}{kA}, \quad \frac{\ln(r_o/r_i)}{2\pi k \ell}, \quad \frac{r_o - r_i}{4\pi k\, r_o r_i}$$

2. Thermal Radiation

   a. Black body emission behavior

      i. Hemispherical radiation intensity--Stefan Boltzmann law

$$W_B = \sigma T^4 \quad \sigma = 0.171 \times 10^{-8} \frac{\text{Btu}}{\text{hrft}^2{}^\circ\text{R}^4}$$

ii. Wave length intensity--Planck's distribution law

$$W_{B,\lambda} = \frac{c_1/\lambda^5}{e^{\frac{c_2}{\lambda T}} - 1} \qquad c_1 = 1.187 \times 10^{-8} \frac{\text{Btu } \mu^4}{\text{ft}^2\text{hr}}$$

$$c_2 = 2.5896 \times 10^4 \text{°R } \mu$$

iii. Wave length of most intense radiation--Wein's displacement law

$$\lambda_{max} T = c_3 \qquad c_3 = 5215.6 \ \mu\text{°R}$$

iv. Directional radiation intensity--Lambert's cosine law

$$\frac{dE}{d\tau} = \frac{\sigma}{\pi} T_1^4 \cos \phi \ dA$$

v. Radiation exchange between black, isothermal surfaces separated by a non-absorbing medium

$$q_{1\to2} = \frac{\sigma}{\pi} (T_1^4 - T_2^4) \int_{A_1} \int_{A_2} \frac{\cos \phi_1 \cos \phi_2}{r^2} dA_1 \ dA_2$$

b. Grey body behavior

i. Definition

$$\alpha = \alpha \ (T)$$

ii. Fractional distribution of incident radiant energy

$$\alpha + r + tr = 1$$

iii. Circuit network for grey body enclosure composed of isothermal surfaces

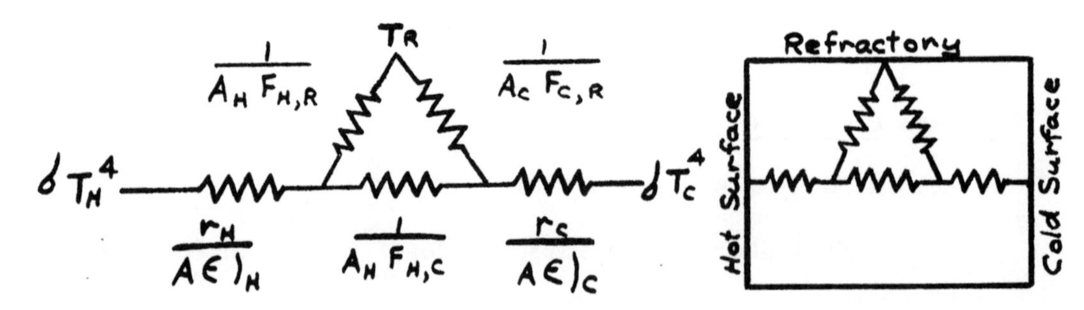

3. Convective Heat Transfer

a. Defining equation

$$dq = h\,dA\,(t_{surface} - t_{fluid})$$

b. Natural convection, t_{fluid} is defined as $t_{surrounding\ bath}$

$$h = h\,(L, \beta, \Delta t, \rho, \mu, g, k, c_p, g_c)$$

Resulting groups are

$$\frac{hL}{k} = f\left(\frac{c_p\,\mu\,g_c}{k}, \frac{L^3\rho^2 g^2 \beta\,\Delta t}{u^2\,g_c^2}\right)$$

c. Unestablished, forced convective heat transfer, t_{fluid} is defined as $t_{free\ stream}$

$$h = h\,(V, x, \rho, \mu, c_p, k, g_c)$$

$$\frac{hx}{k} = f\left(\frac{Vx}{\mu g_c}, \frac{c_p\mu}{kg_c}\right)$$

d. Established forced convection in closed conduits, t_{fluid} is defined as

$$\bar{t}_f = \frac{\int_A t\,V\,dA}{Q}$$

$$h = h\,(V, d, \rho, \mu, c_p, k, g_c) \quad \frac{hd}{k} = f\left(\frac{V\,d\rho}{g_c}, \frac{c_p\,\mu\,g_c}{k}\right)$$

e. Convective resistance for uniform convective coefficient

$$R = \frac{1}{hA}$$

4. Heat flow rate for parallel or counter flow heat exchangers assuming uniform fluid properties and heat transfer coefficients and transverse heat transfer only

$$q = \frac{\Delta t_m}{\Sigma R} = UA\,\Delta t_m \qquad \Delta t_m = \frac{\Delta t_a - \Delta t_b}{\ln \dfrac{\Delta t_a}{\Delta t_b}}$$

where Δt_a and Δt_b are temperature differences at opposite ends of the exchanger.

If Δt_a, Δt_b are not known for exchanger, use heat exchanger effectiveness concept

$$\xi = \frac{q_{actual}}{C_{min}(T_{h,in} - T_{c,in})} = f\left(\frac{C_{min}}{C_{max}}, \frac{AU}{C_{min}} = \frac{1}{\Sigma R C_{min}}\right)$$

(See reference for graphs of ξ versus $\frac{AU}{C_{min}}$)

Definition of Terms:

A	area
$C_{min,max}$	thermal capacitance rate
d,L,ℓ	length
f	function
F	radiation geometry factor
g	acceleration of gravity
g_c	proportionality factor in Newton's 2nd law
h	connective heat transfer coefficient
k	thermal conductivity
n	length in direction of heat flow
p	heat generation per unit volume
Q	volumetric flow rate
q	heat transfer rate per unit area
R,ΣR	thermal resistance, composite resistance
r	reflectivity or radius
t,T	temperature, absolute
tr	transmissivity
U	
V	velocity
w	thickness in direction of heat flow
x,y,z,r,θ	coordinate lengths
τ	time
α	absorbtivity or diffusivity
φ	radiation angle relative to surface normal
ρ	density
μ	viscosity
β	coefficient of cubical expansion
λ	wave length
ξ	heat exchange effectiveness

Reference: Kreith, F., Principles of Heat Transfer, International Textbook Company.

HEAT TRANSFER 1

The steady state heat leakage through a large composite wall is desired. The wall and accompanying heat transfer data are shown in the sketch below. You are asked to find two approximate solutions which lie on opposite sides of the true solution and hence predict the true value by averaging the results. Report the leakage in Btu/hr-ft^2 of the face area.

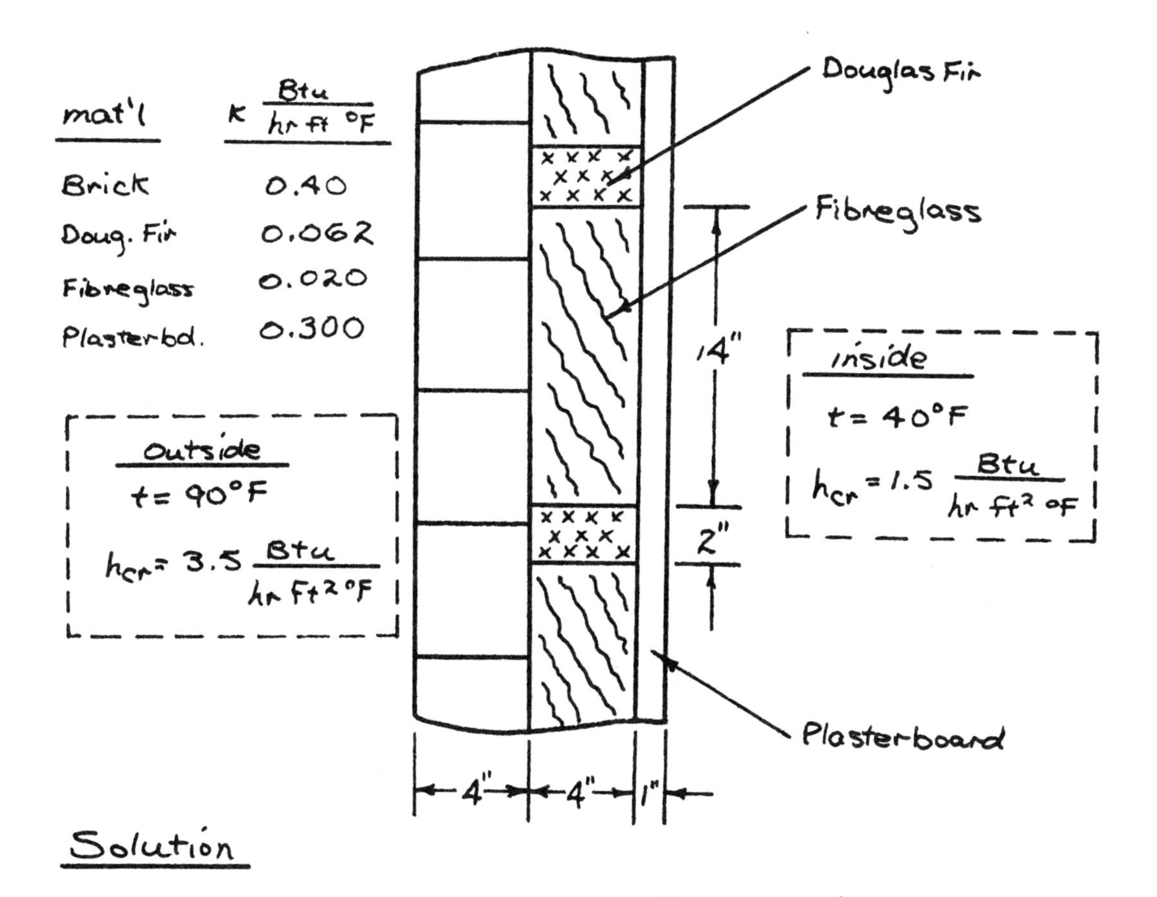

Solution

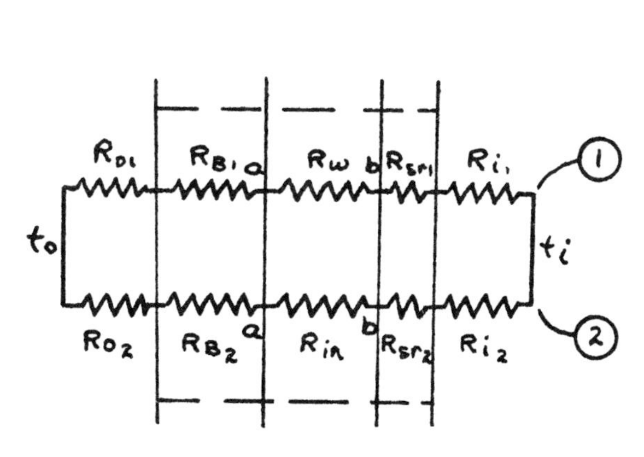

Resistances/unit area

$R_o = \frac{1}{h} = 0.286 \quad \frac{ft^2\,hr\,°F}{Btu}$

$R_B = t/k = \frac{0.333}{0.40} = 0.833$ "

$R_w = \frac{0.333}{0.062} = 5.36$ "

$R_{in} = \frac{0.333}{0.020} = 16.65$ "

$R_{sr} = \frac{1}{12 \times 0.30} = 0.278$ "

$R_i = \frac{1}{1.5} = 0.667$ "

Consider circuits ① + ② as separate, i.e., infinite transverse resistance

1 ft² face

	①	②
	$R_o = 0.286$	$R_o = 0.286$
	$R_B = 0.833$	$R_B = 0.833$
	$R_w = 5.36$	$R_{in} = 16.65$
	$R_{sr} = 0.278$	$R_{sr} = 0.278$
	$R_i = 0.667$	$R_i = 0.667$
	$R_{t①} = 7.424$	$R_{t②} = 18.714$

Composite area

1/8 is circuit ①, 7/8 is circuit ②

$R_{①} = 8 \times 7.42 = 59.4 \; \frac{hr \; ^oF}{Btu}$

$R_{②} = \frac{8}{7} \times 18.71 = 21.4$ "

$$R_{combined} = \frac{21.4 \times 59.4}{21.4 + 59.4} = 15.7 \; \frac{hr \; ^oF}{Btu}$$

$$q/A = \Delta t / R_{comb.} = 50/15.7 = 3.18 \; Btu/hr \; ft^2$$

Consider transverse resistance as zero --- opposite side of reality from first case. This is achieved by imagining circuits are shorted at a-a and b-b.

$R_{in} = 8/7 \; 16.65 = 19.0 \qquad R_w = 8(5.36) = 42.9$

$$R_{combined} = \frac{19.0 \times 42.9}{19.0 + 42.9} = 13.2 \; \frac{hr \; ft^2 \; ^oF}{Btu}$$

$R_o = 0.286$

$R_B = 0.833$

$R_{comb} = 13.2$

$R_{sr} = 0.278$

$R_i = 0.667$

$R_{t\,comb} = 15.3$

$$q/A = \frac{50}{15.3} = 3.27 \ \frac{\text{Btu}}{\text{hr ft}^2}$$

$$(q/A)_{actual} = \frac{3.18 + 3.27}{2}$$

$$= 3.23 \ \frac{\text{Btu}}{\text{hr ft}^2}$$

$$= \left(10.2 \ \text{WATTS}/\text{M}^2\right)$$

HEAT TRANSFER 2

A properly designed steam-heated, tubular preheater, with gases passing through the tubes, is heating 45,000 pounds per hour of air from 70°F to 170°F when using saturated steam at 5 psig. It is proposed to double the air flow rate through the heater and yet heat the air from 70°F to 170°F; this is to be accomplished by increasing the steam pressure. State reasonable assumptions, and determine the new steam pressure required to meet the changed conditions. Express answer in psig.

Solution

Assume:

1) air side heat transfer resistance dominates — turbulent flow

2) steady state

3) no heat leakage

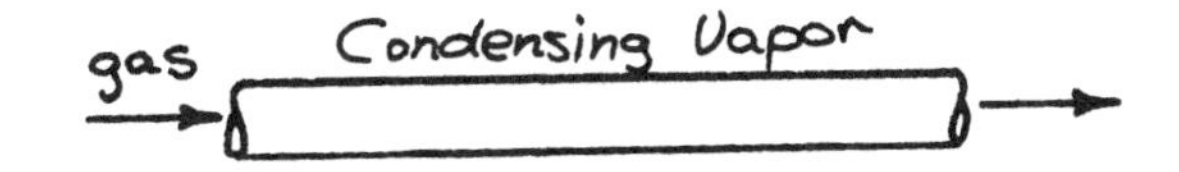

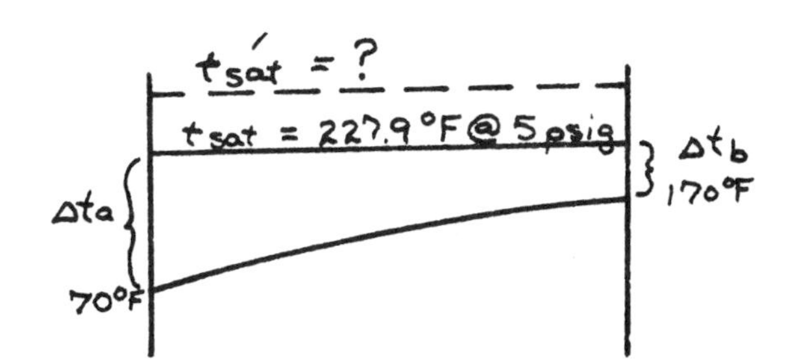

$$q = \frac{\Delta t_m}{R_t} = \frac{\Delta t_m}{\frac{1}{h_a \pi d_a l} + \frac{\ln r_o/r_i}{2\pi k l} + \frac{1}{h_s \pi d_s l}} \simeq \frac{\Delta t_m}{\frac{1}{h_a \pi d_a l}}$$

Let q' = new heat transfer rate

$$\frac{q'}{q} = \frac{h_a'}{h_a}\,\frac{\Delta t_m'}{\Delta t_m} \qquad \Delta t_m = \frac{\Delta t_a - \Delta t_b}{\ln \Delta t_a/\Delta t_b} = 100°F$$

for turbulent flow heat transfer in pipes

$$\frac{hd}{k} = 0.023\left(\frac{vd\rho}{\mu g_c}\right)^{0.80}\left(\frac{c_p \mu g_c}{k}\right)_f^{1/3}$$

$$\therefore \frac{h_a'}{h_a} \simeq \frac{(v\rho)'^{0.80}}{(v\rho)^{0.80}} \simeq (2)^{0.80}$$

Substituting $q'/q = 2$ and $h_a'/h_a = (2)^{0.80}$

$$\Delta t_m' = \Delta t_m \frac{(2)}{(2)^{0.80}} = 100(2)^{0.20} = 114.8°F$$

Guess $t'_{sat} = 241°F$

$$\Delta t_m' = \frac{(241-70)-(241-170)}{\ln \frac{241-70}{241-170}} = 114.8°F$$

$$p'_{steam} = \underline{\underline{25 \text{ psia}}}$$

$$\left(170 \frac{KN}{M^2}\right)$$

HEAT TRANSFER 3

From the data given in the figure, calculate the temperature of the earth. Assume earth to be a perfect absorber.

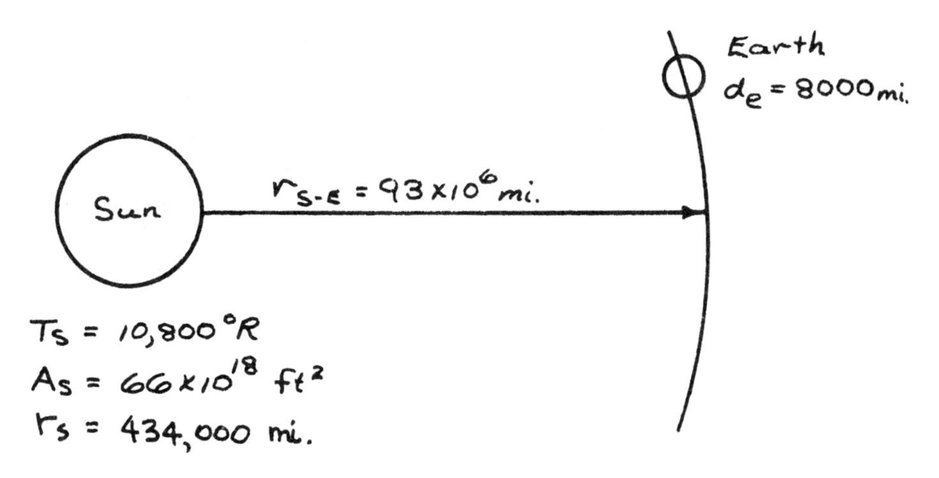

Solution

Assume sun is black radiator, and emits as much as it absorbs.

Radiation rate from sun $\frac{dE}{d\tau} = \sigma A_S T_S^4$

Fraction of this absorbed by Earth $= \left(\frac{\pi d_E^2}{16\pi r_{S-E}^2}\right)\sigma A_S T_S^4$

Radiation rate from earth $= \sigma \pi d_e^2 T_e^4$

$$\therefore \quad \sigma \pi d_e^2 T_e^4 = \left(\frac{\pi d_e^2}{16\pi r_{S-E}^2}\right)\sigma A_S T_S^4$$

$$T_e = \sqrt[4]{\frac{A_S}{16\pi r_{S-E}^2}}\; T_S$$

$$T_e = \underline{\underline{520°R = 60°F}}$$

$$(288.9°K)$$

HEAT TRANSFER 4

Temperature readings taken at the same time at various points in a slab of homogeneous material were plotted as shown below.

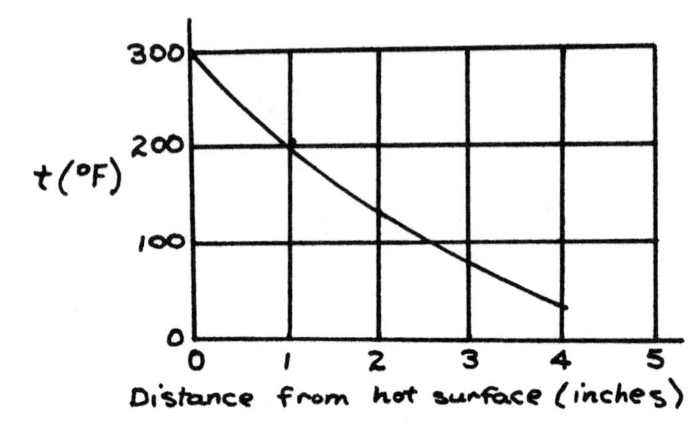

$$c = 0.40 \frac{\text{Btu}}{\text{lbm °F}}$$

$$k = 10.0 \frac{\text{Btu in}}{\text{hr ft}^2 \text{ °F}}$$

Assume the temperature readings are correct, what is the exact rate of heat flow per square foot past the plane two inches from the hot surface? Is temperature rising or falling?

Solution

a) Determine gradient at 2" position

$$q/A = -k \frac{dt}{dn} = 530 \frac{\text{Btu}}{\text{hr ft}^2}$$

$$(1671 \text{ WATTS/m}^2)$$

b) Temperature is rising because inflow of heat (steeper gradient) on left of the 2" plane is greater than outflow on the right.

HEAT TRANSFER 5

Find the area in square feet required for a heating coil to heat 5000 pounds of air per hour from 70°F to 200°F using steam at 250°F.

Material of the coil shall be aluminum with wall thickness of 0.049 inches and having a thermal conductivity of 1570 Btu/hr/ft^2/°F/inch.

Assume the coil to be clean and clear. The conductance of the air surface film is 30 Btu/hr/ft^2/°F. Specific heat of air is assumed constant at 0.24 Btu/pound/°F.

Assume d_i is approximately equal to d_o for the coil.

h_{steam} is approximately equal to 1000 Btu/hr ft^2 °F

The logarithmic mean temperature difference shall be used.

Solution

$$q = \dot{m} c_p \Delta T_{air} = 5\times10^3 \times 0.24 \times 1.3\times10^2 = 1.56\times10^5 \ \frac{\text{Btu}}{\text{hr}}$$

$$q = \frac{\Delta T_{mean}}{R_{air} + R_{wall} + R_{steam}}$$

$$\Delta T_{mean} = \frac{(250-70)-(250-200)}{\ln \frac{(250-70)}{(250-200)}} = 101\ ^\circ\text{F}$$

$$R_{air} = \frac{1}{hA} = \frac{1}{30\times A} = 3.33\times10^{-2}\ \frac{1}{A}$$

$$R_{wall} = \frac{w}{kA} = \frac{.049}{1.57\times10^3 A} = .003\times10^{-2}\ \frac{1}{A}\ \text{(neglect)}$$

$$R_{steam} = \frac{1}{hA} = \frac{1}{1\times10^3 A} = .10\times10^{-2}\ \frac{1}{A}$$

$$\Sigma R = 3.43\times10^{-2}\ \frac{1}{A(\text{ft}^2)}$$

$$A = \frac{1.56\times10^5\ \text{Btu/hr}}{101\ ^\circ\text{F}} \times 3.43\times10^{-2}\ \frac{\text{hr}\ ^\circ\text{F ft}^2}{\text{Btu}}$$

$$= \frac{1.56\times3.43\times10}{1.01} = \frac{52.5\ \text{ft}^2}{(4.87\ \text{m}^2)}$$

HEAT TRANSFER 6

A cylindrical fuel element in a nuclear reactor is shown below:

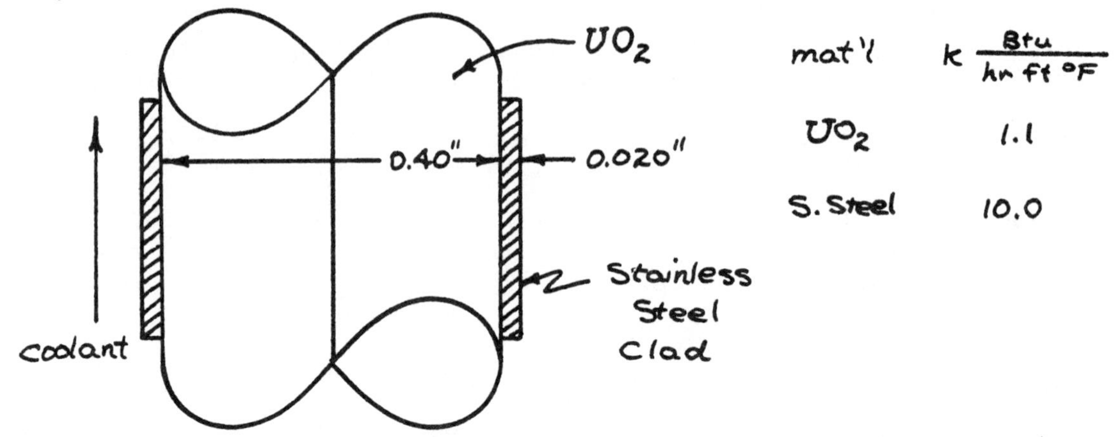

What is the center-line temperature in the UO_2 when the fuel element is producing 40 x 10^6 Btu/ft^{3}hr in the UO_2. Assume the temperature of the coolant is 500°F; and that the heat transfer coefficient for the coolant (h) = 10,000 Btu/ft^{2}hr°F.

Solution

For radial heat flow with heat generation

$$\frac{d^2t}{dr^2} + \frac{1}{r}\frac{dt}{dr} = -\frac{p}{k}$$

$$\frac{d}{dr}\left(r\frac{dt}{dr}\right) = -\frac{p}{k}r$$

$$r\frac{dt}{dr} = -\frac{p}{2k}r^2 + C_1 \qquad @\ r=0\ \frac{dt}{dr}=0 \quad \therefore C_1 = 0$$

$$t = -\frac{p}{4k}r^2 + C_2 \qquad @\ r=0 \quad t = t_{max}$$

$$t_{max} - t_r = \frac{pr^2}{4k}$$

at $r_o = 0.20''$

$$t_{max} - t_{r_o} = \frac{pr_o^2}{4k} = \frac{40\times10^6 \times 0.04}{4\times1.1\times144} = 2520\,°F$$

$(1382.2\,°K)$

–continued–

$$q/ft = p\pi r_o^2 = \frac{40 \times 10^6 \times \pi \times 0.04}{144} = 3.49 \times 10^4 \frac{Btu}{hr\ ft.}$$

$$q/ft = \frac{t_{r_o} - t_f}{R_{clad} + R_{film}} = \frac{t_{r_o} - t_f}{\frac{w}{kA_m} + \frac{1}{hA}}$$

$$t_{r_o} - t_f = 3.49 \times 10^4 \left(\frac{0.02}{10 \times \pi \times 0.42} + \frac{12}{10{,}000 \times \pi \times 0.44} \right)$$

$$= 83°F$$

$$t_{max} = t_f + (t_{r_o} - t_f) + (t_{max} - t_{r_o})$$

$$= 500°F + 83°F + 2520°F = \underline{\underline{3103°F}}$$

HEAT TRANSFER 7

A two-pass surface condenser is to be designed using an overall heat transfer coefficient of 480 Btu/hr-sq ft-°F with reference to the square feet of outside tube surface. The tubes are to be 1 inch O.D. with 1/16 inch walls.

Entering circulating water velocity is to be 6 fps.

Steam enters the condensers at a rate of 1,000,000 lb/hr at a pressure of 1 psia and an enthalpy of 1090 Btu/lb.

Condensate leaves as saturated liquid at 1 psia.

Circulating water enters the condenser at 85°F and leaves at 95°F.

Calculate the required number and length of condenser tubes.

Solution

$$\dot{m}(h_i - h_o)_{s(steam)} = \dot{m}c(t_o - t_i)_{w(water)}$$

$$\dot{m}_w = \frac{1\times10^6(1090-69.7)}{1\times10} = 1.02\times10^8 \ \frac{lbm}{hr}$$

$\dot{m}_s$, h_i, h_o, t_i, $\dot{m}_w$

$t_s = 101.7°F$

95°F

85°F

Let n = no. of tubes/pass

$$n V \rho A = m_w$$

$$n = \frac{1.02\times10^8 \times 4 \times 144}{6\times 3.6\times 0.624\times \pi \times (0.875)^2 \times 10^5}$$

$$= \underline{\underline{18{,}100 \text{ tubes/pass}}}$$

$$q = UA\,\Delta t_m \qquad \Delta t_m = \frac{(101.7-85)-(101.7-95)}{\ln\frac{16.7}{6.7}}$$

$$A = \frac{q}{U\,\Delta t_m} = \frac{\dot{m}_w c(t_o - t_i)}{U\,\Delta t_m}$$

$$= \frac{1.02\times10^9}{480\times10.9} = 1.94\times10^5 \ ft^2 \text{ of outside surface}$$

- continued -

$$2n\pi d_o \ell = 1.94 \times 10^5$$

$$\ell = \frac{1.94 \times 10^5}{2\pi n d_o}$$

$$= \frac{1.94 \times 10^5 \times 12}{1.81 \times 10^4 \times \pi \times 1 \times 2} = \underline{\underline{20.5 \text{ ft}}}$$

$$(6.25 \text{ m})$$

HEAT TRANSFER 8

Assume you are the designer of a line of heat exchangers in which warm water flowing inside of copper tubes is used to heat air outside the tubes. Because of size limitations you find it necessary to use finned surface to achieve a rating of 50,000 Btu/hr for the exchanger. An inventor claims to have a new technique for inserting longitudinal fins inside the tubes at a cost only one quarter that of placing fins outside the tubes for the same net increase in capacity over bare tubes. The water velocity in the tubes is 6 ft per second and the water temperatures are 190°F inlet and 160°F outlet. The air is warmed from 50°F to 110°F. Is the inventor's claim worth pursuing? Justify your answer.

Convective coefficients for air commonly are an order of magnitude smaller than water. Since the copper wall will offer negligable resistance, 90% of the resistance is on the air side. If the water side resistance were completely eliminated, only a 10% improvement in heat exchange rate would be possible.

∴ Adding fins to water side is not worth it.

HEAT TRANSFER 9

The heat capacity of the interior structure and contents of a building is 100,000 Btu/°F. The conductance (net effect of walls, roof, windows, etc.) is 6500 Btu/hr°F. At 5:00 P.M., the heat supply is stopped and the interior temperature is 70°F. The outdoor temperature is constant at 40°F. Estimate, by appropriate calculations, the interior temperature at 1:00 A.M.

Solution

Let c = capacitance of bldg.

Let K = overall conductance

$$-c\frac{dt}{d\tau} = K(t - t_o)$$

$$\int_{t_i}^{t_f} \frac{dt}{t - t_o} = -\frac{K}{C}\int_0^{\tau} d\tau$$

$$\ln\frac{t_f - t_o}{t_i - t_o} = \frac{K\tau}{C} \qquad \text{or} \quad t_f = t_o + (t_i - t_o)e^{-\frac{K\tau}{c}}$$

$$t_f = 40°F + 30°F\left(e^{-\frac{6500 \times 8}{1 \times 10^5}}\right)$$

$$= \underline{\underline{57.9°F}}$$

$$(14.4°C)$$

HEAT TRANSFER 10

A 5 inch diameter polished copper sphere is heated electrically to maintain a surface temperature of 300°F. If the sphere is suspended in a large room where the air and wall temperatures are 75°F, how many Btu per hour must be supplied to the sphere to compensate for the radiation loss?

Solution

$$q_{rad.} = \delta A \epsilon \left(T_s^4 - T_r^4 \right)$$

$$= 0.171 \times \pi D^2 \times 0.05 \left(\overline{7.6}^4 - \overline{5.35}^4 \right)$$

$$= 0.171 \times \pi \left(\frac{5}{12} \right)^2 \times 0.05 \left(\overline{7.6}^4 - \overline{5.35}^4 \right)$$

$$= 12.2 \ \text{Btu/hr}$$

(3.57 WATTS)

HEAT TRANSFER 11

The figure below describes the core geometry of a gas turbine regenerator for a 5,500 shp open cycle gas turbine plant. Operating conditions and the heat transfer surfaces are specified as follows:

Operating Conditions

Air flow rate, humid	200,000 lbs/hr
Air humidity	0.015 lbs/H_2O/lb dry air
Fuel-air ratio	0.015 lbs/lb
Fuel: fuel oil	H/C = 0.15, lbs/lb

Air Side

Louvered plate-fin surface	3/8 - 6.06
Entering pressure	130 lbs/in^2 abs
Entering temperature	350°F = 810°R

Gas Side

Plain plate-fin surface	11.1
Entering pressure	15 lbs/in^2 abs
Entering temperature	800°F = 1260°R

The objective of this problem is to determine for the specified conditions and the basic heat transfer and flow friction characteristics of the surface, (1) the regenerator effectiveness, and (2) the pressure drops for both the air and gas sides.

The steps in the general analysis require the determination of eleven specific factors. Name at least 8 of the factors.

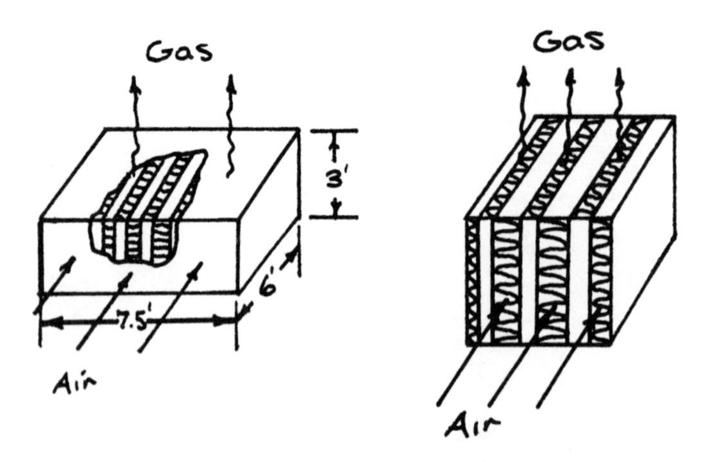

Air Side - Surface Louvered - Plate - Fin 3/8-6.06
Air Side - Surface Plain - Plate - Fin 11.1

Solution

Given:

1) Heat exchanger size & form
2) Inlet gas states
3) Mass flow rates

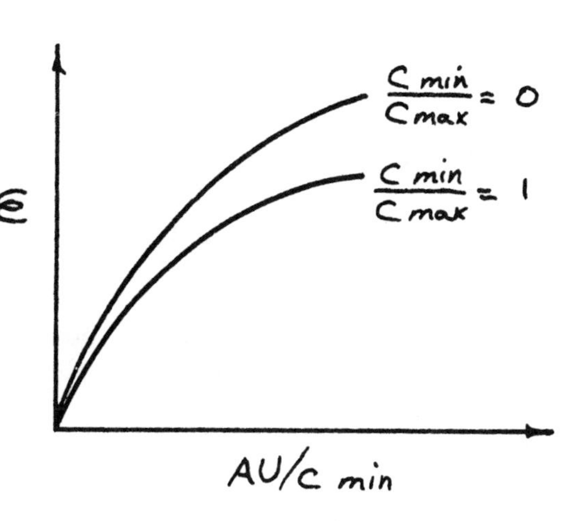

The following must be determined to find ϵ :

1. C_{min}
2. C_{max}
3. C_{min}/C_{max}
4. h_{hot}
5. h_{cold}
6. $\eta_{fin,cold}$
7. $\eta_{fin,hot}$
8. U
9. AU/C_{min}

The following must be determined to find Δp :

10. $\bar{\rho}_{hot}$
11. $\bar{\rho}_{cold}$
12. f_{hot}
13. f_{cold}

HEAT TRANSFER 12

The partial differential equation for one-dimensional heat conduction is:

$$\frac{\partial T}{\partial t} = \alpha \frac{\partial^2 T}{\partial x^2}$$

where T is temperature, t is time, α is thermal diffusivity and x is material thickness.

Convert this equation in a step-by-step procedure into a finite-difference equation that includes the heat balance.

Make a sketch illustrating the terms of the final equation.

Solution

Consider application of equation to plane-x

$$\frac{\partial T}{\partial \tau} = \frac{k}{\rho c} \frac{\partial^2 T}{\partial x^2}$$

in difference form,

$$\frac{\Delta T}{\Delta \tau} = \frac{k}{\rho c} \frac{\Delta}{\Delta x} \frac{\Delta T}{\Delta x}$$

by rearrangement,

$$\underbrace{\rho c \Delta x \frac{\Delta T}{\Delta \tau}}_{\text{rate of energy storage}} = \underbrace{\Delta \left(k \frac{\Delta T}{\Delta x} \right)}_{\text{inflow heat} \; - \; \text{outflow heat}}$$

inflow heat: $-k \left.\frac{\Delta T}{\Delta x}\right)_{x-\frac{\Delta x}{2}}$ − outflow heat: $-k \left.\frac{\Delta T}{\Delta x}\right)_{x+\frac{\Delta x}{2}}$

8

Gas Dynamics and Combustion

MICHEL A. SAAD

GAS DYNAMICS

While problems can be solved by use of the Mach function relationships which follow, it is expeditious to use Mach function tables and the problem solutions which are presented utilize the tables. For reference see "The Dynamics and Thermodynamics of Compressible Flow" by A. H. Shapiro.

Fundamental Relations:

Speed of sound $c = \sqrt{g_c \left(\frac{\partial p}{\partial \rho}\right)_s}$

For Perfect Gas $c = \sqrt{k g_c R T}$

For Air $(k = 1.4,\ R = 53.34 \text{ ft-lbf/lbm°R})$

$c = 49 \sqrt{T}$ ft/sec (T°R)

Mach Number $M = \frac{V}{c}$

Energy Equation $h_o = h + \frac{V^2}{2g_c J}$

$$T_o = T + \frac{V^2}{2g_c J c_p} \qquad \text{(perfect gas)}$$

Note: subscript o refers to stagnation state; superscript * refers to Mach 1 condition in the relations that follow.

Isentropic Relations for Perfect Gas:

$$\frac{p}{p_o} = \left(\frac{\rho}{\rho_o}\right)^k = \left(\frac{T}{T_o}\right)^{\frac{k}{k-1}} = \left(\frac{c}{c_o}\right)^{\frac{2k}{k-1}}$$

(Adiabatic Perfect Gas)

$$\frac{T_o}{T} = 1 + \frac{k-1}{2} M^2$$

(Isentropic Perfect Gas)

$$\frac{p_o}{p} = \left(1 + \frac{k-1}{2} M^2\right)^{\frac{k}{k-1}}$$

(Isentropic Perfect Gas)

$$\frac{\rho_o}{\rho} = \left(1 + \frac{k-1}{2} M^2\right)^{\frac{1}{k-1}}$$

(Adiabatic Perfect Gas)

$$\frac{c_o}{c} = \left(1 + \frac{k-1}{2} M^2\right)^{1/2}$$

At M = 1 (and k = 1.4):

$$\frac{T^*}{T_o} = \frac{c^{*2}}{c_o^2} = \frac{2}{k+1} = 0.8333$$

$$\frac{\rho^*}{\rho_o} = \left(\frac{2}{k+1}\right)^{\frac{1}{k-1}} = 0.6339$$

$$\frac{p^*}{p_o} = \left(\frac{2}{k+1}\right)^{\frac{k}{k-1}} = 0.5283$$

Mass rate of flow per unit area:

$$\frac{\dot{m}}{A} = \sqrt{\frac{kg_c}{R}} \frac{p_o}{\sqrt{T_o}} \frac{M}{\left(1 + \frac{k-1}{2} M^2\right)^{\frac{k+1}{2(k-1)}}}$$

$$\left(\frac{\dot{m}}{A}\right)_{max} = \frac{\dot{m}}{A^*} = \sqrt{\frac{kg_c}{R} \left(\frac{2}{k+1}\right)^{\frac{k+1}{k-1}}} \frac{p_o}{\sqrt{T_o}}$$

For air:

$$\frac{\dot{m}}{A^*} = 0.532 \frac{p_o}{\sqrt{T_o}}$$

where $\dot{m}$ is in lbm/sec, A^* is ft^2, p_o in lbf/ft^2 and T_o in °R.

<u>Isentropic Flow through a Nozzle:</u>

$$\frac{dA}{A} = \frac{dV}{V}(M^2-1) = g_c \frac{(1-M^2)}{\rho V^2} dp = \frac{1-M^2}{M^2} \frac{d\rho}{\rho}$$

	M < 1			M > 1		
	dV	dp	dρ	dV	dp	dρ
dA +	-	+	+	+	-	-
dA -	+	-	-	-	+	+

$$\frac{A}{A^*} = \frac{\dot{m}/A^*}{\dot{m}/A} = \frac{1}{M}\left[\frac{2 + (k-1)\,M^2}{k+1}\right]^{\frac{k+1}{2(k-1)}}$$

Frictional Flow Relations (Constant Area, Adiabatic Flow):

$$\frac{V}{V^*} = M\sqrt{\frac{k+1}{2 + (k-1)M^2}}$$

$$\frac{p}{p^*} = \frac{1}{M}\sqrt{\frac{k+1}{2 + (k-1)M^2}}$$

$$\frac{T}{T^*} = \frac{k+1}{2 + (k-1)M^2}$$

$$\frac{\rho}{\rho^*} = \frac{V^*}{V} = \frac{1}{M}\sqrt{\frac{2 + (k-1)M^2}{k+1}}$$

$$\frac{p_o}{p_o^*} = \frac{1}{M}\sqrt{\left(\frac{2 + (k-1)M^2}{k+1}\right)^{\frac{k+1}{(k-1)}}}$$

$$\frac{4\bar{f}L^*}{D} = \frac{1-M^2}{kM^2} + \frac{k+1}{2k}\ln\left[\frac{(k+1)M^2}{2 + (k-1)M^2}\right]$$

$$\frac{4\bar{f}L_{1-2}}{D} = \left(\frac{4\bar{f}L^*}{D}\right)_{M_1} - \left(\frac{4\bar{f}L^*}{D}\right)_{M_2}$$

Simple Heating - Rayleigh Line:

$$\frac{T}{T^*} = \frac{(1+k)^2M^2}{(1+kM^2)^2}$$

$$\frac{p}{p^*} = \frac{1+k}{1+kM^2}$$

$$\frac{V}{V^*} = \frac{\rho^*}{\rho} = \frac{(k+1)M^2}{1+kM^2}$$

$$\frac{T_o}{T_o^*} = \frac{(k+1)M^2(2+(k-1)M^2}{(1+kM^2)^2}$$

$$\frac{p_o}{p_o^*} = \frac{k+1}{1+kM^2}\left[\frac{(2+(k-1)M^2}{k+1}\right]^{\frac{k}{k-1}}$$

COMBUSTION

Fundamental Concepts:

Air-Fuel Ratio (A/F) -
Mass or molal ratio of air to fuel in a combustion process.

Stoichiometric -
A/F such that sufficient oxygen is present to fully oxidize the fuel with no excess oxygen.

Heating Value (HV) or Heat of Reaction -
The amount of energy released by the chemical reaction being carried out either at constant pressure and temperature or constant volume and temperature.

Higher Heating Value (HHV) -
The water formed by combustion is assumed condensed.

Lower Heating Value (LHV) -
The water formed by combustion is assumed to remain in vapor phase.

Combustion Example Problems:

Problems in stoichiometry are divided into two main categories, depending on whether the composition of the fuel or the composition of the reaction products is known.

Example 1: Determine the stoichiometric air required for the complete combustion of 1 lbm of normal heptane C_7H_{16}. What is the percentage analysis of the products on a mass and a mole basis?

Solution: The chemical equation is

$$C_7H_{16}(\ell) + 11\ O_2(g) + 11(3.76)\ N_2(g) \rightarrow 7\ CO_2(g) + 8\ H_2O(g) + 11(3.76)\ N_2(g)$$

The air fuel ratio on a mass basis is

$$\frac{(11 \times 32) + (11 \times 3.76 \times 28)}{(7 \times 12) + (16 \times 1.008)} = 15.12 \text{ lbm air/lbm fuel}$$

Alternately, since there are (11 + 11 x 3.76) moles of air, the air fuel ratio is

$$\frac{(11 + 11 \times 3.76)(28.97)}{(100.205)} = 15.12 \text{ lbm air/lbm fuel}$$

The air fuel ratio on a mole basis is

$$\frac{11 + 11 \times 3.76}{1} = 52.3 \text{ moles air/mole fuel}$$

The analysis of the products on a mass and molal basis are as follows:

	By Mole		By Mass	
CO_2	8	12.42%	7 x 44 = 308	19.11%
H_2O	7	14.20%	8 x 18 = 144	8.94%
N_2	41.36	73.38%	41.36 x 28 = 1160	71.95%
Totals	56.36	100.00%	1612	100.00%

Example 2: Five moles of propane C_3H_8 are completely burned at atmospheric pressure in the theoretical amount of air. Determine

(a) The volume of air used in the combustion process measured at 14.7 psia and 77°F.

(b) The partial pressure of each constituent of the products.

(c) The volumetric analysis of the dry products.

Solution: (a) The chemical equation is

$$5\ C_3H_8(g) + 25\ O_2(g) + 25(3.76)\ N_2(g) \rightarrow 15\ CO_2(g) + 20\ H_2O(g) + 25(3.76\ N_2(g)$$

or

$$5\ C_3H_8(g) + 25.0_2(g) + 94.0\ N_2(g) \rightarrow 15\ CO_2(g) + 20\ H_2O(g) + 94.0\ N_2(g)$$

Thus, 25 moles of O_2 are required or 25 (1/0.2099) = 119 moles of air (25 moles of O_2 plus 94 moles of N_2).

If air is assumed a perfect gas, then

$$V = \frac{nRT}{p} = \frac{119 \times 1545(77 + 460)}{14.7 \times 144} = 46{,}700 \text{ ft}^3$$

(b) The partial pressures of the product constituents are proportional to the mole fractions. Therefore,

$$p_{CO_2} = \frac{15}{15 + 20 + 94} \times 14.7 = 1.71 \text{ psia}$$

$$p_{H_2O} = \frac{20}{129} \times 14.7 = 2.28 \text{ psia}$$

$$p_{N_2} = \frac{94}{129} \times 14.7 = \underline{10.71 \text{ psia}}$$

Total pressure = 14.7 psia

(c) $$\frac{V_{CO_2}}{V_{CO_2} + V_{N_2}} = \frac{15}{15 + 94} = 13.75 \text{ per cent}$$

$$\frac{V_{N_2}}{V_{CO_2} + V_{N_2}} = 1 - 0.1375 = 86.25 \text{ per cent}$$

Example 3: The analysis of a sample of coal gives the following values by weight (the remainder is ash):

Carbon	80.7%
Hydrogen	4.9%
Sulfur	1.8%
Oxygen	5.3%
Nitrogen	1.1%

What is the air fuel ratio by weight if 20 per cent excess air is used in the combustion process?

Solution: For 1 lbm of fuel, the following table can be formulated:

	Number of Moles per lbm of Fuel	Moles of O_2 Required for Complete Combustion of 1bm of Fuel	Reaction Equation
C	$\frac{0.807}{12} = 0.0673$	0.0673	$0.0673\ C(s) + 0.0673\ O_2(g) \rightarrow 0.0673\ CO_2(g)$
H_2	$\frac{0.049}{2} = 0.0245$	0.01225	$0.0245\ H_2(g) + 0.01225\ O_2(g) \rightarrow 0.0245\ H_2O(g)$
S	$\frac{0.018}{32} = 0.000563$	0.000563	$0.000563\ S(s) + 0.000563\ O_2(g) \rightarrow 0.000563\ SO_2(g)$
O_2	$\frac{0.053}{32} = 0.001655$	0	
N_2	$\frac{0.011}{28} = 0.000393$	0	

O_2 required = 0.080113 mole; O_2 in fuel = 0.001655 mole;
Difference = 0.078458 mole O_2 per lbm of fuel

Stoichiometric air fuel ratio = 0.078458(100/20.99) x 28.97

= 10.83 lbm air/lbm fuel

With 20 per cent excess air, the air fuel ratio is

$$10.83 \times 1.2 = 13 \text{ lbm air/lbm fuel}$$

Example 4: The composition of a hydrocarbon fuel by mass is 85 per cent carbon, 13 per cent hydrogen, and 2 per cent oxygen. Determine

(a) The chemically correct mass of air required for the complete combustion of 1 lbm of fuel.

(b) The volumetric analysis of the products of combustion and the dew point if the total pressure is 14.7 psia.

(c) If the products of combustion are cooled to 60°F, what is the mass of the water vapor that condensed?

Solution: (a) Consider the combustion of 1 lbm of fuel composed of 0.85 lbm of carbon, 0.13 lbm of hydrogen, and 0.02 lbm of oxygen. In the complete combustion process the carbon oxidizes to CO_2 and the hydrogen to H_2O according to the equations

$$C(s) + O_2(g) \rightarrow CO_2(g)$$

$$H_2(g) + \tfrac{1}{2}\, O_2(g) \rightarrow H_2O(g)$$

The oxygen requirement as dictated by these two equations is, however, reduced by the amount of oxygen already existing in the fuel. The number of moles of

$$O_2 \text{ required} = \frac{0.85}{12.01} + \frac{0.13}{2.016 \times 2} - \frac{0.02}{32} = 0.07075 + 0.03225 - 0.000625$$

$$= 0.102375 \text{ mole of } O_2/\text{lbm of fuel}$$

$$\text{mass of air} = \frac{0.102375}{0.2099} \times 28.97 = 14.1 \text{ lbm air/lbm fuel}$$

(b) The analysis of the products of combustion by mass and volume is given in the accompanying table:

	Moles per lbm Fuel	lbm per lbm Fuel	Molal Analysis
CO_2	$\frac{0.85}{12.01} = 0.07075$	3.12	0.136
H_2O	$\frac{0.13}{2.016} = 0.0645$	1.16	0.124
N_2	(3.76 x 0.102375) = 0.385	10.82	0.740
Totals	0.52025	15.1	1.000

The dew point temperature may be found from steam tables corresponding to the pressure of the water vapor in the combustion products:

Pressure of H_2O = 0.124 x 14.7 = 1.82 psia

Dew point = 122.5°F

(c) Since 60°F is below the dew point temperature, some water vapor will condense. The vapor pressure is then equal to the saturation vapor pressure corresponding to 60°F.

From steam tables, $p_{g,sat}$ at 60°F = 0.2563 psia and v_g = 1206.7 ft^3/lbm. The volume occupied by the water vapor is equal to the volume occupied by the CO_2 and the N_2 at their partial pressure of 14.7 - 0.2563 = 14.4437 psia. The volume according to perfect gas law is

$$V = \frac{nRT}{p} = \frac{(0.07075 + 0.385)\ 1545 \times 520}{14.4437 \times 144} = 176.3\ ft^3$$

$$\text{Mass of vapor in products} = \frac{V}{v_g} = \frac{176.3}{1206.7} = 0.1462\ lbm$$

Mass of vapor that condensed = 1.16 - 0.1462 = 1.0138 lbm

Example 5: A hydrocarbon fuel in the vapor state is burned with atmospheric air at 14.7 psia, 80°F, and 60 per cent relative humidity. The volumetric analysis of the dry products of combustion is

Product	Percentage
CO_2	10.00
O_2	2.37
CO	0.53
N_2	87.10 (by difference from 100)
	100.00

Calculate

(a) Ratio of hydrogen to carbon by mass for the fuel

(b) Air-fuel ratio by mass

(c) Air used as percentage of the stoichiometric value

(d) Volume of the humid air supplied per lbm of fuel

Solution: (a) The following table gives the number of atoms of carbon, oxygen, and nitrogen per mole of dry products:

	Moles (per Mole of Dry Products)	Carbon	Oxygen	Nitrogen
CO_2	0.10	0.10	0.20	--
O_2	0.0237	--	0.0474	--
CO	0.0053	0.0053	0.0053	--
N_2	0.871	--	--	1.742
Totals	1.0000	0.1053	0.2527	1.742

Assuming that H_2O is the only remaining product of combustion, the amount of hydrogen in the fuel may be determined by finding the amount of oxygen used to form the H_2O.

Total moles of O_2 supplied with 0.871 mole of N_2 $= \frac{0.871}{3.76} = 0.231.$

Moles of O_2 accounted for in the reactions considered $= 0.12635.$

Hence the difference used in formation of H_2O $= 0.10465$ moles of O_2

The formation of H_2O can thus be expressed as

$$0.2093\ H_2(g) + 0.10465\ O_2(g) \rightarrow 0.2093\ H_2O(g)$$

which indicates that 0.2093 mole of H_2O is formed per mole of dry products. The hydrogen-carbon ratio in the fuel is

$$\frac{H}{C} = \frac{2 \times 0.2093}{0.1053} = 3.97 \quad \text{(by atoms)}$$

$$= \frac{(2 \times 0.2093) \times 1.008}{(0.1053) \times 12.01} = 0.333 \quad \text{(by weight)}$$

(b) The air-fuel ratio $= \frac{[0.871 \times (1.0/0.79)]\ 28.97}{(0.1053 \times 12.01) + (0.2093) \times 2.016}$

$$= \frac{31.95}{1.687} = 18.9 \text{ lbm air/lbm fuel}$$

(c) The number of moles of oxygen required for stoichiometric combustion is

$$0.1053\ C(s) + 0.1053\ O_2(g) \rightarrow 0.1053\ CO_2(g)$$

$$0.2093\ H_2(g) + \underline{0.10465}\ O_2(g) \rightarrow 0.2093\ H_2O(g)$$

0.20995 moles of O_2 per mole of dry products

Since the oxygen-air ratio is constant in atmospheric air, the percentage of the air supplied is the same as that of the oxygen. Hence the percentage of the stoichiometric air used is

$$\frac{0.231}{0.20995} = 1.1 \text{ or } 110 \text{ per cent}$$

(d) The number of moles of dry air is

$0.87\left(\frac{1}{0.79}\right) = 1.102$ moles air/mole dry products, but the mass of fuel = 1.687 lbm fuel/mole dry products. Therefore, the number of moles of air per lbm of fuel = 1.102/1.687 = 0.653.

At 80°F

$$p_{g,sat} = 0.5069 \text{ psia}$$

and

$$p_g = 0.6 \times 0.5069 = 0.30414 \text{ psia}$$

Hence

$$V = \frac{nRT}{p} = \frac{0.653 \times 1545.33 \times 540}{(14.7 - 0.30414) \times 144} = 263 \text{ ft}^3\text{/lbm of fuel}$$

GAS DYNAMICS AND COMBUSTION 1

Air flows through a convergent-divergent nozzle with inlet area 0.8 sq. in., minimum area 0.5 sq. in., and exit area 0.60 sq. in. At the inlet the air velocity is 344 ft/sec, stream pressure 100 psia and stream temperature 160°F.

(a) Calculate the mass rate of flow through the nozzle in lbm/hr.

(b) Determine the Mach number at the minimum area.

(c) Determine the stream velocity and pressure at the exit.

Solution

*Note: It is presumed that mach function tables are available.

Assume isentropic, steady flow.

a) $\dot{m} = \rho_1 A_1 V_1$

$$\rho_1 = \frac{P_1}{RT_1} = \frac{100 \times 144}{53.34 \times 620} = 0.435 \ \text{lbm/ft}^3$$

$$\dot{m} = (0.435)(344)\left(\frac{0.8}{144}\right) = \underline{\underline{0.832 \ \text{lbm/sec.}}}$$

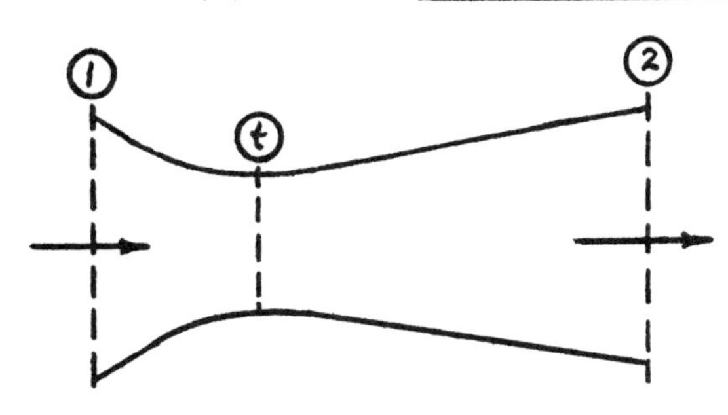

$A_1 = .8 \ in^2$ $A_t = 0.5 \ in^2$ $A_2 = 0.6 \ in^2$

$U_1 = 344$ ft/sec.

$P_1 = 100$ psia

$T_1 = 160°F$

-continued-

b) $M_1 = \dfrac{344}{49\sqrt{620}} = 0.282$

$\dfrac{A^*}{A_1} = 0.467$ but $\dfrac{A_t}{A_1} = \dfrac{0.5}{0.8} = 0.625$

Therefore, $\dfrac{A^*}{A_t} = \dfrac{A^*/A_1}{A_t/A_1} = \dfrac{0.467}{0.625} = 0.747$

from tables, $M_t = \underline{\underline{0.5}}$

c) $\dfrac{A_2}{A_1} = \dfrac{0.6}{0.8} = 0.75$, $\dfrac{A^*}{A_2} = \dfrac{A^*/A_1}{A_2/A_1} = \dfrac{0.467}{0.75} = 0.6227$

from tables $M_2 = 0.395$ and,

$$\left.\begin{aligned} \frac{P_2}{P_0} &= 0.898 \\ \frac{P_1}{P_0} &= 0.946 \end{aligned}\right\} \quad P_2 = P_1 \frac{0.898}{0.946} = 100 \times \frac{0.898}{0.946}$$

$$= \underline{\underline{94.9 \text{ psia}}}$$

$$(644\ KN/m^2)$$

$$\left.\begin{aligned} \frac{T_2}{T_0} &= 0.9696 \\ \frac{T_1}{T_0} &= 0.9843 \end{aligned}\right\} \quad T_2 = T_1 \frac{0.9696}{0.9843} = 620 \times \frac{0.9696}{0.9843}$$

$$= 611\ ^\circ R$$

$V_2 = c_2 M_2 = 49\sqrt{611} \times 0.395 = \underline{\underline{478 \text{ ft./sec.}}}$

$(145.7\ m/sec)$

GAS DYNAMICS AND COMBUSTION 2

Dry air at 100 psia and 1500°R enters a converging-diverging nozzle through a line of 0.05 ft^2 area and expands to a delivery region pressure of 5 psia. Assuming isentropic expansion and a mass rate of flow of 2 lbm/sec, find:

(a) The stagnation enthalpy.

(b) The temperature and enthalpy at discharge.

(c) The Mach number and velocity of the air stream at discharge.

(d) The mass flow rate per unit area.

Solution

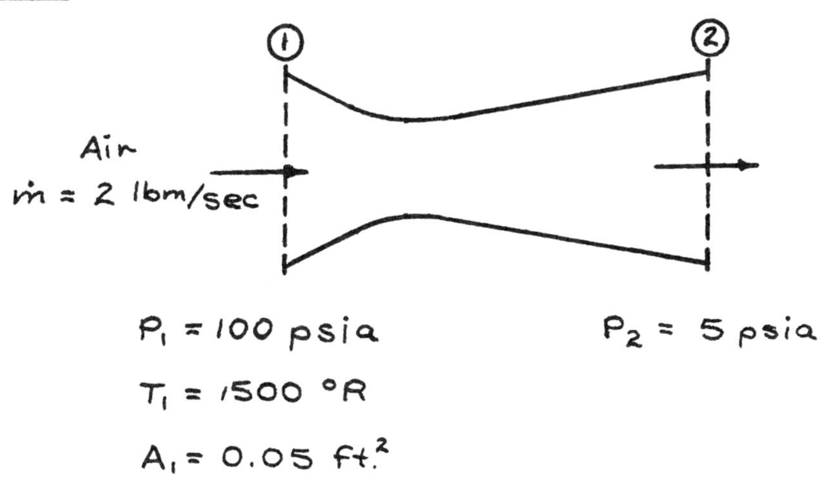

$P_1 = 100$ psia $\qquad P_2 = 5$ psia

$T_1 = 1500$ °R

$A_1 = 0.05$ ft.^{2}

$$\rho_1 = \frac{P_1}{RT_1} = \frac{100 \times 144}{53.34 \times 1500} = 0.180 \text{ lbm/ft.}^3$$

$$V_1 = \frac{\dot{m}}{\rho_1 A_1} = \frac{2}{0.180 \times .05} = 222 \text{ ft./sec.}$$

$$M_1 = \frac{222}{49\sqrt{1500}} = 0.117$$

therefore $\frac{T_1}{T_o} = 0.997$, $T_o = 1505$ °R

$\frac{P_1}{P_o} = 0.990$, $P_o = 101$ psia

-continued-

a) $h_o = h_1 + \frac{V_1^2}{2g_c} = c_p T_o = 0.241 \times 1505 = 362 \text{ Btu/lbm}$

b) + c)

$\frac{P_2}{P_o} = \frac{5}{101} = 0.0495 \qquad M_2 = \underline{\underline{2.61}}$

$\frac{T_2}{T_o} = 0.423 \qquad T_2 = \underline{\underline{636.2\ °R}}$

$V_2 = M_2 c_2 = 2.61 \times 49\sqrt{636.2} = \underline{\underline{3226 \text{ ft/sec.}}}$ (983 m/sec)

$h_2 = h_o - \frac{V_2^2}{2g_c} = 362 - \frac{(3226)^2}{2 \times 32.2 \times 778} = \underline{\underline{154 \text{ Btu/lbm}}}$ (358 KJ/Kg)

d) Not clear which mass flow rate per unit area is sought.

$G_1 = \frac{\dot{m}}{A_1} = \frac{2}{0.05} = 40 \text{ lbm/ft}^2 \cdot \text{sec.}$

$G^* = \frac{\dot{m}}{A^*} = 0.532 \frac{P_o}{\sqrt{T_o}} = 0.532 \frac{101 \times 144}{38.8}$

$= 200 \text{ lbm/ft}^2 \cdot \text{sec}$

GAS DYNAMICS AND COMBUSTION 3

Air is flowing in a supersonic wind tunnel. At a given point, the area is 1 ft^2 and Mach number is 2.5. The reservoir conditions are 2116 lb/ft^2 and 59°F. Find the temperature pressure and density at the point where Mach number equals 2.5 and the area at the throat of the tunnel.

Solution

Assume isentropic, steady flow.

$P_o = 2116$ psf

$T_o = 519$ °R

$P = ?$

$T = ?$

$\rho = ?$

$M = 2.5$

$A = 1$ ft.²

$$\frac{T}{T_o} = 0.44444 \qquad T = \underline{230.6°R} \quad (128.1\ °K)$$

$$\rho_o = \frac{P_o}{RT_o} = 0.0764 \ lbm/ft^3$$

$$\frac{P}{P_o} = 0.05853 \qquad P = \underline{123.3 \ psf} \quad (5.90 \ KN/m^2)$$

$$\frac{\rho}{\rho_o} = 0.13169 \qquad \rho = \underline{0.01006 \ lbm/ft.^3} \quad (1.58 \ N/m^3)$$

$$\frac{A}{A^*} = 2.6367 \qquad A^* = \underline{0.3793 \ ft^2} \quad (0.0352 \ m^2)$$

GAS DYNAMICS AND COMBUSTION 4

A 5 ft^3 tank of compressed air discharges through a 0.500 in. diameter converging nozzle which is located in the side of the tank. If the mass flow coefficient of the nozzle, based on isentropic flow through it, is 0.95 and the gas within the tank expands isothermally, find the time for the tank to discharge from 150 psia to 50 psia if the temperature of the tank is 70°F and the surrounding pressure is 14.7 psia.

Solution

For the tank:

$$p_o V = m R T_o \qquad V, T_o = \text{constant}$$

$$\frac{dp_o}{d\tau} = \frac{dm}{d\tau}\frac{RT_o}{V}$$

For sonic flow through nozzle, note $\frac{P_{exit}}{P_o} < 0.528$, $M_e = 1$

$$\frac{dm}{d\tau} = -\frac{p_o A}{\sqrt{T_o}} C_m \sqrt{\frac{Kg_c}{R}\left(\frac{2}{K+1}\right)^{\frac{K+1}{K-1}}}$$

By elimination of $\frac{dm}{d\tau}$,

$$\frac{dp_o}{d\tau} = -\frac{p_o A}{\sqrt{T_o}} C_m \sqrt{\frac{kg_c}{R}\left(\frac{2}{K+1}\right)^{\frac{K+1}{K-1}}}\frac{RT_o}{V} = \text{constant}\,(C)\,p_o$$

Therefore,

$$\int_{p_i}^{p_f}\frac{dp_o}{p_o} = C\int_0^{\tau} d\tau$$

or

$$\tau = \frac{1}{C}\ln\frac{p_f}{p_i} = -\frac{V}{C_m A\sqrt{Kg_c R T_o\left(\frac{2}{K+1}\right)^{\frac{K+1}{K-1}}}}\ln\frac{p_f}{p_i}$$

-continued-

for air,

$K = 1.4$

$R = 53.34 \ ft\text{-}lbf/lbm°R$

$$\tau = \frac{-0.0353V \ln \frac{P_f}{P_i}}{C_m A \sqrt{T_i}} = \underline{\underline{6.49 \text{ sec.}}}$$

GAS DYNAMICS AND COMBUSTION 5

An attitude control system for a space vehicle uses nitrogen expanding through a nozzle to provide thrust vector control. At maximum thrust nitrogen is used at a rate of 360 lb/hr from an initial pressure of 500 psia and 0°F.

(a) What is the thrust of the nozzle?

(b) Calculate the throat area of the nozzle.

Solution

$\dot{m} = 360 \ lbm/hr = 0.1 \ lbm/sec.$

$P_o = 500 \ psia \qquad C_p = 0.25 \ Btu/lbm°F$

$T_o = 460°R \qquad K = 1.4 \qquad R = 55.14 \ ft.lbf/lbm°R$

for maximum thrust $P_{exit} = P_{atm}$

a) $$F = \frac{\dot{m}}{g_c} V_e + \underbrace{A_e (p_e - p_a)}_{0 \text{ for max. thrust}}$$

but, $$V_e = \sqrt{2 g_c C_p (T_o - T_e)} = \sqrt{2 g_c C_p T_o \left(1 - \left(\frac{P_e}{P_o}\right)^{\frac{K-1}{K}}\right)}, \quad C_p = \frac{KR}{K-1}$$

-continued-

$$\therefore F_{max} = \frac{0.1}{32.2}\sqrt{\frac{2 \times 1.4 \times 1545.44 \times 460 \times 32.2}{28 \times 0.4}\left(1 - \left(\frac{14.7}{500}\right)^{0.285}\right)}$$

$$= \underline{\underline{5.91 \text{ lbf}}}$$

b) $$\frac{\dot{m}}{A_t} = \sqrt{\frac{K g_c}{R/M_{(molecular\ wt.)}}\left(\frac{2}{K+1}\right)^{\frac{K+1}{K-1}}}\ \frac{P_o}{\sqrt{T_o}} \approx 0.523 \frac{P_o}{\sqrt{T_o}}$$

$$= 0.523 \frac{500 \times 144}{\sqrt{460}} = 1757 \text{ lbm/ft}^2\text{sec}$$

$$\therefore A_t = \frac{0.1 \times 144}{1757} = \underline{\underline{0.00820 \text{ in.}^2}} \quad (5.29 \text{ mm}^2)$$

GAS DYNAMICS AND COMBUSTION 6

A turbine test area hopes to obtain 1/4 lb/sec of cooling air through a 29 ft 1/2 in. I.D. pipe.

Assuming no heat transfer, what is the available maximum flow rate from a 100°F 200 psia source if the outlet is maintained at 50 psia?

Friction factor in pipe is f = 0.01 $\left(\Delta P = \frac{4fl}{D}\frac{\rho V^2}{2g}\right)$

List all assumptions and references.

Solution

Assume:

a) fanno line solution for a perfect gas

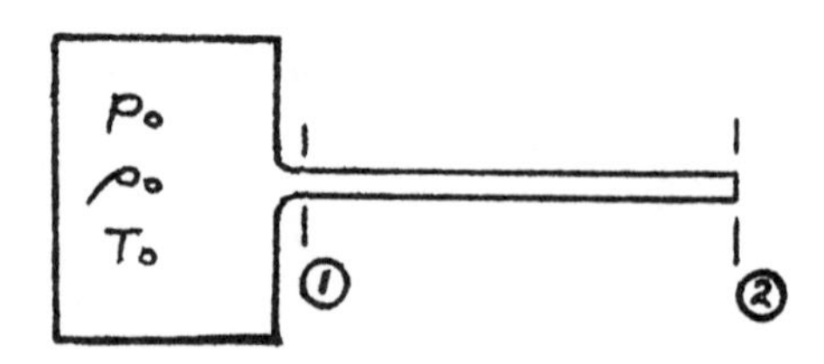

b) pipe entrance velocity low, such that $T_o \simeq T_1$, $\rho_o \simeq \rho_1$ (check subsequently)

-continued-

For desired flow rate,

$$V_1 = \frac{\dot{m}}{\rho_o A_1} = \frac{0.25 \times 53.34 \times 560 \times 4}{200 \times 144 \times \pi \times 0.25} = 190 \text{ ft/sec.}$$

For this velocity, assumption b) is not seriously violated.

$$\therefore \quad M_1 = \frac{190}{49\sqrt{560}} = 0.16$$

$$\frac{4 f L_{1-2}}{D} = \frac{0.04 \times 29}{1/24} = 27.84$$

For $M_1 = 0.16$ from tables $\frac{4 f L_{max}}{D} = 22.96$

Then conclude M_1 is too large.

Further $\frac{p_1}{p^*} = 6.6$ from tables, but $\frac{p_1}{p_2} = 4$ from problem gives additional support

$M_1 < 0.164$

<u>This means $\dot{m}$ will be below desired flow rate</u>

To estimate $\dot{m}$, assume $M_1 = 0.15$ Then $\left.\frac{4 f L_{max}}{D}\right)_1 = 27.93$

$$\left.\frac{4 f L_{max}}{D}\right)_2 = \left.\frac{4 f L_{max}}{D}\right)_1 - 27.84 = 0.09$$

$$\therefore M_2 = 0.78 \quad \text{from tables,} \quad \frac{p_1/p^*}{p_2/p^*} = \frac{p_1}{p_2} = \frac{7.29}{1.326} = 5.5$$

$\therefore$ M_1 is slightly lower than 0.15 to give $p_1/p_2 = 4$

$$V_1 \simeq 190 \times \frac{0.15}{0.16} = 174 \text{ ft/sec.}$$

$$\dot{m} = 0.25 \times \frac{0.15}{0.164} = \underline{\underline{0.229 \text{ lbm/sec.}}} \ (0.104 \text{ Kg/sec})$$

GAS DYNAMICS AND COMBUSTION 7

Gas flows through a duct that extends from station "a" through station "d" and it has a cross-sectional area of 100 in^2. At station "a" the gas has a velocity of 880 ft/sec, a static pressure of 20.0 psia and a total temperature of 670°R. Between stations "b" and "c" heat is transferred to/from the flow. If the gas has a specific heat at constant pressure of 0.275 Btu/lb - °R and a gas constant of 70 ft-lbf/lbm°R, what heat flow rate is necessary for a Mach number of 0.300 at station "d"? Is the heat added or removed? Neglect velocity distribution and all real gas effects.

Solution

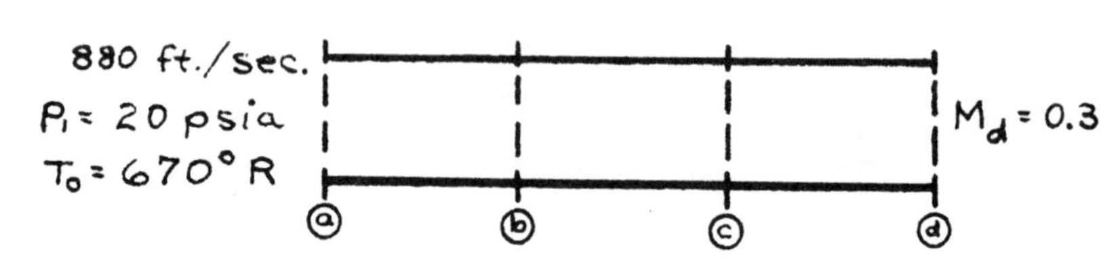

$A = 100 \text{ in}^2$

$C_p = 0.275$ Btu/lbm °F

$R = 70$ ft·lbf/lbm°R

$$M_a = \frac{880}{\sqrt{K g_c R T_a}}$$

$C_p - C_v = R$

$$C_v = 0.275 - \frac{70}{778} = .185$$

$$K = \frac{C_p}{C_v} = \frac{.275}{.185} = 1.49$$

$$T_a = T_0 - \frac{V^2}{2 g_c C_p} = 670 - \frac{(880)^2}{2 \times .275 \times 778 g_c} = 670 - 56.2 = 613.8°R$$

Therefore,

$$M_a = \frac{880}{\sqrt{1.49 \times 32.2 \times 70 \times 613.8}} = 0.613$$

Since $M_a > M_d$ and both are subsonic

Therefore, heat is removed.

$$\frac{T_{o_2}}{T_{o_1}} = \frac{M_2^2(1 + k M_1^2)^2 (1 + \frac{K-1}{2} M_2^2)}{M_1^2(1 + K M_2^2)^2 (1 + \frac{K-1}{2} M_1^2)}$$

$$= \frac{.09(1 + 1.49 \times .376)^2 (1 + .245 \times .09)}{.376(1 + 1.49 \times .09)^2 (1 + .245 \times .376)} = 0.447$$

-continued-

Therefore, $T_{o2} = 0.447 \times 670 = 300°R$

Heat removed $= c_p(T_{o1} - T_{o2}) = .275(670-300) = 101.9$ Btu/lbm

Total $Q = \left(\frac{20 \times 144}{70 \times 613.8} \times \frac{100}{144} \times 880\right)(101.9) = 4167$ Btu/sec

GAS DYNAMICS AND COMBUSTION 8

A 16-inch pipe line is transporting natural gas across the desert. The pressure and temperature as it leaves pumping station A is 400 psia and 97°F. One hundred miles further along the pipe line at B, measurements reveal a pressure of 175 psia and a temperature of 110°F.

How much could the temperature be reduced at B by insulating the 100 miles of pipe?

Solution

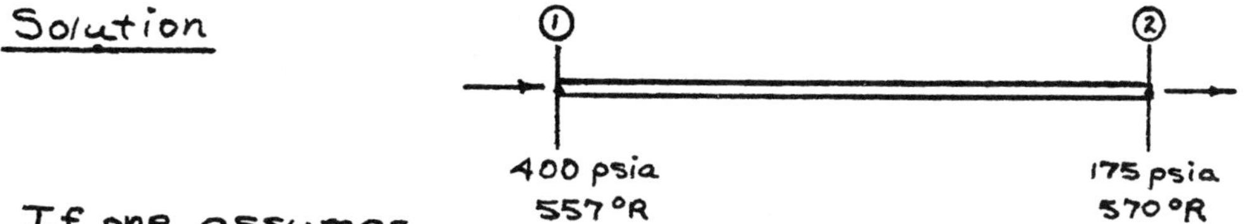

If one assumes no heat transfer, the fanno line would represent the solution.

For $M<1$, $T_2 > T_1$

∴ There is heat supplied to pipe.

From 1st Law,

$$h_1 + \frac{V_1^2}{2g_c} + q = h_2 + \frac{V_2^2}{2g_c}$$

$$c_p(T_2 - T_1) = q + \frac{V_1^2 - V_2^2}{2g_c}$$

-continued-

Check to see if $\Delta\left(\frac{V^2}{2g_c}\right)$ is negligible

$$\Delta p \simeq f \frac{\ell}{D} \frac{\bar{V}^2}{2g_c} \bar{\rho} \simeq .01 \times \frac{5.28 \times 10^5}{1.25} \frac{\bar{V}^2}{64.4} \frac{\bar{p} M}{R_o \bar{T}}$$

* for check, use arithmetic average pressure and temp.

Then,

$$\bar{V}^2 = \frac{225 \times 144 \times 1.25 \times 1544 \times 563 \times 64.4}{.01 \times 5.25 \times 10^5 \times 288 \times 144 \times 16}$$

$$\bar{V} = 25.5 \text{ ft./sec.}$$

∴ neglect vel. effect.

By perfect insulation $\underline{\underline{T_2 \simeq T_1 \ (97°F)}}$ (36.1°C)

GAS DYNAMICS AND COMBUSTION 9

A residual fuel oil used in central station steam power plant has the following ultimate analysis:

Constituent	Per Cent by Weight
Carbon	84.00
Hydrogen	13.10
Oxygen	.75
Nitrogen	1.00
Sulfur	1.10
Ash	.05

Assume that this fuel is burned completely with nine per cent excess air by volume.

The regulations of the air pollution authorities require that sulfur products calculated as sulfur dioxide not exceed 0.2 per cent by volume of the exhaust products, and that ash particles discharged not exceed 0.2 grain per cubic foot of exhaust gas at 60°F and atmospheric pressure. Assume that only the original fuel ash remains in the solid state after combustion.

Will the regulations of the authorities be complied with?

Show data and calculations necessary to support your answer.

Solution

Moles/lbm of fuel		Moles of O_2 required for complete combustion
C	$\frac{0.84}{12} = 0.07$	0.07
H_2	$\frac{0.13}{2} = 0.065$	0.0325
O_2	$\frac{.0075}{32} = 0.000234$	—
N_2	$\frac{0.01}{28} = 0.000357$	—
S	$\frac{0.011}{32} = 0.000344$	0.000344
Ash	—	—

0.102844 O_2 required

0.000234 O_2 in fuel

0.102610 (moles of O_2/lbm fuel)

$$\text{Air-fuel ratio} = 0.102610 \times \frac{100\%}{20.99\%}(1.09)$$

$$= 0.53 \text{ moles of air/lbm fuel}$$

$$= 15.35 \text{ lbm air/lbm fuel}$$

-continued-

Product Analysis

	Moles/lbm fuel	
CO_2	0.07	= 0.07
H_2O	0.065	= 0.065 Assume all H_2O in vapor state
O_2	0.102610 x 0.09	= 0.00922
N_2	3.76 x .102610 x 1.09	= 0.420
SO_2	0.000344	= 0.000344
		0.565

$$SO_2 = \frac{.000344}{.565} = \underline{\underline{0.06\% < 0.2\%}} \quad (OK)$$

mass of ash = 0.0005 x 7000 = 3.5 grains/lbm fuel

$$V_{products} = \frac{nRT}{P} = 214 \ ft^3/lbm \ fuel,$$

$$ash = \frac{3.5}{214} = \underline{0.0163} \ grains/ft.^3 \ product \quad (OK)$$

$$(37.3 \times 10^{-6} \ Kg/m^3)$$

GAS DYNAMICS AND COMBUSTION 10

An air pollution regulation restricts the discharge to atmosphere of sulfur compounds calculated as sulfur dioxide (SO_2) to 0.2 per cent by volume.

Fuel having the following analysis on a weight basis is available as boiler fuel:

Ash - 0.10, Carbon - 85.5, Hydrogen - 9.50
Nitrogen - 1.10, Oxygen - 0.80, Sulfur - 3.0
Dry bulb temperature - 80°F
Web bulb temperature - 70°F
Exit flue gas temperature - 300°F
Atmospheric pressure - 14.7 psi

(a) What would the sulfur dioxide (SO_2) content be in the flue gas when burning this fuel with 20 per cent excess air?

(b) Would this fuel meet requirements of sulfur discharge?

Solution

	moles / pound fuel	moles O_2 required
C	0.855/12 = 0.0713	0.0713
H_2	0.095/2 = 0.0475	0.0238
N_2	0.0110/28 = 0.0039	—
O_2	0.008/32 = 0.0003	—
S	0.030/32 = 0.0009	0.0009
	(less O_2 in fuel)	− 0.0003
		Σ = 0.0957

moles air = 4.76 x 0.0957 x 1.2 = 0.5466

$$\text{moles } N_2 = \frac{3.76}{4.76} \times 0.5466 = 0.4318$$

-continued-

moles exhaust:

CO_2 = 0.0713

H_2O = 0.0475

N_2 = 0.0039 + 0.4318 = 0.4357

O_2 = .0957 x 0.20 = 0.0191

SO_2 = 0.0009

Σ = 0.5745

since water will be in vapor form.

$$\frac{n_{SO_2}}{n_{total}} = \frac{V_{SO_2}}{V_{total}} = \frac{0.0009}{0.5745} = 0.16\%$$ level is OK

GAS DYNAMICS AND COMBUSTION 11

Determine the horsepower required to drive the forced draft fans on a pressurized furnace type boiler unit with output at 500,000 pounds of steam per hour at 1250 psig and 1000°F at superheater outlet. Efficiency of the boiler unit is 86.0 per cent when burning fuel oil of the following ultimate analysis:

Carbon	- 85.5 per cent by weight	
Hydrogen	- 9.5 per cent by weight	
Oxygen	- 0.80 per cent by weight	
Nitrogen	- 1.10 " " " "	
Sulfur	- 3.00 " " " "	
Ash	- 0.10 " " " "	Btu/pound 18,500

Firing condition is 20% excess air and 21" water total pressure difference.

Solution

assume excess air is on a mass base

at 1250 psig + 1000°F, h = 1497 Btu/lbm

at 410 °F + 1250 psig, h_f = 374.97 + 1.13 = 376.1 Btu/lbm

-continued-

$$Q = 500{,}000\,(1497 - 376.1) = 560.45 \times 10^6 \text{ Btu/hr.}$$

$$Q_{fuel} = \frac{560.45 \times 10^6}{0.86} = 652 \times 10^6 \text{ Btu/hr.}$$

$$\text{Mass of fuel} = \frac{652 \times 10^6}{18500} = 35200 \text{ lbm/hr.}$$

To determine air required/lbm of fuel :

	Moles/lbm of fuel		Moles of O_2 required
C	$\frac{.855}{12}$	= .0713	.0713
H_2	$\frac{.095}{2}$	= .0475	.0238
O_2	$\frac{.008}{32}$	= .0003	—
N_2	$\frac{.011}{28}$	= .0004	—
S	$\frac{.03}{32}$	= .0009	.0009
			.0960
			−.0003 O_2 in fuel
			.0957

$$A/F \text{ ratio} = .0957 \times \frac{1.20}{0.21} \times 28.97 = 15.84 \text{ lbm air/lbm fuel}$$

$$\text{Total mass of air} = 35200 \times 15.84 = 5.58 \times 10^5 \text{ lbm/hr}$$

$$H_a\text{, head in ft. of same fluid} = H_w \frac{\rho_w}{\rho_a}$$

$$\rho_a = \frac{14.7 \times 144}{53.34 \times 540} = .0735 \text{ lbm/ft}^3$$

Therefore,

$$HP = \frac{\dot{m}_a H_a}{33000\,\eta}$$

$$= \frac{5.58 \times 10^5}{60}\ \frac{21}{12} \times \frac{62.4}{.0735} \times \frac{1}{33000 \times 0.7}$$

$$= \underline{\underline{598 \text{ HP}}}$$

$$(445.9 \text{ KW})$$

GAS DYNAMICS AND COMBUSTION 12

A tube solid propellant grain is bonded to its case and has inhibitor on the grain ends. The case I.D. is 40.0 inches in diameter. The grain I.D. is 26.0 inches in diameter and has a 70.0 inch length. The propellant density is 0.062 lbs/cu.in. and its burning rate is characterized by the following equation:

$b = ap^n$, in/sec, where a = 0.001 and n = 0.75

p - chamber pressure, psia

The products of combustion have a constant specific heat ratio of 1.25, a gas constant of 70.0 ft-lbf/lbm°R, and a flame temperature of 3950°F. At zero time the chamber pressure is 190 psia. What is the chamber pressure at the end of 40.0 seconds? Neglect grain erosion effects. State all assumptions.

Solution

* Assume: Stagnation Conditions in combustion zone radial burning only.

Initial burning rate

$$b_i = 0.001 \times 190^{0.75} = 0.0512 \text{ in/sec}$$

$$\therefore \dot{m}_i = \pi D \ell b_i \rho = \pi \times 26 \times 70 \times 0.0512 \times 0.062$$

$$= 18.15 \text{ lbm/sec.}$$

* Assume: Sonic Flow of perfect gas at nozzle throat.

$$\dot{m}/A^* = \left[\frac{K g_c}{R}\left(\frac{2}{K+1}\right)^{\frac{K+1}{K-1}}\right]^{1/2} \frac{p_{o,i}}{T_{o,i}^{1/2}}$$

$$A^* = \frac{18.15}{190}\left[\frac{4410}{\frac{1.25 \times 32.2}{70}\left(\frac{2}{2.25}\right)^{\frac{2.25}{0.25}}}\right]^{1/2}$$

$$= 14.2 \text{ in}^2$$

—continued—

*Assume: burning rate remains constant for 40 sec.

$$\therefore \dot{m}_{40\,sec.} = \pi (D_i + 2 b_i t) \ell b_i \rho$$

$$= \pi (26 + 2 \times 0.0512 \times 40) 70 \times .0512 \times .062$$

$$= 21.0 \text{ lbm/sec}$$

*Assume: $T_{o,i} = T_{o,40\,sec.}$

Then, $$P_{o,40\,sec.} = \frac{\dot{m}}{A} \frac{T_o^{1/2}}{\left[\frac{K g_c}{\bar{R}} \left(\frac{2}{K+1}\right)^{\frac{K+1}{K-1}}\right]^{1/2}}$$

$$= \underline{\underline{220 \text{ psia}}}$$
$$(1.52 \times 10^3 \text{ KN/m}^2)$$

For more accurate value, use $\bar{p}_o$ to get new $\bar{b}$ & hence a better value of $P_{o,40\,sec.}$

9

Hydraulic Machines

R. IAN MURRAY

In this chapter the performance characteristics of pumps, fans and turbines of the hydrodynamic type are reviewed. Emphasis is placed on the determination of the operating point for specific conditions and on the selection of appropriate machines.

Dimensional analysis leads to the definition of the following dimensionless variables:

Capacity Coefficient	$C_Q = \frac{Q}{ND^3}$
Head Coefficient	$C_H = \frac{Hg}{N^2D^2}$
Power Coefficient	$C_P = \frac{P}{\rho N^3 D^5}$
Efficiency	$\frac{Q\gamma H}{P}$
Machine Reynolds Number	$\frac{ND^2}{\nu}$

N is the machine speed in revolutions per unit of time; D is a characteristic dimension of the machine, such as the impeller or runner diameter. H is the change of total head across the machine. For fans, H is generally replaced by $\frac{\Delta p}{\gamma}$ where Δp is the increase in total pressure produced by the fan. Other symbols are as defined in Chapter 6.

When dealing with low viscosity fluids such as water and air, the effects of changes in the machine Reynolds number are negligible for the machine sizes and speeds generally encountered. For this reason the machine Reynolds number is omitted from the following relationships.

PUMPS

For geometrically similar pumps operating under conditions which ensure that the pressure is everywhere greater than the vapor pressure of the fluid, the head coefficient is a function of the capacity coefficient only. The significance of this statement is that the relationship among H, Q, N, and D for an entire family of geometrically similar pumps operating at different speeds can be represented by a single curve of C_H vs. C_Q. This knowledge makes it possible, for example, to predict the H vs. Q characteristic at speed N_2 from test data at speed N_1.

Corresponding points are points that plot as a single point on the C_H vs. C_Q curve. Thus for corresponding points

$$C_Q = \frac{Q_1}{N_1 D_1^3} = \frac{Q_2}{N_2 D_2^3}$$

$$C_H = \frac{H_1 g}{N_1^2 D_2^2} = \frac{H_2 g}{N_2^2 D_2^2}$$

Furthermore,

$$\frac{C_Q^2}{C_H} = \frac{Q^2}{H\ gD^4} = \text{constant}$$

is the locus of corresponding points on an H vs. Q plot. The constant, of course, is different for each point on the C_H vs. C_Q curve.

From a hydraulic point of view alone, and under the same assumptions as above, the power coefficient and the efficiency are each functions of the capacity coefficient only. However, since the hydraulic analysis does not take mechanical friction into account, the above statement for power and efficiency is only a first approximation.

The above relationships are used to predict one set of performance characteristics from another. They should not be used directly to predict one operating point from another, for in general two operating points are not corresponding points. An operating point is determined by the intersection of a "head rise required" vs. flow rate curve with the H vs. Q characteristic curve of the pump. The "head rise required" (or system) curve is determined by the principles reviewed in Chapter 6. It depends on the system the pump must supply, not on the pump.

All of the above is predicated on the assumption that the pressure is everywhere greater than the vapor pressure of the liquid being pumped, i.e., that the pump does not cavitate. To ensure this, the "net positive suction head" (NPSH) must be greater than a critical value that is a function of the design, the size, the flow rate and the speed of the pump. The NPSH is defined as the total head at inlet in feet of liquid absolute, less the vapor pressure of the liquid in feet of liquid. The critical value of NPSH scales in accordance with the head coefficient.

To select the proper type of pump for a given application, the notion of "specific speed" (N_S) is introduced. If C_Q and C_H are combined in such a way as to eliminate the size factor D, the following dimensionless variable is obtained:

$$\frac{NQ^{1/2}}{(Hg)^{3/4}}$$

When this is evaluated at the point of maximum efficiency, it becomes a useful parameter to identify the type or design of pump. In the United States it is common practice to omit g and to specify the units of H in feet, the units of Q in gpm and the units of N in rpm. The parameter is then known as specific speed. Its value ranges from 500 for large centrifugal pumps to 15,000 for propeller pumps.

TURBINES

Speed, head and flow rate are independent variables in the performance of turbines. Thus, under the same assumptions as for pumps, the power coefficient and the efficiency are each functions of the capacity coefficient and the head coefficient. In the operation of a turbine, however, the speed is generally fixed, and the head is approximately constant or a function of the flow rate. Under these conditions the power is a function of the flow rate only (density also being fixed) and the efficiency can be expressed as a function of the power only.

Specific speed is again a useful parameter for selecting the proper type of turbine for a given application. The definition, however, is different than that for a pump:

$$N_S = \frac{NP^{1/2}}{H^{5/4}} \quad \text{(P is in hp)}$$

Small values of N_S are associated with impulse turbines; large values, with axial flow.

Energy is extracted from a fluid as it passes through the rotor of a turbine. Corresponding to this energy extraction either the velocity or the pressure of the fluid decreases, or both may decrease. The ratio of the energy extraction associated with a pressure drop to the total energy extraction is called "reaction". An "impulse" turbine is one in which the "reaction" is zero.

HYDRAULIC MACHINES 1

You are to select a boiler-feed pump for condensate at a maximum temperature of 130°F and maximum vacuum of 25 inches of mercury. Proposed are a horizontal centrifugal pump and a vertical turbine pump. Each requires 8 feet of net positive suction head at its suction inlet for water at 72°F. Your design has the suction inlet 4 feet below the condensate surface. It is suggested the suction condition is improved by lowering the suction inlets 10 feet. This would be done by lengthening the horizontal pump's suction pipe (friction = 10 feet per 100 feet) and lengthening the vertical pump's column pipe (friction = 15 feet per 100 feet).

REQUIRED: (a) Are the inlets all right as is?

(b) What do you think of the suggestion?

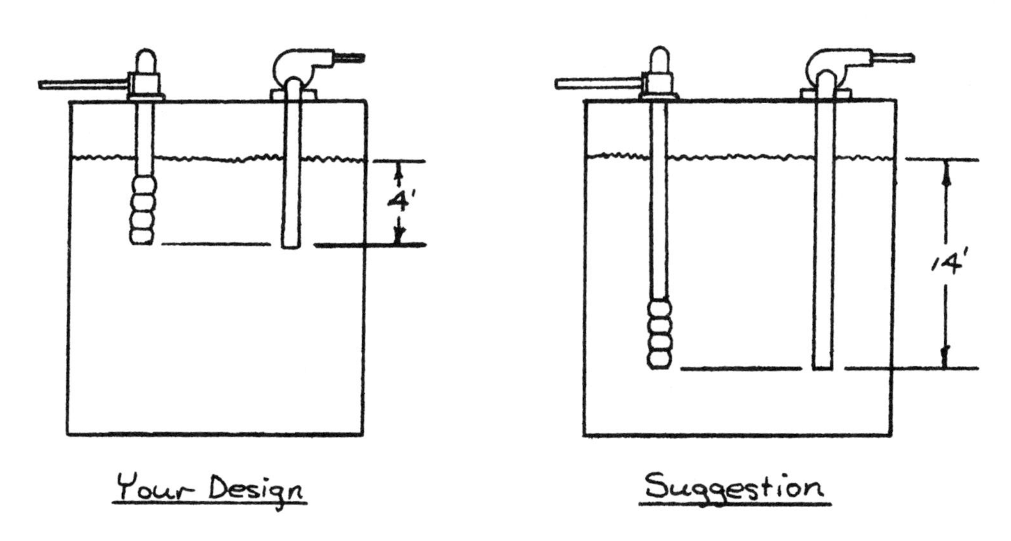

Solution

a) The design is not satisfactory. The pressure at the surface of the condensate is equal to the vapor pressure. Thus the net positive suction head (NPSH) on the verticle turbine pump is approximately 4 ft.; on the horizontal centrifugal pump it is zero.

b) The suggestion for the horizontal centrifugal pump is useless, for the NPSH would remain zero.

—Continued—

The suggestion for the verticle turbine pump is excessive, for the NPSH would then be approximately 14 feet.

HYDRAULIC MACHINES 2

Water at the boiling point of 228°F is pumped at a rate of 120 gallons per minute from a deaerator into a boiler. Pressure in the deaerator is 5.3 psig. The water level in the tank is controlled by a high-low device. The water level varies in the tank from a maximum height of 17.0 feet to a minimum height of 15.0 feet above the centrifugal pump center line. The piping is 2-1/2" steel pipe, Sched. 40, 18 feet long. There is one 2-1/2" gate valve, and one 2-1/2" long radius elbow. The pump operates at sea level. Pump impeller is 8-31/32" in diameter. At 228°F, viscosity for water is 0.26 centipoises and specific weight is 59.43 pounds per cubic foot. K = 0.5 for square edge inlet; K = 0.22 for flanged wedge disc gate valve; K = 0.27 for 90 degree large radius elbow; pipe roughness $\frac{\varepsilon}{D}$ = 0.000729. The pump curves and schematic are shown below.

REQUIRED: Determine each of the following:

(a) Net positive suction head available.
(b) Net positive suction head required.
(c) Will the centrifugal pump cavitate under these operating conditions?

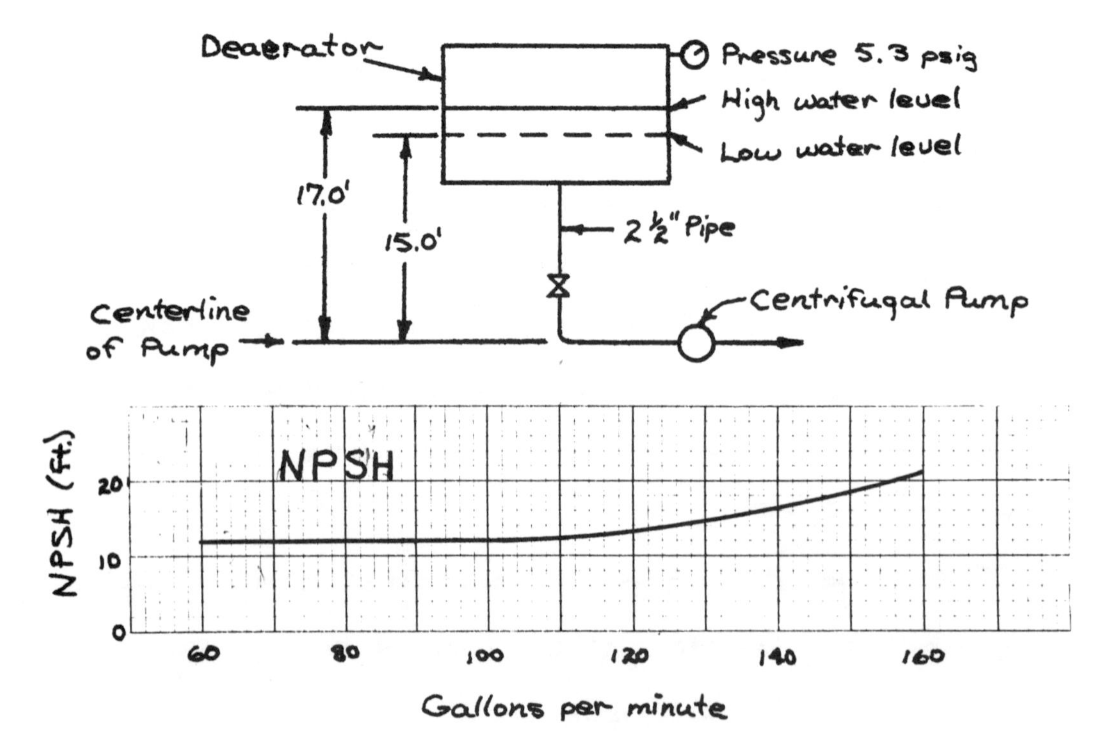

Solution

a) $NPSH = \begin{Bmatrix} 17' \\ 15' \end{Bmatrix} - H_L$

$$V = \frac{Q}{A} = \frac{120\ \text{gal./min.} \times \frac{\text{ft}^3}{7.48\ \text{gal.}} \times \frac{\text{min.}}{60\ \text{sec.}}}{\frac{\pi}{4}\left(\frac{2.47}{12}\right)^2\ \text{ft}^2} = 8.04\ \text{ft/sec.}$$

$$\frac{V^2}{2g} = 1\ \text{ft.}$$

$$N_R = \frac{v d \rho}{\mu} = \frac{(8.04\ \text{ft./sec.})\left(\frac{2.47}{12}\ \text{ft.}\right)\left(\frac{59.4}{32.2}\ \frac{\text{lb}\cdot\text{sec}^2}{\text{ft.}^4}\right)}{(0.26\ \text{centipoise})\left(\frac{\text{lb. sec.}}{4.79 \times 10^4\ \text{ft.}^2\ \text{centipoise}}\right)}$$

$$N_R = 5.6 \times 10^5$$

$$f = 0.0185 \quad (\text{from Moody Diagram})$$

$$H_L = \left(0.0185 \times \frac{18 \times 12}{2.47} + 0.50 + 0.22 + 0.27\right)\frac{V^2}{2g} = 2.6'$$

NPSH ranges from <u>12.4' to 14.4'</u> ⟵

b) NPSH required = 13' (from graph)

c) Pump will cavitate

HYDRAULIC MACHINES 3

You need to increase the pressure of a hydrocarbon fuel 125 psi at 450 gallons per minute. The hydrocarbon has a specific gravity of 0.72 and a vapor pressure of 35 psia, and a temperature of 90°F. You are considering a centrifugal pump whose characteristics are shown by a 1770 rpm test with water at 72°F to be:

G.P.M.	Head (feet)	N.P.S.H. Required (feet)
0	436	
100	400	3.9
200	377	5.0
300	346	6.5
400	287	8.3
500	200	11.0

REQUIRED: (a) What speed must the pump be operated for your needs?

(b) What suction pressure must be supplied the pump?

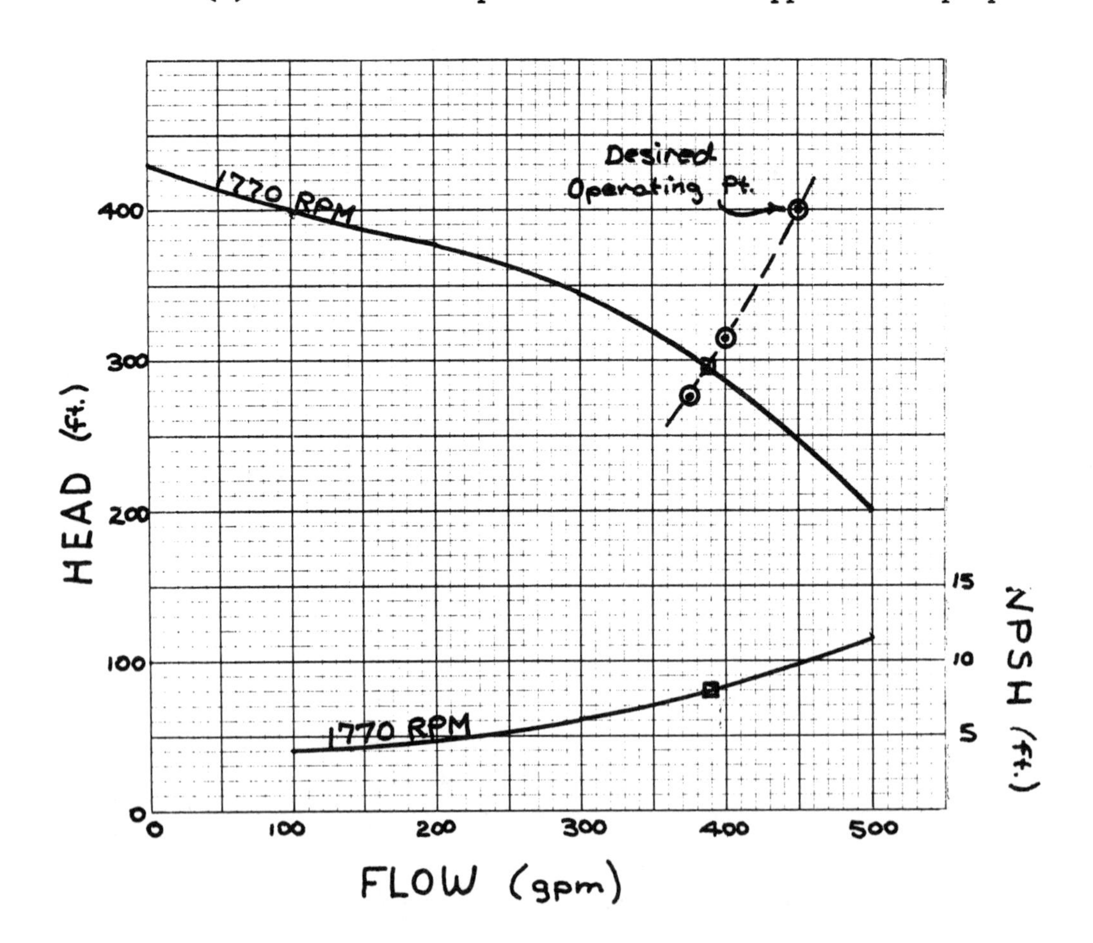

Solution

$$H = \frac{\Delta P}{\gamma} = \frac{125 \times 144}{0.72 \times 62.4} = 400 \text{ ft.} \quad @ \ 450 \text{ gpm}$$

* For corresponding points to desired operating points,

$$\frac{Q^2}{H} = \text{constant}$$

$$= \frac{(450)^2}{400} = \frac{(400)^2}{316} = \frac{(375)^2}{277}$$

* Corresponding point at 1770 RPM,

$$H_1 = 295 \text{ ft.}$$

$$Q_1 = 385 \text{ gpm}$$

$$NPSH_1 = 8 \text{ ft.}$$

$$N_2 = N_1 \frac{Q_2}{Q_1}$$

$$N_2 = 1770 \times {}^{450}\!/_{385} = \underline{\underline{2070 \text{ rpm}}} \longleftarrow$$

* Since the NPSH scale's in accordance with the head coeficient,

$$NPSH_2 = 8 \text{ ft.} \times \left({}^{450}\!/_{385}\right)^2 = 10.9 \text{ ft} = \frac{P_s + P_a - P_v}{\gamma}$$

$$P_s = \frac{0.72 \times 62.4 \times 10.9}{144} + 35 - 14.7 = 3.4 + 35 - 14.7$$

$$= \underline{\underline{23.7 \text{ psig}}} \longleftarrow$$

P_s is the total suction pressure.

HYDRAULIC MACHINES 4

An electric-motor-driven (1770 rpm) pump is used to raise water from one open tank to another, as shown. The pump data presented to you is:

Flow (gpm)	Head (feet)	Overall Efficiency
0	635	0
500	570	41.5
1000	474	66.2
1250	415	71.0
1500	334	69.5
1750	235	65.0
2000	120	49.5

The system curve, before installing the automatic flow control valve, is:

Flow (gpm)	Head (feet)
0	127
400	133
800	151
1200	180
1600	222
2000	275

The valve allows a flow of 1400 gpm. What is the pressure differential across the valve?

If another pump is available with the same head-capacity curve and 10% less power consumption, how much could economically be paid for it, assuming pumping to be 1000 acre-feet per year, for 5 years, power cost 1.8¢ per kwh, and money worth 7%?

Calculate the pump cost with and without the flow control valve.

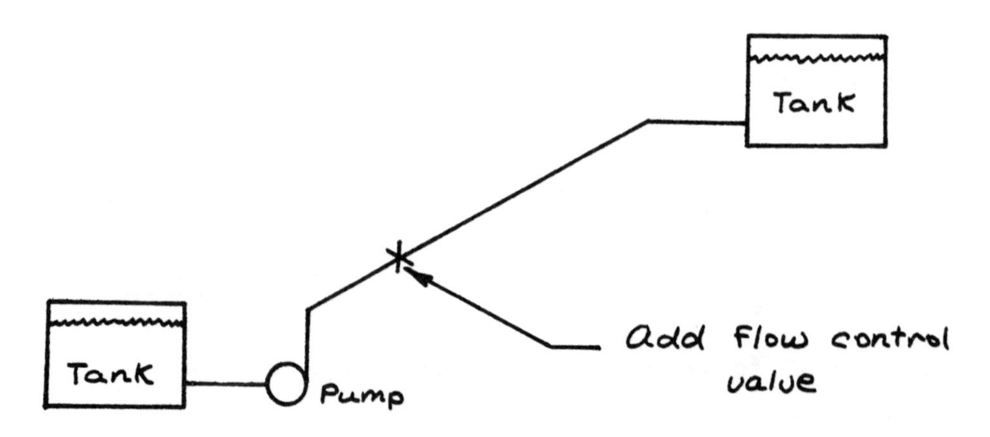

Solution

Plot the given characteristics for both pump and system (see next page for plot). At 1400 gpm, observe the pump head to be 370 ft. and the system head to be 200 ft.. Thus, the head loss at the control value is 170 ft.

$$\Delta p = \gamma \Delta H = \left(62.4 \frac{lb}{ft^3}\right)(170\ ft)\left(\frac{ft^2}{144\ in^2}\right)$$

$$= \underline{\underline{73.7\ psi}}$$

Operation without control value:

$$\left.\begin{array}{l} Q = 1730\ gpm \\ H = 240\ ft. \\ \eta = 66\% \end{array}\right\} P = \frac{Q \gamma H}{\eta} = 118.5\ kw$$

$$P = \left(1730 \frac{gal}{min.}\right)\left(\frac{ft^3}{7.48\ gal}\right)\left(62.4 \frac{lb}{ft^3}\right)(240\ ft)\left(\frac{1}{.66}\right)\left(\frac{kw \cdot min.}{44{,}254\ ft \cdot lb}\right)$$

$$\underset{(per\ year)}{\text{Pumping time}} = \frac{1000\ acre\ ft.}{1730\ gpm} \times \frac{326 \times 10^3\ gal.}{acre \cdot ft} \times \frac{hr}{60\ min} = 3140\ hr.$$

$$\text{Energy Cost per year} = \frac{\$0.018}{Kwh} \times 118.5\ kw \times 3140\ hr$$

$$= \$6700$$

Annual Savings with other pump = \$670

Present value @ 7% for 5 yrs. = 4.1 x \$670 = $\underline{\underline{\$2740}}$

Operation with control value:

$$\left.\begin{array}{l} Q = 1400\ gpm \\ H = 370\ ft \\ \eta = 71\% \end{array}\right\} P = 1.16 \times 118.5\ kw$$

$$\underset{(per\ year)}{\text{Pumping time}} = \frac{1730}{1400} \times 3140\ hr = 1.235 \times 3140\ hr.$$

Energy cost per year = 1.16 x 1.235 x \$6700 = \$9600

Annual savings with other pump = \$960

Present value* @ 7% for 5 yrs. = 4.1 x \$960 = \$3940

*See Chapter 12.

However, control valve should not be used unless it is needed for reasons not indicated in the problem statement.

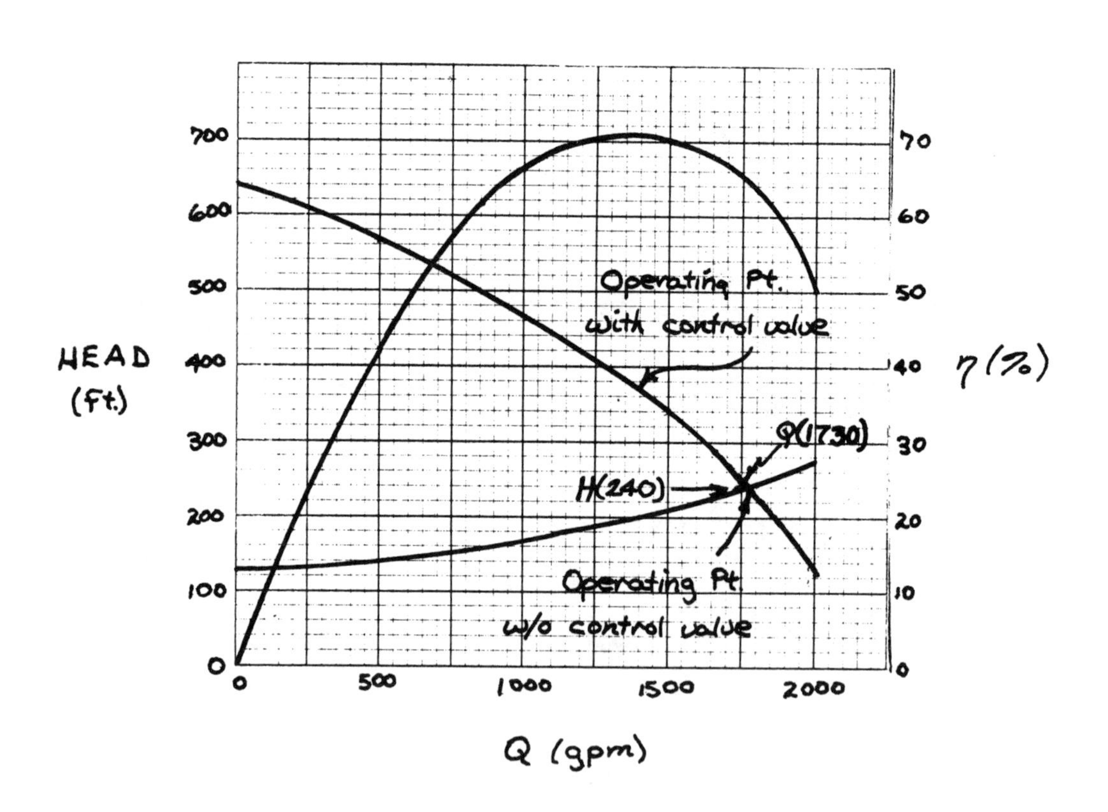

HYDRAULIC MACHINES 5

A variable-speed centrifugal water pump has the following H - Q curve at its maximum speed of 1680 rpm:

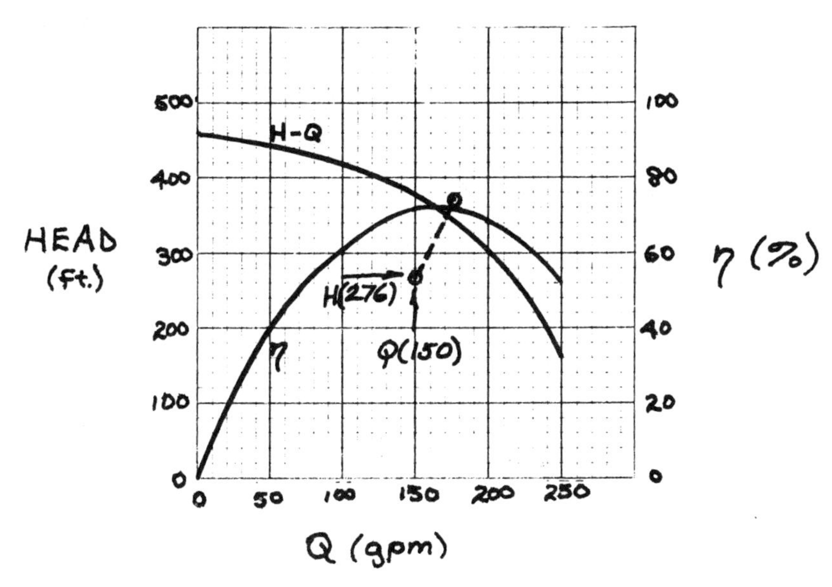

Suction pressure varies from 23 to 40 psig. The system demand varies from 0 to 200 gpm.

(a) What speed range must the pump have to maintain a constant discharge pressure of 150 psig?

(b) What speed would the pump be running when pumping 150 gpm with a discharge pressure of 150 psig and suction pressure of 30 psig?

Solution

Suction pressure is minimum at maximum flow.

$$H = \frac{(150-40) \times 144}{62.4} = 253 \text{ ft.} \quad @\ Q = 0 \quad N = ?$$

$$H = \frac{(150-23) \times 144}{62.4} = 293 \text{ ft} \quad @\ Q = 205 \text{ gpm} \quad N = 1680 \text{ rpm}$$

$$H = \frac{(150-30) \times 144}{62.4} = 276 \text{ ft} \quad @\ Q = 150 \text{ gpm} \quad N = ?$$

- Continued -

a) At shut off :

$$N_2 = \sqrt{\frac{H_2}{H_1}}\, N_1 = \sqrt{\frac{253}{460}} \times 1680 \text{ rpm} = 1240 \text{ rpm}$$

Thus, the speed range is from 1240 rpm to 1680 rpm.

b) For corresponding points :

$$\frac{Q^2}{H} = \frac{(150\text{ gpm})^2}{276 \text{ ft.}} = \frac{(175 \text{ gpm})^2}{H_3}$$

$$H_3 = 375 \text{ ft.}$$

Sketch curve of corresponding points on graph and estimate that @ 1680 rpm , $Q_5 = 168$ gpm.

Then,

$$N_4 = \frac{Q_4}{Q_5} N_5 = \frac{150}{168} \times 1680$$

$$= \underline{\underline{1500 \text{ rpm}}}$$

HYDRAULIC MACHINES 6

A pump takes water from the Mad River and delivers it through 4000 feet of pipe line to a paper plant water tank. The static lift is 200 feet. The pump when new had the following performance:

Flow in gpm	Head in feet	Efficiency of Pump and Electric Motor Combined
0	785	0
1000	685	43.5
1500	620	57.8
2000	579	67.1
2500	539	73.0
3000	476	75.0
3500	389	71.3
4000	260	61.9

The pump has worn so that it presently pumps to the tank 2000 gpm at a head of 323 feet, while using power at the rate of 328 kilowatts. You are to decide whether to leave the pump as is for awhile, repair it to original condition, or replace it. Power costs 1.5¢ per kwh. Repair of the present pump to new condition would cost $3000. Replacement would cost $40.00 per nominal motor h.p. The plant uses 1.1 billion gallons annually, operating 363 days. The pump should deliver 20% more than the annual average, to allow for future wear.

What is the economical solution?

Solution

$$\text{Use rate} = \frac{1.1 \times 10^9 \text{ gal.}}{363 \text{ days}} \times \frac{\text{day}}{24 \text{ hrs.}} \times \frac{\text{hr}}{60 \text{ min}}$$

$$= 2100 \text{ gpm}$$

Thus, one alternative is eliminated; the present pump will not do the required job.

The pump should deliver $1.2 \times 2100 \text{ gpm} = 2530$ gpm. Assume that the head loss in the system is proportional to the square of the flow rate.

$$H = 200 \text{ ft} + KQ^2$$

$$H = 200\text{ ft} + KQ^2$$

$$323 = 200 + K(2000)^2 \rightarrow K = \frac{123}{4\times10^6}$$

$$H = 200 + \frac{123}{4\times10^6}(2530)^2 = 396\text{ ft.}$$

$$P_{new\ pump} = \frac{Q\gamma H}{\eta} = \frac{2530\text{ gpm}}{7.48\text{ gal/ft}^3}\cdot\frac{62.4\text{ lb/ft}^3\ 396\text{ ft}}{33{,}000\ \frac{\text{ft}\cdot\text{lb}}{\text{hp}\cdot\text{min.}}}\cdot\frac{1}{\eta}$$

$$= \frac{253.5}{\eta}\text{ hp}$$

Assume $\eta = 75\%$, then $P_{new\ pump} = 338\text{ hp} = 252\text{ Kw}$

$\therefore$ minimum replacement cost $= 40 \times 338 = \$13{,}500$

$$\underset{(\text{new pump})}{\text{Power cost}} = 252\text{ Kw}\left(363 \times 24 \times \frac{2100}{2530}\text{ hr}\right)\$0.015/\text{kwh}$$

$$= \$27{,}300/\text{yr.}$$

Operating point if present pump were repaired would be 3000 gpm @ 476 ft. head and 75% efficiency. This is obtained by inspection of the pump characteristic data and trying $Q = 3000$ gpm in the head required equation developed above. It could also be obtained by plotting the pump characteristic data and the head required equation and observing their intersection.

$$\$27{,}300 \times \frac{476\text{ ft.}}{396\text{ ft.}} = \$32{,}800$$

Additional annual power cost for repaired pump is <u>$5,500</u>. Thus, the incremental cost of a new pump would be recoved in two years. Buy a new pump!

HYDRAULIC MACHINES 7

A water purveyor pumps from the Sacramento River, as shown, schematically.

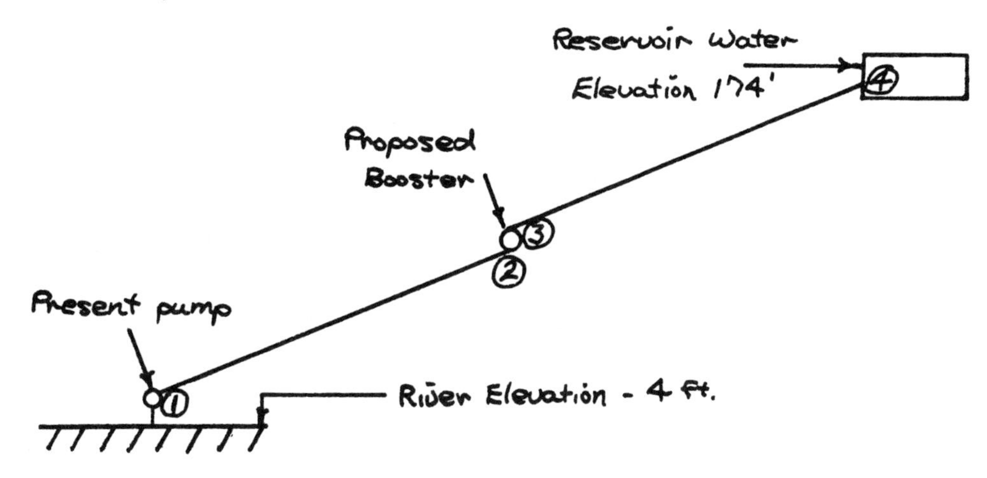

The existing booster is a pump that has the following performance:

Flow - gpm	Head in feet	Efficiency of Pump and Electric Motor Combined
0	615	0
1000	590	27.3
2000	555	49.8
3000	500	66.5
4000	465	75.8
5000	415	81.7
6000	345	81.4
7000	265	75.3
8000	155	56.1

This pump presently pumps 5,500 gpm through 16,500 feet of 16" I.D. concrete-lined steel pipe from elevation 4' to elevation 174'. To increase the system flow, the addition of a booster is proposed at elevation 94', 8,000 feet from the existing booster. The proposed booster is to take suction directly from the pipe line.

REQUIRED: (a) What is the maximum flow for which the new booster can be designed?

(b) A second proposal is to construct a parallel pipe line and booster to give the same increase in flow as the first proposal. Assume friction in the new pipe line to be 1.5 feet per 100 feet, and velocity to be 6.0 feet per second. What would be the annual savings in power cost for 2,000 hours operation at 1.4¢ per kilowatt hour?

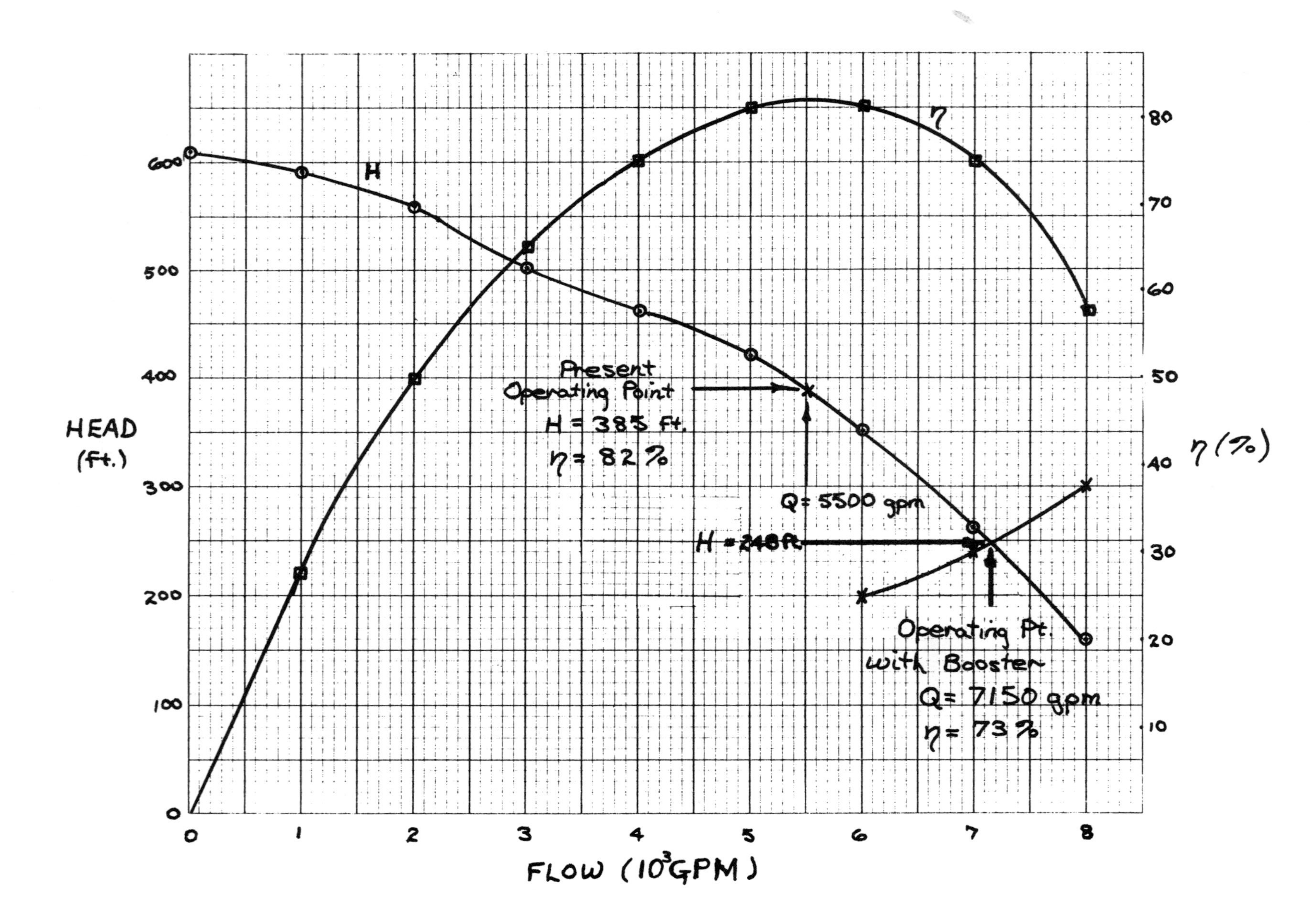
HEAD
(Ft.)
600
500
400
300
200
100
0
η (%)
80
70
60
50
40
30
20
10
FLOW (10³GPM)
0
1
2
3
4
5
6
7
8
H
η
Present
Operating Point
H = 385 ft.
η = 82%
Q = 5500 gpm
Operating Pt.
with Booster
Q = 7150 gpm
η = 73%

Solution

Present Operation:

$$H_1 + H_p = H_4 + H_{L_{1-4}}$$

$$4' + 385' = 174' + H_{L_{1-4}} \longrightarrow H_{L_{1-4}} = 215 \text{ ft.}$$

Assume $H_L = K\ell Q^2 \longrightarrow K = 0.43 \times 10^{-9} (\text{gpm})^{-2}$

Operation with Booster:

$$\text{For } \ell = 8000', \ K\ell = 3.44 \times 10^{-6} \frac{\text{ft}}{(\text{gpm})^2}$$

$$\text{Thus, } H_{L_{1-2}} = 3.44 \times 10^{-6} \frac{\text{ft}}{(\text{gpm})^2} Q^2$$

Since no information is given about NPSH, assume a suction head of −15 ft. at the inlet to the booster.

* Note: a zero suction head would be a better assumption; but the problem asks for the maximum flow, and −15 ft. is a common rule of thumb.

Then, $H_2 = 94' - 15' = 79 \text{ ft.}$

$$H_1 + H_p = H_2 + H_{L_{1-2}}$$

$$4' + H_p = 79' + 3.44 \times 10^{-6} Q^2$$

$$\longrightarrow H_p = 75 + 3.44 \times 10^{-6} Q^2$$

Plot the curve of H vs. Q for the pump and the curve of H_p vs. Q. for the system. From the intersection of these curves obtain:

$Q = 7150$ gpm ← max. flow

$H_p = 248$ ft.

$\eta = 73\%$

<u>Booster requirements</u>:

$$\text{For } \ell = 8{,}500' \; , \; K\ell = 3.65 \times 10^{-6} \frac{ft}{(gpm)^2}$$

$$H_{L_{3-4}} = (3.65 \times 10^{-6})(7150)^2 = 186.5 \text{ ft.}$$

$$H_2 + H_B = H_4 + H_{L_{3-4}}$$

$$79' + H_B = 174' + 186.5' \longrightarrow H_B = 282 \text{ ft.}$$

<u>Power Requirements</u>

$$P = \frac{Q\gamma H}{\eta} = \frac{QH}{\eta}\left(62.4 \frac{lb}{ft^3} \times \frac{ft^3}{7.48\,gal} \times \frac{2.26 \times 10^{-5}\,kw\text{-}min}{ft \cdot lb}\right)$$

$$= \frac{QH}{\eta}\left(1.88 \times 10^{-4} \frac{kw \cdot min}{ft \cdot gal}\right)$$

<u>Present Operation</u>:

$$P = \frac{5500 \times 385}{0.82}(1.88 \times 10^{-4}) = 485 \text{ kw}$$

<u>Pump (Original) when Booster is used</u>:

$$P = \frac{7150 \times 248}{0.73}(1.88 \times 10^{-4}) = 457 \text{ kw}$$

<u>Booster (assuming $\eta = 80\%$)</u>:

$$P = \frac{7150 \times 282}{0.80}(1.88 \times 10^{-4}) = 474 \text{ kw}$$

Total Power 931 kw

<u>Parallel installation (assuming $\eta = 80\%$)</u>:

$$Q = 7150 - 5500 = 1650 \text{ gpm}$$

$$H = 170 + 1.5 \times 165 = 418 \text{ ft}$$

$$P = \frac{1650 \times 418}{0.80}(1.88 \times 10^{-4}) = 162 \text{ kw}$$

Original Inst. → 485

647 kw Total

$$\Delta P = 931 - 647 = 284 \text{ kw}$$

—continued—

Annual power cost savings for parrallel installation:

$$284 \text{ Kw} \times 2000 \text{ hr.} \times \frac{\$0.014}{\text{Kw-hr.}} = \underline{\underline{\$7950}}$$

HYDRAULIC MACHINES 8

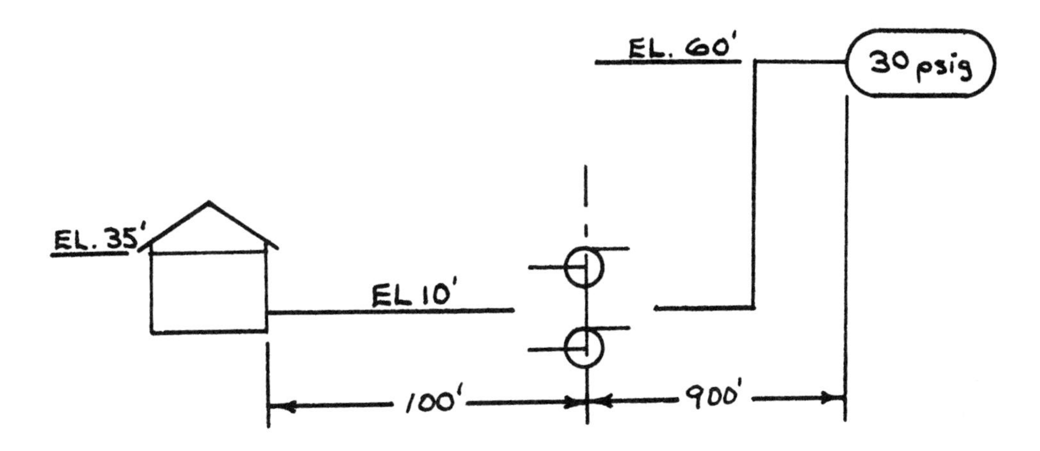

Gasoline is to be pumped from a feed tank whose liquid surface is at elevation 35' to a closed vessel operating at 30 psig pressure and located at elevation 60'. Two identical centrifugal pumps are available, located at elevation 10'. They can be operated either singly, in series, or in parallel to pump over a pipe line from the feed tank to the vessel. The pipe line consists of 100 equivalent feet of 4 inch pipe from the feed tank to the pumps, and 900 equivalent feet of 4 inch pipe from the pumps to the vessel. At the pumping temperature, the gasoline has a specific gravity of 0.72 and a vapor pressure of 6 psia.

Data for constructing the pump head capacity curves and the pipe line friction curve are given in the following tabulation:

G.P.M.	Pump Head Feet of Liquid	Friction Drop in Pipe Feet of Liquid per 100 Feet of Pipe
0	252	0
200	252	4.29
400	235	15.5
600	187	32.8
800		55.8
1000		84.3

(a) Find the maximum pumping rate.

(b) Find the available NPSH at this pumping rate.

Solution

To obtain the H vs. Q characteristic curve for pumps in series, add the heads corresponding to common flow rates; for pumps in parallel, add the flow rates corresponding to common heads.

To obtain the piping system curve:

$$H_p = H_2 - H_1 + H_L = (60 - 35)\,\text{ft.} + \frac{30 \times 144}{0.72 \times 62.4}\,\text{ft.} + H_L$$

$$= 121\ \text{ft} + H_L \quad \left(\text{neglecting velocity head at discharge}\right)$$

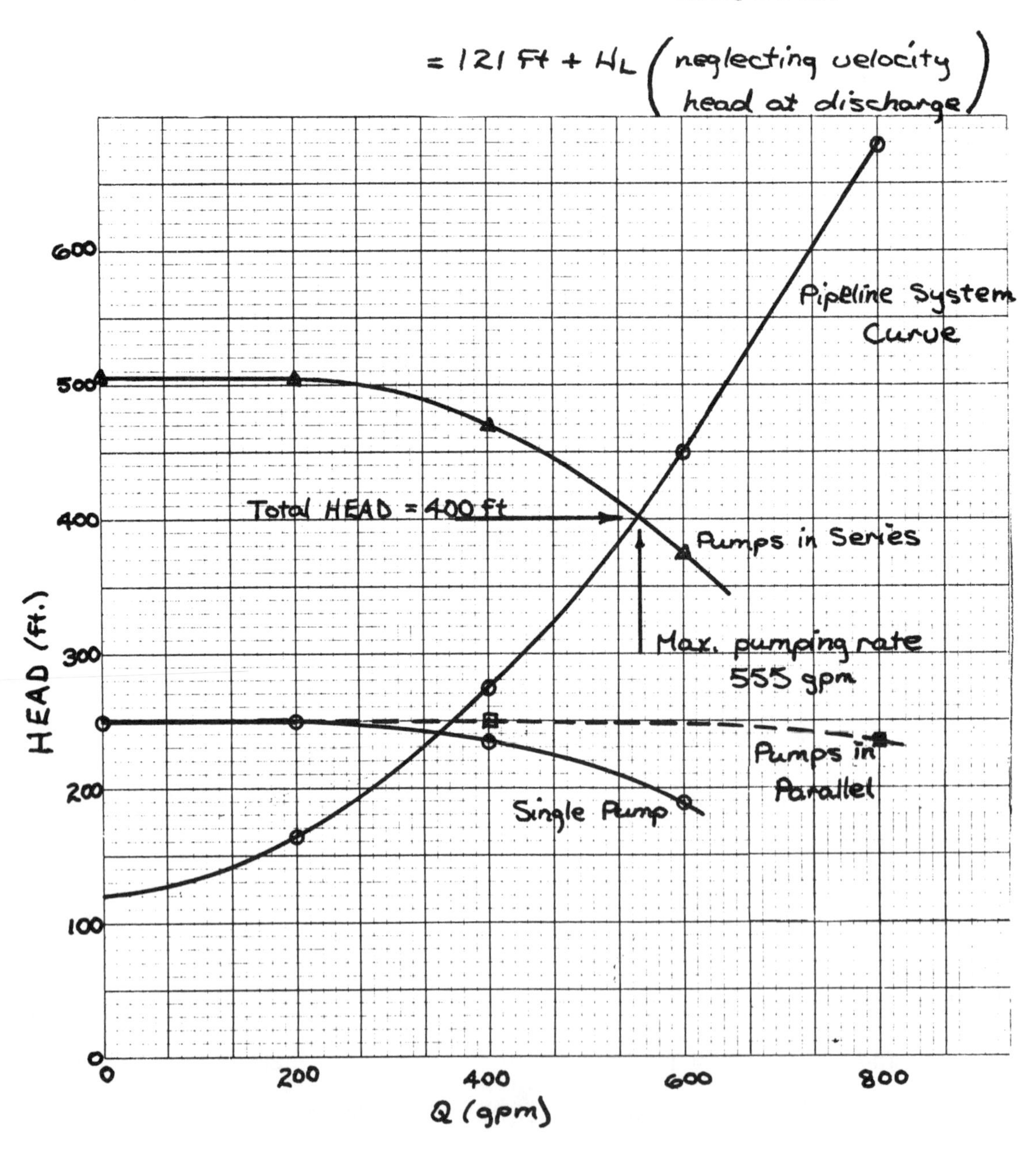

From the intersection of the pumps in series curve and the piping system curve the maximum pumping rate is 555 gpm at a total pumping head of 400 ft.

Thus, the head loss is $400' - 121' = 279'$

$\longrightarrow 27.9 \text{ ft}/100 \text{ ft}.$

$$NPSH = (35' - 10') - 27.9' + \frac{(14.7 - 6.0) \times 144}{0.72 \times 62.4} \text{ ft}$$

$$= 25 \text{ ft.}$$

HYDRAULIC MACHINES 9

An engine-driven deep well pump takes water from a well, and puts it through 1300 feet of pipe of unknown size to irrigate a field of sugar beets. The level beet field is 75 feet higher than ground level at the well. The water is dispersed through a number of sprinkler heads. Flow is 700 gallons per minute and pressure at the pump is 115 psig.

The water level in the well is 175 feet below ground level when pumping 700 gpm and 140 feet when not pumping. Assume the well water level to be a linear relationship with the flow up to 2000 gpm.

The pump and engine are running at 1500 rpm, which is the limit of the governor. The engine at 1500 rpm is rated at 150 continuous bhp and 190 bhp intermittent. At 1800 rpm, it is rated at 175 bhp continuous and 225 bhp intermittent.

The farmer needs a greater flow. You are an engineer for a pump company with pumps for all conditions. Assume any pump you have has an efficiency of 75%.

(a) What is the efficiency of the existing pump? Should it be pumping more?

(b) What would be the flow with your best pump for the existing engine, pipe, and sprinkler system? Run the engine at 1800 rpm.

(c) What would you recommend be done?

Solution

a) Assume the velocity head to be negligible.

$$\text{Pump head } H = 175\text{ ft.} + \frac{115 \times 144}{62.4} = 175 + 265 = 440\text{ ft.}$$

$$\text{Power supplied to Flow} = Q\gamma H$$

$$= 700\text{ gpm} \times \frac{\text{ft}^3}{7.48\text{ gal}} \times 62.4\frac{\text{lb}}{\text{ft}^3} \times 440\text{ ft.} \times \frac{\text{hp}\cdot\text{min}}{33{,}000\text{ ft}\cdot\text{lb}}$$

$$= 77.9\text{ hp}$$

Efficiency cannot be determined since the actual Engine power is not known. Correspondingly, pumping performance cannot be evaluated.

b) Assume the head loss in the pipe and sprinkler heads is proportional to Q^2.

* System head curve equation : H is in Ft.
Q is in GPM

$$H = 140' + \frac{35'}{700\,gpm} Q + 75' + \frac{265' - 75'}{(700\,gpm)^2} Q^2$$

$$H = 215 + 5Q + 3.88Q^2 \quad (Q \text{ in } 100 \text{ GPM units})$$

* System power curve equation : P in hp
H in ft.
Q in 100 GPM

$$P = \frac{Q\gamma H}{\eta} = \frac{62.4\,QH}{7.48 \times 330 \times 0.75} = 0.03375\,QH$$

Q (100 gpm)	H (ft.)	P (Hp)
7	440	104
8	503	136
9	574	174

New Pump @1800 rpm
900 gpm (continuous)

Computed Values from equations above

c) Present pump @ 1800 rpm.

Corresponding point with present operating point :

$$Q_2 = \frac{N_2}{N_1} Q_1 = \frac{1800}{1500} \times 700 = 840\,gpm$$

$$H_2 = \left(\frac{N_2}{N_1}\right)^2 H_1 = \left(\frac{1800}{1500}\right)^2 \times 440 = 634 \text{ ft.}$$

The new operating point can not be determined without knowing the pump characteristic curve, but it can be approximated by sketching typical curve segments through known points.

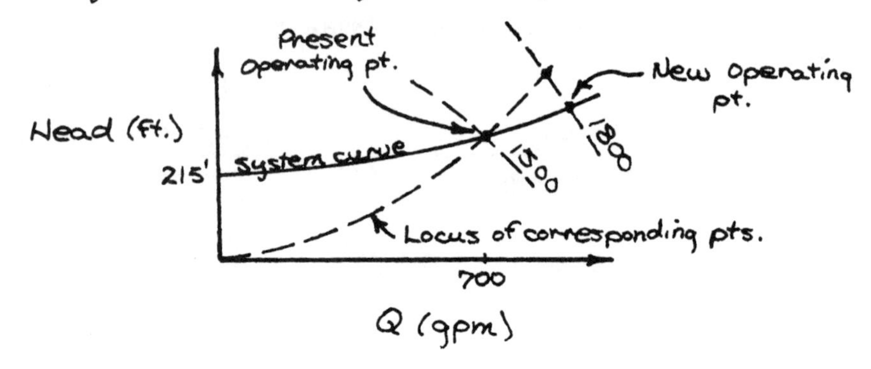

By comparing approximate new operating point with what a new pump would do, it appears that the present pump is still in good condition.

Recommendation :

1) Keep Present Pump

2) Reset governor for 1800 rpm.

HYDRAULIC MACHINES 10

A Francis-type hydroelectric turbine and generator unit is to be installed at a plant where the operating head is 200 feet. The unit is to have a rating of 10,000 hp and the power system operates at 60 cps. Assume the plant elevation is 500 feet above sea level and ambient temperature is 70°F.

REQUIRED: Answer the following questions, giving full reference for all formulas, charts, grpahs, etc., you use:

(a) What specific speed would you recommend for the runner?

(b) What is the maximum recommended runner speed in rpm?

(c) What is the closest synchronous speed to the maximum recommended runner speed that is still less than the maximum recommended runner speed?

(d) How many pole pieces (not pairs) will the generator have at this synchronous speed?

(e) What runner discharge diameter would be recommended for the given head and power rating?

(f) What setting (vertical distance between the bottom of the runner and tail water elevation) would you recommend?

(g) What maximum efficiency would you expect the turbine (exclusive of the generator) to have?

(h) Across what range of percentages of full load will this peak efficiency be expected to occur?

Solution

Ref: Baumeister + Marks, "Std. Handbook for M.E.", seventh ed., pp 9-(183-200)

a) $n_s = 675/\sqrt{H} = 675/\sqrt{200} = \underline{\underline{47.7}}$

b) $n_s = n\sqrt{P}/H^{5/4} \quad \therefore n = n_s H^{5/4}/\sqrt{P}$

$$n = \frac{47.7(200)^{5/4}}{\sqrt{10^4}} = \underline{\underline{359 \text{ rpm}}}$$

n_s - specific speed; H in ft., P in hp., n in rpm

c) 360 and $\underline{\underline{327 \text{ rpm}}}$ are synchronus speeds.

d) $n = 120 \times f/p \qquad p = \frac{120 \times 60}{327} = \underline{\underline{22}}$

* Note: the above answers the question as worded, but does not represent good design. The 360 rpm synchronus speed should be chosen which would mean 20 poles and a specific speed of 47.9. These values will be used.

e) $\varphi = \frac{\pi D n}{720\sqrt{2gH}} = 0.73$ $\quad \begin{cases} D \text{ in inches} \\ n \text{ in rpm} \\ H \text{ in ft.} \end{cases}$

$$D = \frac{0.73 \times 720 \times 8.03\sqrt{200}}{\pi \times 360} = \underline{\underline{52.8 \text{ in.}}}$$

f) $\sigma = \frac{H_b - H_v - H_s}{H} \geq \frac{n_s^{3/2}}{2000} = 0.1658$ (if model test data not available)

$H_s \leq H_b - H_v - 33.16$ Ft.

Assume $p_{atm} = 14.7 - 0.3 = 14.4$ psia

$p_v = 0.4$ psia

$$H_b - H_v = \frac{14.0 \times 144}{62.4} = \underline{32.31 \text{ Ft}}$$

$H_s \leq \underline{\underline{-0.85 \text{ Ft.}}}$ (Negative value means Submergence)

g) & h) $\underline{\underline{90-92\%}}$ $\underline{\underline{\text{From } 60-100\% \text{ of rated load}}}$

HYDRAULIC MACHINES 11

A steam turbine stage has a blade root velocity of 800 ft/sec and a root radius of 2.0 feet. Swirl velocity is to be zero at exit to blading along its total length (6") while axial velocity remains constant along length.

Assume free vortex swirl velocity distribution, "0" reaction at root, and blade entrance angle = blade exit angle = 45° at root.

REQUIRED: Find:

(a) The blade entrance and exit angles at tip.

(b) The amount of reaction at tip.

Solution

For the conditions of this problem the energy transfer associated with a pressure difference is equal to the change of kinetic energy associated with the relative velocity of the fluid with respect to the blade. Thus,

$$\text{Reaction} = \frac{V_{R_2}^2 - V_{R_1}^2}{V_1^2 - V_2^2 + V_{R_2}^2 - V_{R_1}^2}$$

where: V_R - relative velocity

V - absolute velocity

V_B - blade velocity

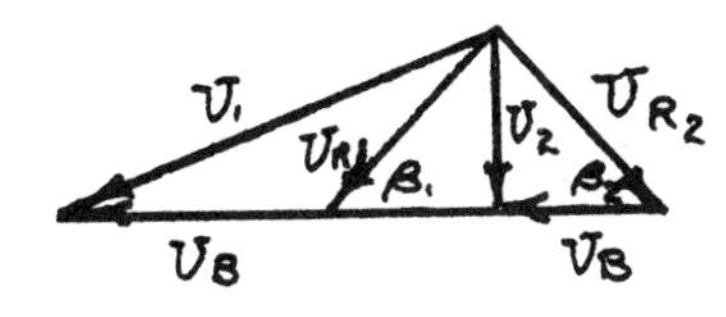

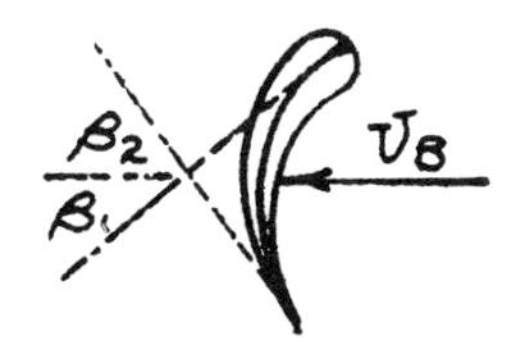

Condition at root:

$V_{R_1} = V_{R_2}$ for zero reaction

$\beta_1 = \beta_2 = 45°$

$V_2 = V_B = 800$ ft/sec

V_s - tangential component of $V_1 = 2V_B = 1600$ ft/sec

$V_s r = \text{const.} = 3200$ ft²/sec

$\omega = \frac{V_B}{r} = 400\ \text{sec}^{-1}$

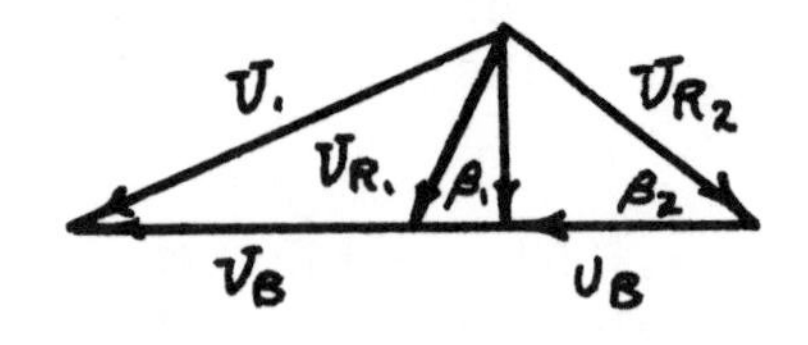

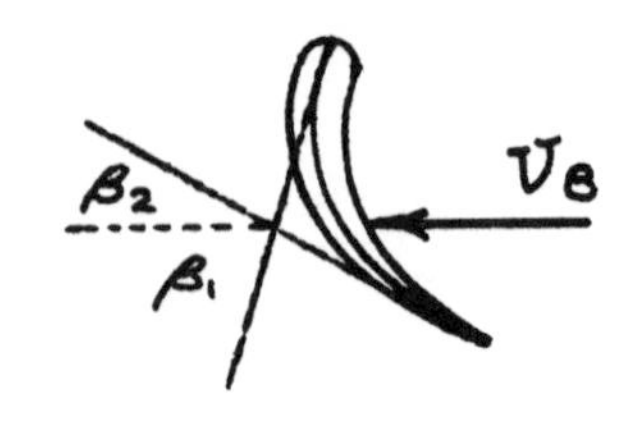

Condition at tip:

$$V_B = r\omega = 2.5 \times 400 = 1000 \text{ ft/sec}$$

$$V_s = \frac{3200}{2.5} = 1280 \text{ ft/sec.}$$

$$\beta_1 = \tan^{-1}\frac{V_2}{V_s - V_B} = \tan^{-1}\frac{800}{1280-1000}$$

$$\beta_1 = \underline{\underline{70.7^\circ}}$$

$$\beta_2 = \tan^{-1}\frac{V_2}{V_B} = \tan^{-1}\frac{800}{1000} = \underline{\underline{38.7^\circ}}$$

$$V_{R_2} = \frac{V_2}{\sin\beta_2} = 1280 \text{ ft/sec}$$

$$V_{R_1} = \frac{V_2}{\sin\beta_1} = 848 \text{ ft/sec}$$

$$V_1 = \sqrt{V_2^2 + V_s^2} = 1510 \text{ ft/sec}$$

$$V_1^2 - V_2^2 = V_s^2 = 1.64 \times 10^6 \text{ ft}^2/\text{sec}^2$$

$$V_{R_2}^2 - V_{R_1}^2 = 0.919 \times 10^6 \text{ ft}^2/\text{sec}^2$$

$$\text{Reaction} = \frac{0.919}{1.64 + 0.919} = \underline{\underline{35.9\ \%}}$$

10

Power Plants

RICHARD K. PEFLEY

Reciprocating Engine Closed Cycle Models:

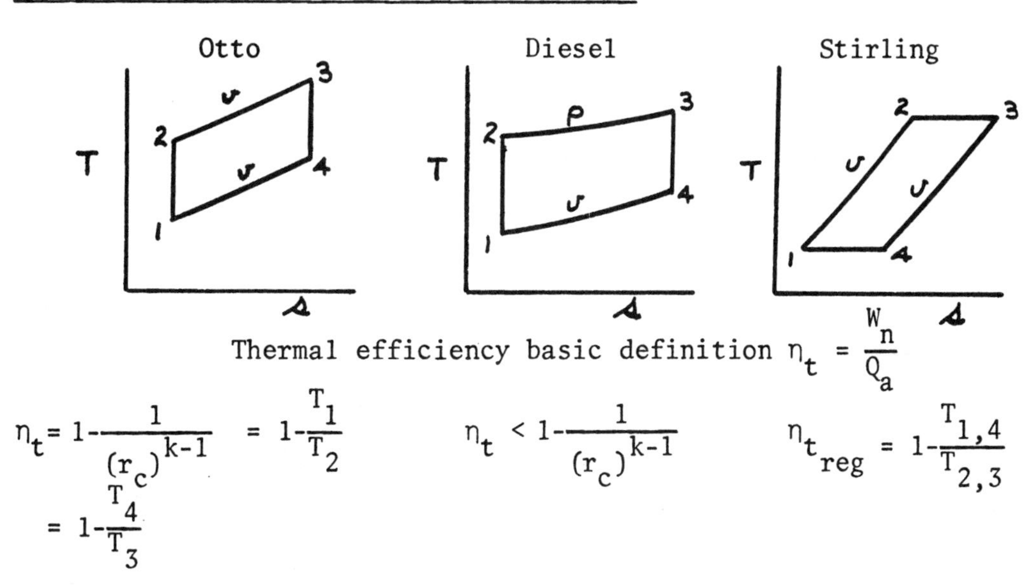

Thermal efficiency basic definition $\eta_t = \frac{W_n}{Q_a}$

$$\eta_t = 1 - \frac{1}{(r_c)^{k-1}} = 1 - \frac{T_1}{T_2} = 1 - \frac{T_4}{T_3} \qquad \eta_t < 1 - \frac{1}{(r_c)^{k-1}} \qquad \eta_{t_{reg}} = 1 - \frac{T_{1,4}}{T_{2,3}}$$

Other terms that are frequently encountered:

Volumetric efficiency: $\eta_v = \left.\frac{\nu_a}{\nu_d}\right]_{p,T}$

Compression ratio: $r_c = \frac{\nu_d + \nu_c}{\nu_c}$

Mean effective pressure (brake, indicated)

$$\text{bmep} = \frac{w_{b,cc}}{A\ell} \qquad \text{imep} = \frac{w_{i,cc}}{A\ell}$$

Specific fuel consumption (brake, indicated)

$$\text{bsfc} = \frac{\dot{m}_f}{\text{bhp}} \qquad \text{isfc} = \frac{\dot{m}_f}{\text{ihp}}$$

Horsepower (brake, indicated)

$$bhp = \frac{2\pi NT}{33,000} = \frac{(bmep)\ell AnN}{33,000}$$

$$ihp = bhp + fhp = \frac{(imep)\ell AnN}{33,000}$$

Gas Turbines - Brayton Cycle:

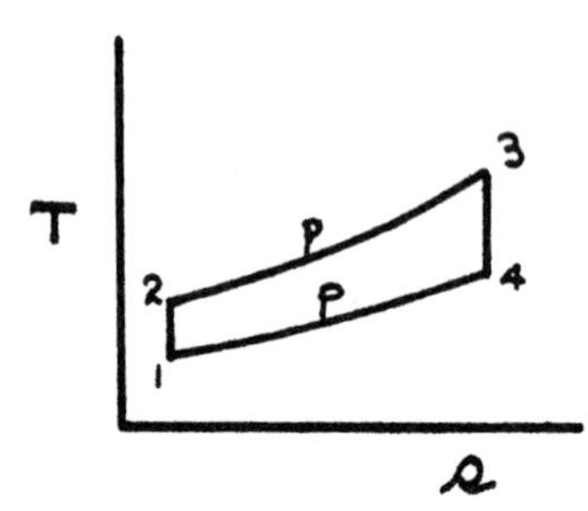

$$\eta_t = 1 - \frac{1}{(r_p)^{\frac{k-1}{k}}} = 1 - \frac{T_1}{T_2} = 1 - \frac{T_4}{T_3}$$

Other terms that are frequently encountered:

Isentropic efficiency (compressor, turbine)

$$\eta_{s,c} = \frac{W_s}{W_a} \qquad \eta_{s,tu} = \frac{W_a}{W_s}$$

Work ratio: $\frac{W_{tu}}{W_c}$

Regenerative ideal heat exchange: $\eta_t = 1 - \frac{T_2}{T_3}$

Rankine Vapor Power Cycle:

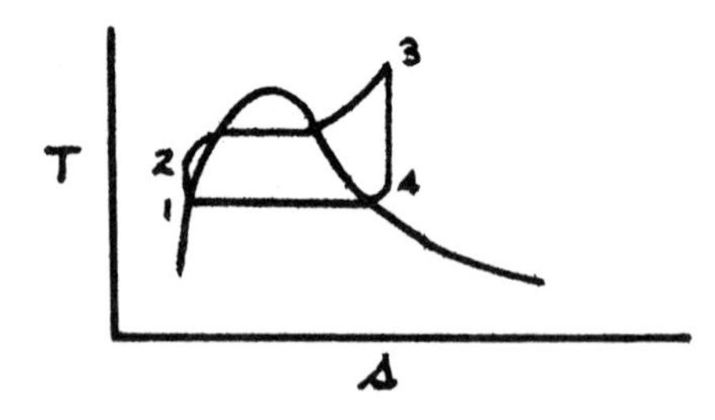

$$\eta_t = \frac{(h_3 \ h_4) - (h_2 - h_1)}{(h_3 - h_2}$$

Other terms that are frequently encountered:

Heat rate: $HR = \frac{Q_a\,(Btu)}{W_n\,(kw\text{-}hr)}$

Boiler hp = h_{fg} x 34.5 lbm/hr at 212°F = 33,475 Btu/hr

Terminology for Power Plants Review

Symbols:

A	piston area
bhp	brake horsepower
bmep	brake mean effective pressure
bsfc	brake specific fuel consumption
fhp	friction horsepower
HR	heat rate
ihp	indicated horsepower
imep	indicated mean effective pressure
isfc	indicated specific fuel consumption
k	specific heat ratio
ℓ	stroke length
$\dot{m}$	mass flow rate
N	revolutions per minute
n	number of cylinders
p	pressure
r	ratio
T,t	temperature
ν	specific volume
W	work
η	efficiency

Subscripts:

a	actual, added
b	brake
c	clearance, compressor, compression
cc	cylinder-cycle
d	displacement
i	indicated
n	net
t	thermal
tu	turbine
reg	regeneration
v	volume
f	fuel

NEW VEHICLE EMISSIONS STANDARDS SUMMARY
(Taken from California Air Resources Board Fact Sheet 72-6 5000 6-72)
LIGHT-DUTY VEHICLES UNDER 6,000 lbs

Year	Standard	Cold Start Test	Hydrocarbons	Carbon Monoxide	Oxides of Nitrogen
Prior to Controls			850 ppm (11 gm/mi)*	3.4% (80 gm/mi)	1000 ppm (4 gm/mi)
1966-67	State	7-mode	275 ppm	1.5%	no std.
1968-69	State & Federal	7-mode			
		50-100 CID	410 ppm	2.3%	no std.
		101-140 CID	350 ppm	2.0%	no std.
		over 140 CID	275 ppm	1.5%	no std.
1970	State & Federal	7-mode	2.2 gm/mi	23 gm/mi	no std.
1971	State	7-mode	2.2 gm/mi	23 gm/mi	4 gm/mi
	Federal	7-mode	2.2 gm/mi	23 gm/mi	-
1972	State	7-mode	1.5 gm/mi	23 gm/mi	3 gm/mi
		or			
		CVS-1	3.2 gm/mi	39 gm/mi	3.2 gm/mi**
	Federal	CVS-1	3.4 gm/mi	39 gm/mi	-
1973	State	CVS-1	3.2 gm/mi	39 gm/mi	3 gm/mi
	Federal	CVS-1	3.4 gm/mi	39 gm/mi	3 gm/mi
1974	State	CVS-1	3.2 gm/mi	39 gm/mi	2 gm/mi
	Federal	CVS-1	3.4 gm/mi	39 gm/mi	3 gm/mi
1975	State	CVS-1	1 gm/mi	24 gm/mi	1.5 gm/mi
	Federal	CVS-2	0.41 gm/mi	3.4 gm/mi	3 gm/mi
1976	State	CVS-1	1 gm/mi	24 gm/mi	1.5 gm/mi
	Federal	CVS-2	0.41 gm/mi	3.4 gm/mi	0.4 gm/mi

ppm parts per million concentration

gm/mi grams per mile

7-mode is a 137 second driving cycle test

CVS-1 is a constant volume sample cold start test

CVS-2 is a constant volume sample cold start test average with a constant volume sample hot start test, both with the Federal 22-minute driving cycle

* The values in parentheses are approximately equivalent values

NEW VEHICLE EMISSIONS STANDARDS SUMMARY
(Taken from California Air Resources Board Fact Sheet 72-6 5000 6-72)

HEAVY-DUTY VEHICLES OVER 6,000 lbs

Year	Standard	Hydrocarbons	Carbon Monoxide	Oxides of Nitrogen
1969-71	State-gasoline	275 ppm	1.5%	no std.
1972	State-gasoline	180 ppm	1.0%	no std.
1973-74	State-gasoline & diesel	$HC + NO_X$ = 16 gm/BHP hr. CO = 40 gm/BHP hr.		
1975 & later	State-gasoline & diesel	$HC + NO_X$ = 5 gm/BHP hr. CO = 25 gm/BHP hr.		

gm/BHP hr. grams per brake horsepower-hour

POWER PLANTS 1

A space vehicle in an earth distance orbit about the sun has a solar panel array which is its electrical power generator. The solar cells are of the P-N type with a "red-blue" filter and have 50 mil. thick protective glass covers. The solar panels are sun oriented. If the radiant solar energy in this orbit is 1400 watts/m^2, and the array is 10% efficient and develops 460 watts, what power will it develop in a Mars distance orbit about the sun? Assume that the earth and Mars orbits are 1.495×10^8 km and 2.276×10^8 km, respectively, from the sun.

Solution

* Assume temperature of panels and hence efficiency are constant.

* Note that radiation intensity varies inversely as the square of the distance from the sun.

Then, $P \propto \frac{1}{r^2}$ $\quad \therefore \frac{P_e}{P_m} = \frac{r_m^2}{r_e^2}$

$$P_m = 460\left(\frac{1.495}{2.278}\right)^2 = \underline{\underline{200 \text{ watts}}}$$

POWER PLANTS 2

From a test of a 7 x 10 inch, single-acting, four-stroke-cycle gas engine, the following data was obtained: speed = 340 rpm, average area of indicator cards = 1.13 sq. in.; length of cards = 2.90 in.; scale of indicator spring = 200 lb/sq.in./in.; net brake load = 37.1 lb.; length of brake arm = 4.0 ft.

Determine:
- (a) The indicated mean effective pressure (mep).
- (b) The indicated horsepower (ihp).
- (c) The brake horsepower (bhp).
- (d) The mechanical efficiency.
- (e) The brake torque.

Solution

* Note: no. of cylinders not specified

a) $\bar{h} = \frac{A}{\ell}\Big)_{card} = \frac{1.13}{2.90} = 0.39 \text{ in}$

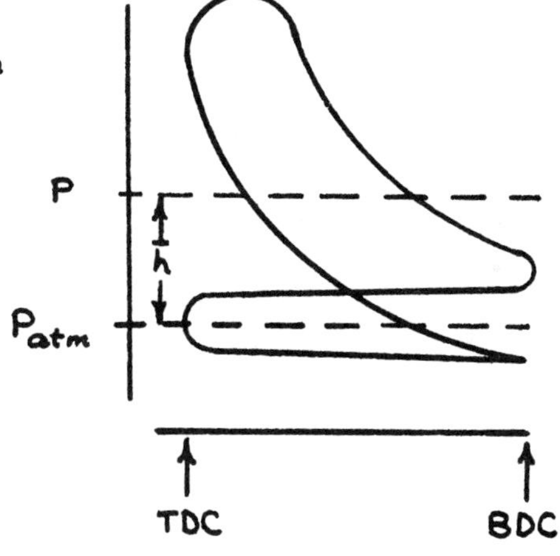

$$imep = \bar{h} k_{spring}$$

$$= 0.39 \times 200$$

$$= \underline{\underline{78 \text{ psi}}}$$

b) $ihp/cyl = imep \times A_p \times stroke \times \frac{rpm}{2}$

$$= \frac{78 \times \pi \times 49 \times 10 \times 170}{4 \times 33{,}000 \times 12} = \underline{\underline{12.9 \text{ hp}}}$$

c) $bhp = \frac{2\pi \times rpm \times T}{33{,}000} = \frac{2\pi \times 340 \times 4 \times 37.1}{33{,}000} = \underline{\underline{9.61 \text{ hp}}}$ (7.17 kW)

∴ single cyl. engine since bhp is a reasonable % of ihp/cyl.

d) $\eta_{mech.} = \frac{9.61}{12.9} = \underline{\underline{74.5\%}}$

e) $T = F \times L = 37.1 \times 4 = \underline{\underline{148.4 \text{ ft}\cdot\text{lbf}}}$

(2.012 Nm)

POWER PLANTS 3

A vertically-launched rocket vehicle has an initial weight of 3,000 lbm. The rocket has 1,500 lbm of first-stage propellant and a variable expansion ratio nozzle with a constant system delivered specific impulse of 240 sec.

If the liftoff thrust-to-weight ratio is 2.00 lbf/lbm, then at first-state burnout:

(a) What is the time elapsed from launch?

(b) What is the thrust-to-weight ratio?

(c) Neglecting aerodynamic drag, what is the velocity?

Solution

$$I_{\text{(specific imp.)}} = \frac{F_t \text{ (thrust)}}{\dot{W} \text{ (prop. flow rate)}}$$

Assume $\dot{w}$ = constant

$$\therefore \dot{w} = \frac{F_t}{I} = \frac{6000 \text{ lbf}}{240 \text{ sec.}} = 25 \text{ lbf/sec.}$$

a) $t = \frac{W}{\dot{w}} = \frac{1500}{25} = \underline{\underline{60 \text{ sec}}}$

b) $\frac{F_t}{W_T \text{(total)}} = \frac{6000}{W_{T,0} - \dot{W}t} = \frac{6000}{1500} = \underline{\underline{4:1}}$

c) at any instant,

$$F_t = \frac{1}{g_c} \frac{d\, m_T V_r \text{(rocket)}}{dt} + W_T$$

-continued-

$$m_T = m_{T,0} - \dot{m}t$$

$$F_t = \frac{1}{g_c}\frac{d[(m_{T,0} - \dot{m}t)V_r]}{dt} + (W_{T,0} - \dot{\omega}t)$$

$$F_t = \frac{1}{g_c}(m_{T,0} - \dot{m}t)\left(\frac{dV_r}{dt} + g\right)$$

$$\int_0^V dV_r = \int_0^t F_t\, g_c \frac{dt}{(m_{T,0} - \dot{m}t)} - \int_0^t g\, dt$$

$$V_r = \frac{F_t g_c}{\dot{m}} \ln \frac{m_{T,0}}{(m_{T,0} - \dot{m}t)} - gt$$

$$= \frac{6000 \times 32.2}{25} \ln 2 - 32.2 \times 60 = 3428 \frac{ft}{sec}$$

(1.045 KM/SEC)

POWER PLANTS 4

Consider the ideal isotope powered space power system shown below.

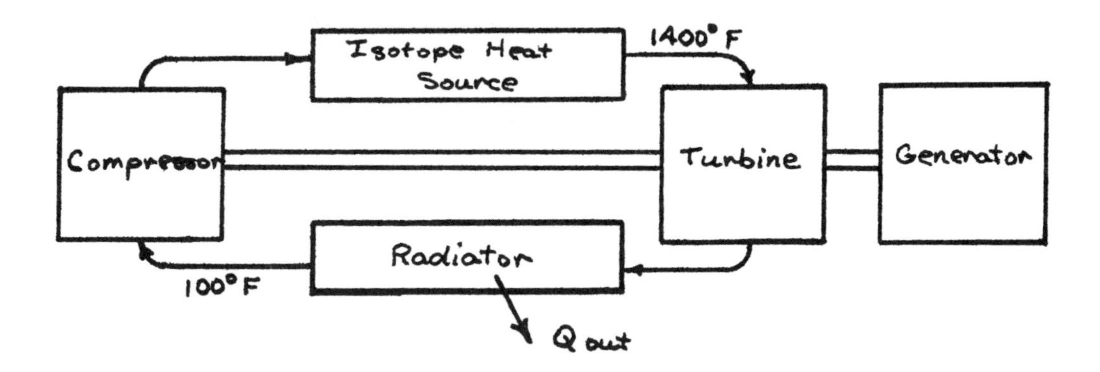

Assume: (1) A simple Brayton Cycle.

(2) There are no losses in the system.

(3) The working fluid is a perfect monatomic gas.

(4) The compressor pressure ratio is two to one.

-continued-

Then:

(a) What is the cycle efficiency?

(b) How does the cycle efficiency compare with that of a Carnot engine operating between the same temperature limits?

(c) What would you do to greatly improve the cycle efficiency of the Brayton Cycle while still operating between the same temperature limits?

(d) What is the cycle efficiency of this improved cycle assuming ideal conditions?

Solution

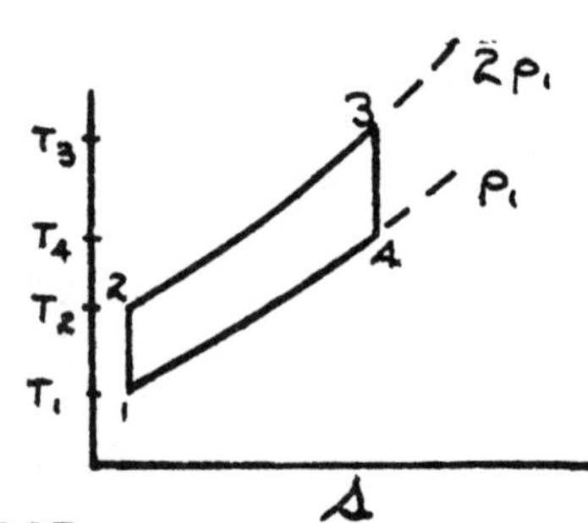

a) For the Brayton Cycle

$$\eta_t = 1 - \frac{1}{r_p^{\,k-1/k}}$$

From kinetic theory for monatomic gas

$C_v = \frac{3}{2}\frac{R_o}{M}$ and $C_p = C_v + \frac{R_o}{M}$ $\therefore k = 5/3 = 1.67$

$$\eta_t = 1 - \frac{1}{2^{0.67/1.67}} = \underline{\underline{24.2\ \%}}$$

b) $\eta_{t\ carnot} = 1 - \frac{T_{min}}{T_{max}} = 1 - \frac{T_1}{T_3} = 1 - \frac{560}{1860} = \underline{\underline{69.9\ \%}}$

c) Use a regenerative heat exchanger which ideally would heat compressed gases from T_2 to T_4. (THEN $\eta_t = 1 - T_2/T_3$)

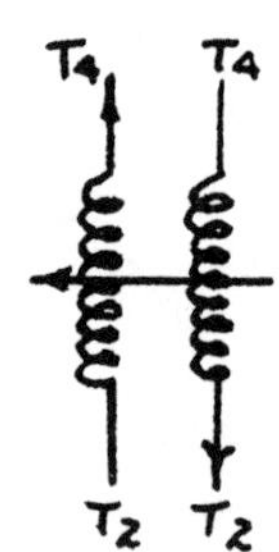

d) From isentropic process

$$T_2 = T_1\, r_p^{\,k-1/k} = 560 \times 1.32 = 740°R$$

$$T_4 = T_3 / r_p^{\,k-1/k} = \frac{1860}{1.32} = 1410°R$$

For original cycle $\eta_{t_o} = \frac{W_{net}}{Q_{add_o}} = \frac{W_{net}}{C_p(T_3 - T_2)}$

For regenerator $\eta_{t_r} = \frac{W_{net}}{Q_{add_r}} = \frac{W_{net}}{C_p(T_3 - T_4)}$

Since W_{net} same for both cases,

$$\eta_{t_r} = \eta_{t_o}\frac{T_3 - T_2}{T_3 - T_4} = 24.2\,\frac{1860 - 740}{1860 - 1410} = \underline{\underline{60.3\ \%}}$$

POWER PLANTS 5

A 55,000 lb truck (loaded weight) has a power train which delivers 200 horsepower at the wheels. When the loaded truck is traveling at 60 mph on a level road, 100 horsepower is required to overcome wind drag and the remaining 100 horsepower is required to overcome rolling resistance.

Can the same truck and load maintain a speed of 20 mph on a 6 per cent uphill grade? Support your answer with calculations. Show all assumptions.

Solution

P_z (Climbing power required) = $W_{(weight)}$ V_z (vert. vel.)

$$V_z = 20 \text{ mph} \times \frac{5280}{3600} \times .06 = 1.76 \frac{\text{ft.}}{\text{sec.}}$$

$$P_z = \frac{55{,}000 \times 1.76}{550} = 176 \text{ hp}$$

P_a (air drag power) $\propto V^3$ (Veh. Vel.) $\quad P_{a_{20mph}} = 100\left(\frac{20}{60}\right)^3 = 3.7 \text{ hp}$

P_r (rolling drag power) $\propto V^*$ $\quad P_{r_{20mph}} = 100\left(\frac{20}{60}\right) = 33.3 \text{ hp}$

$P_z + P_a + P_r = \underline{\underline{213 \text{ hp}}}$ higher than 200 hp available
(159 kW)

* actually rolling drag increases slowly with speed so that $P_r \propto V^{1+}$. This depends on tire pressure, type of road, etc.. Thus, it is possible that the truck could maintain 20 mph on grade.

POWER PLANTS 6

The compressed air motor shown consists of two identical chambers (A) to which nozzles are attached. Air is admitted through expansion valves (B). The chambers are mounted on a hub which rotates freely about the supply line (C). The air in the chambers (A) has negligible velocity.

Pressure in supply line	500 psig
Pressure in chambers (A)	100 psig
Exhaust pressure	14.7 psia

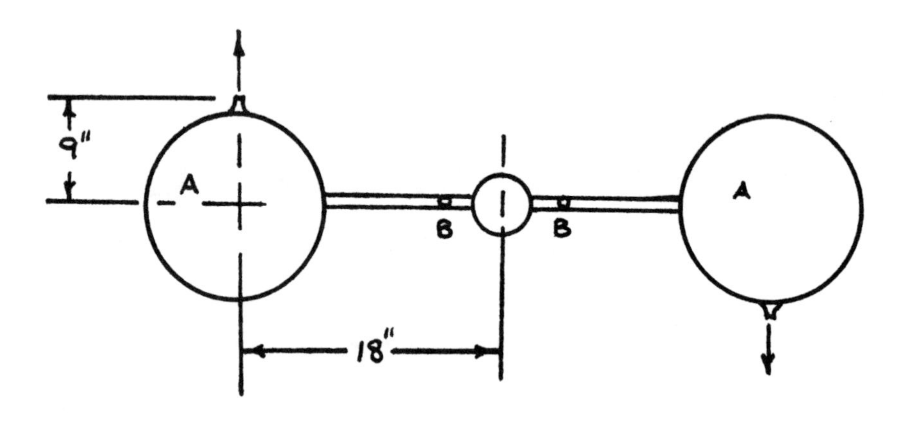

Temperature of supply air	100°F
Temperature of exhaust	20°F
Discharge rate of air	2,000 lb/hr
Distance from center of bearing to center lines of nozzles	18 in
Radius of gyration of motor	15 in
Weight of motor (all rotating parts)	200 lb
Distance from center of chambers (A) to ends of nozzles	9 in

Find: (a) If the motor is originally held by a brake which is suddenly released, what speed will the motor attain in 6 seconds?

(b) What will be the maximum speed of the motor, assuming no frictional or air resistance?

Give both answers in revolutions per minute.

Solution

Assume adiabatic nozzle flow & stagnation in spheres.

From 1st law of thermo,

$$V_r\,(\text{rel. exh.}) = \sqrt{2 g_c C_p (T_0 - T_{exh})}$$

$$= (64.4 \times 0.241 \times 778 \times 80)^{1/2}$$

$$= 980 \text{ ft/sec.}$$

From impulse-momentum principle for angular momentum,

$$\Sigma T + \frac{1}{g_c} \dot{m} (\vec{r} \times \vec{V}_{abs})_{in\ flow} = \frac{1}{g_c} m r_g^2 \frac{d\omega}{dt}$$

for free rotation $\Sigma T = 0$. Let cc. moments be (+).

$$\frac{1}{g_c} \dot{m} (\vec{r} \times \vec{V}_{abs}) = -\frac{1}{32.2}\left(\frac{2000}{3600}\right)(1.5)(-V_r + r\omega)$$

$$= -\frac{1}{32.2}\left(\frac{2000}{3600}\right)(1.5)(-980 + 1.5\omega)$$

$$= 25.9 - .039\,\omega$$

$$\frac{1}{g_c} m r_g^2 \frac{d\omega}{dt} = \frac{200}{32.2}(1.25)^2 \frac{d\omega}{dt} = 9.68 \frac{d\omega}{dt}$$

Assume ω small for first 6 seconds.

$$\therefore 9.68\ d\omega/dt = 25.9 \qquad \frac{d\omega}{dt} = 2.68 \text{ rad/sec}^2$$

a) in six seconds, $\omega = 6 \times \frac{d\omega}{dt} = \underline{\underline{16 \frac{\text{rad}}{\text{sec}} = 2.5 \text{ rps}}}$

b) at runaway $d\omega/dt = 0$

$$.039\,\omega = 25.9 \qquad \omega = \underline{\underline{666 \frac{\text{rad}}{\text{sec}} = 106 \text{ rps}}}$$

POWER PLANTS 7

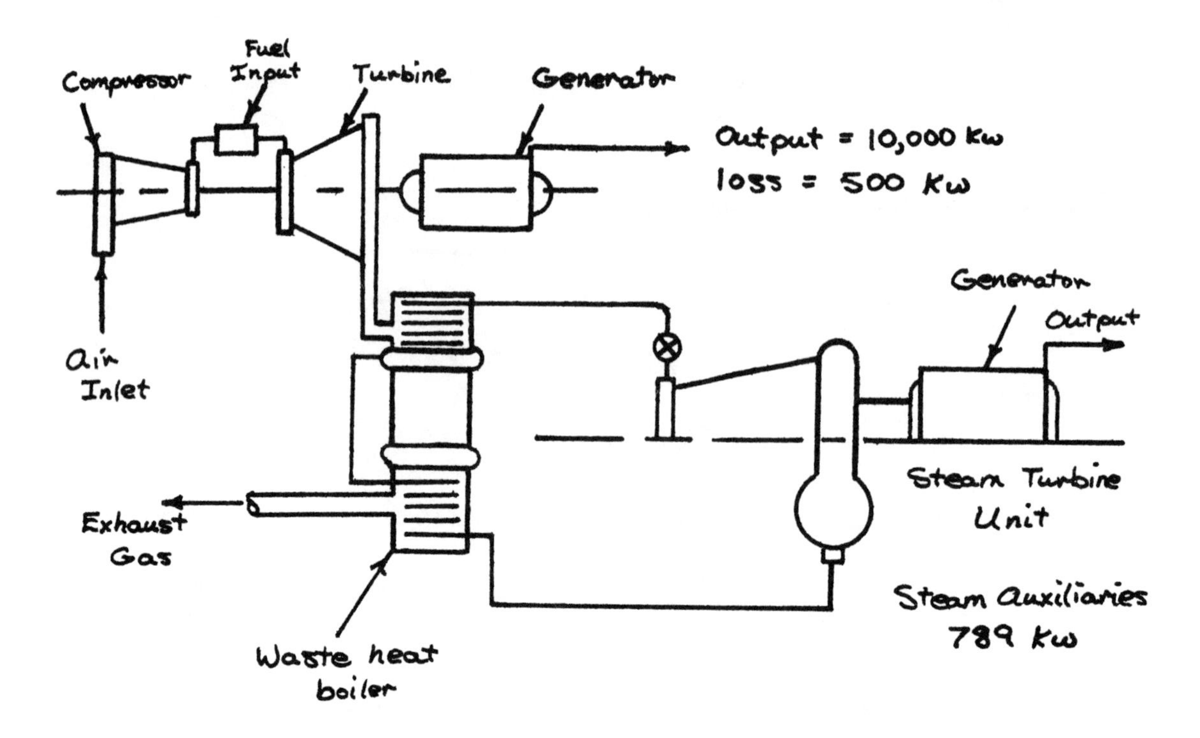

The above diagram shows a combined cycle gas and steam turbine electric generating unit. At 10,000 kw output on the gas turbine unit the electrical and mechanical loss is 500 kw. The overall thermal efficiency of the gas turbine generator unit is 20 per cent. The gas turbine exhausts to a waste heat steam generator which is 75 per cent efficient. The steam from the steam generator drives a steam turbine electric generator which has an overall heat rate of 11,000 Btu per kwh. Steam turbine unit auxiliaries require 789 kw.

Determine: (a) The output of the steam turbine generating unit in kw.

(b) The net heat rate in Btu per kwh of the combined cycle.

(c) The thermal efficiency.

Solution

Assume 10,000 kw output is net output

$$\dot{E}_{\text{supplied to gas turbine}} = \frac{P_{net}}{\eta_t} = \frac{10{,}000}{0.20} = 50{,}000 \text{ kw}$$

$$\dot{E}_{boiler} = \dot{E}_{turbine} - (P_{net} + \text{loss}) = 39{,}500 \text{ kw}$$

$$\dot{E}_{steam} = \dot{E}_{boiler} \times \eta_{boiler} = 29{,}600 \text{ kw}$$

$$\text{boiler heat rate} = \frac{11{,}000 \text{ Btu/kwh}}{3413 \text{ Btu/kwh}} = 3.22 \frac{\text{kw input}}{\text{kw output}}$$

a) Assume boiler heat rate based on $\dot{E}_{boiler}$.

$$\therefore P_{\text{steam turb.}} = \frac{39{,}500}{3.22} = \underline{\underline{12{,}200 \text{ kw}}}$$

b) $$\text{Heat rate} = \frac{50{,}000 \times 3413}{10{,}000 + 12{,}200} = \underline{\underline{7700 \frac{\text{Btu}}{\text{kwh}}}}$$

c) $$\eta_{t,\text{overall}} = \frac{22{,}200}{50{,}000} = \underline{\underline{44.5\ \%}}$$

POWER PLANTS 8

A turbo-generator produces 8500 kw in a refinery power plant. Generator efficiency is 95%. Turbine inlet steam is 815 psia, 750°F; exhaust steam is 145 psia, 440°F at A. If the turbine trips out, steam would be depressed and desuperheated to the 145 psia header. Steam flow rate, pressure and temperature are to be the same at A and B.

(a) What is the turbine engine efficiency?

(b) What is the flow rate of A?

(c) What is the flow rate at C?

(d) What is the steam temperature at C?

Assume line losses and heat losses are negligible.

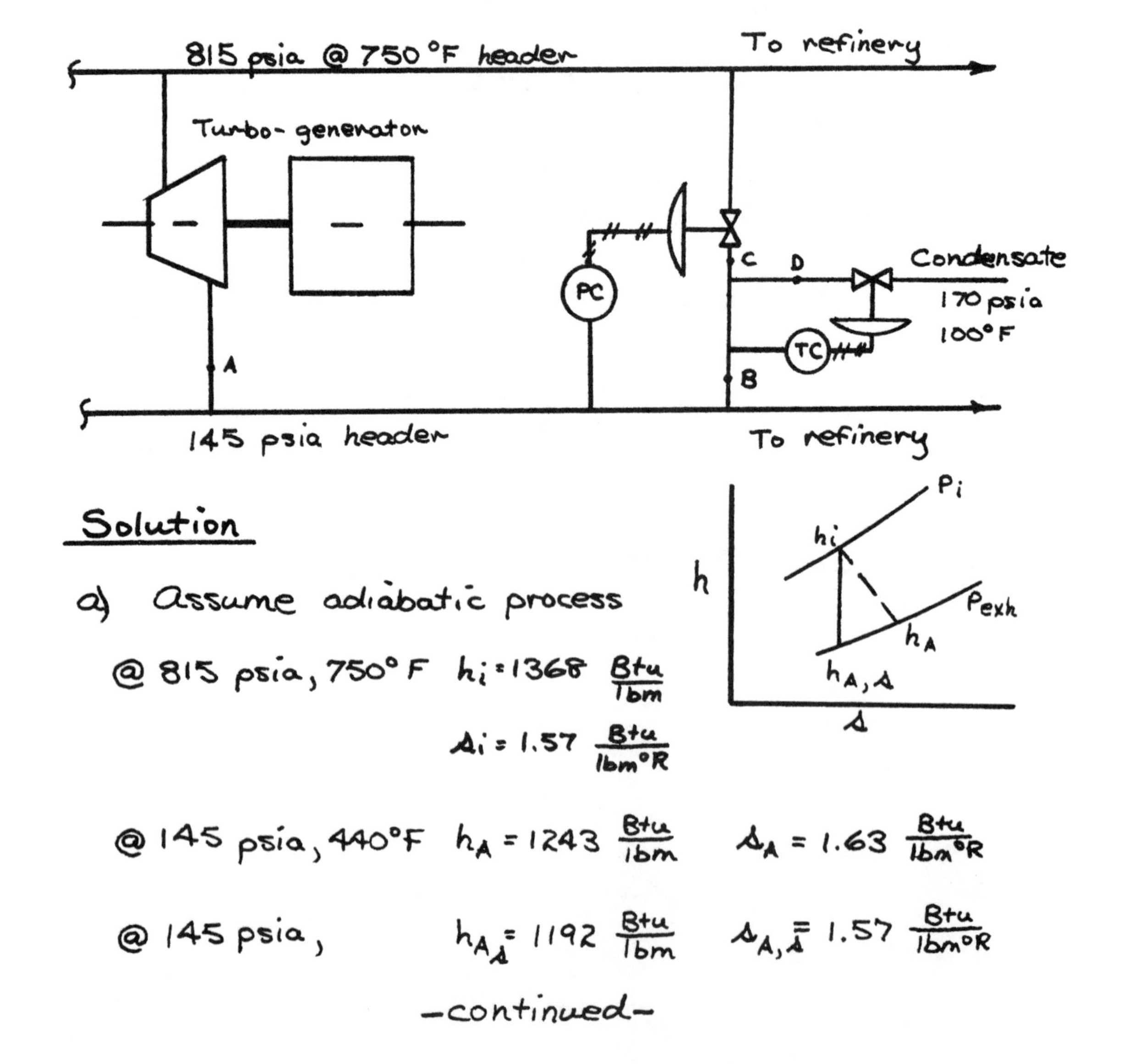

Solution

a) Assume adiabatic process

@ 815 psia, 750° F $h_i = 1368 \frac{Btu}{lbm}$

$s_i = 1.57 \frac{Btu}{lbm°R}$

@ 145 psia, 440°F $h_A = 1243 \frac{Btu}{lbm}$ $s_A = 1.63 \frac{Btu}{lbm°R}$

@ 145 psia, $h_{A_s} = 1192 \frac{Btu}{lbm}$ $s_{A,s} = 1.57 \frac{Btu}{lbm°R}$

–continued–

$$\eta_{engine} = \frac{\Delta h_{actual}}{\Delta h_s} = \frac{1368-1243}{1368-1192} = \underline{\underline{71\%}}$$

b) From 1st Law of Thermo.

$$\dot{m}(h_i - h_A) = P_{turbine} = \frac{P_{gen.}}{\eta_{gen.}} = \frac{8500\,\text{KW} \times 3413\,\frac{\text{Btu}}{\text{KWh}}}{0.95}$$

$$\dot{m}_A = \frac{8500 \times 3413}{0.95 \times 125} = \underline{\underline{2.44 \times 10^5\ \frac{\text{lbm}}{\text{hr}}}}$$

(111 Kg/HR)

c) @ 170 psia, 100°F $h_D = 68\ \frac{\text{Btu}}{\text{lbm}}$

Assume adiabatic mixing and negligible K.E. change of streams.

$$\dot{m}_C(h_i - h_{B=A}) = \dot{m}_D(h_B - h_D) \quad \& \quad \dot{m}_D = \dot{m}_A - \dot{m}_C$$

$$\dot{m}_C\,(125) = (\dot{m}_A - \dot{m}_C)(1243 - 68)$$

$$\dot{m}_C = \frac{1175}{1300}\,\dot{m}_A = \underline{\underline{2.21 \times 10^5\ \frac{\text{lbm}}{\text{hr}}}}$$

(100 Kg/HR)

d) Assume steam is throttled

$h_C = h_i \qquad P_C = 145$ psia

$$\underline{\underline{T = 680\,°\text{F}}}$$

(360°C)

POWER PLANTS 9

A stationary diesel engine generates 2,000 brake horsepower with a specific fuel rate of 0.42 lb/bhp-hr. The heating value of the fuel is 19,000 Btu/lb, and the air-fuel ratio is 16/1. Assume surroundings of 90°F, 14.7 psia and the exhaust temperature 800°F.

Estimate the cooling water flow rate (gpm) necessary if the water cooling apparatus can produce a 30°F temperature drop in the water.

Solution

$$\dot{E}_{shaft} = 2000 \times 2544 = 5.09 \times 10^6 \ \frac{Btu}{lbm}$$

$$\dot{m}_f = bsfc \times bhp = 0.42 \times 2000 = 840 \ lbm/hr.$$

$$\dot{E}_{supplied} = LHV \dot{m}_f = 18{,}000 \frac{Btu}{lbm} \times 840 \frac{lbm}{hr}$$

$$= 15.2 \times 10^6 \ \frac{Btu}{hr}$$

* Note: assume HV given was HHV & LHV ≃ HHV − 1000

$$\dot{E}_{exh} = \dot{m} c_p (T_{exh} - T_{atm}) = (\dot{m}_{air} + \dot{m}_{fuel}) c_p (T_e - T_a)$$

$$= \dot{m}_f (16+1)(0.26)(800-90)$$

$$= 2.63 \times 10^6 \ \frac{Btu}{hr}$$

allow 5% $\dot{E}_{supplied}$ for incomplete comb. & heat loss to surroundings

$$\dot{E}_{coolant} = (\dot{E}_{su} - \dot{E}_{sh} - \dot{E}_{ex} - .05 \dot{E}_{su})$$

$$= 6.73 \times 10^6 \ Btu/hr$$

$$gpm = \frac{6.73 \times 10^6 \ Btu/hr}{60 \frac{min}{hr} \times 30°F \times 1 \frac{Btu}{lbm\,°F} \times 8.35 \frac{lbm}{gal}} = \underline{\underline{445}}$$

$$(0.028 \ m^3/SEC)$$

POWER PLANTS 10

A boat, powered by a normally aspirated diesel engine, has been given a pier (stationary) test in San Francisco Bay. The resulting required engine horsepower curve and the maximum engine horsepower curve versus speed are shown on the graph below. If this boat is moved to the Great Salt Lake in Utah, its performance will not be satisfactory. Explain why (credit will be based on the quality of your analysis using the data supplied) and recommend changes that will improve the performance.

	San Francisco	Great Salt Lake
Elevation - ft	0	4,000
Baro pressure - inches of Hg	29.9	26.6
Air Temp. - °F	65.0	65.0
Water Temp. - °F	55.0	55.0
Density of water - lb/ft^3	65.0	78.0
Viscosity of water - $lb\ sec/ft^2$	2.65×10^{-5}	4.15×10^{-5}

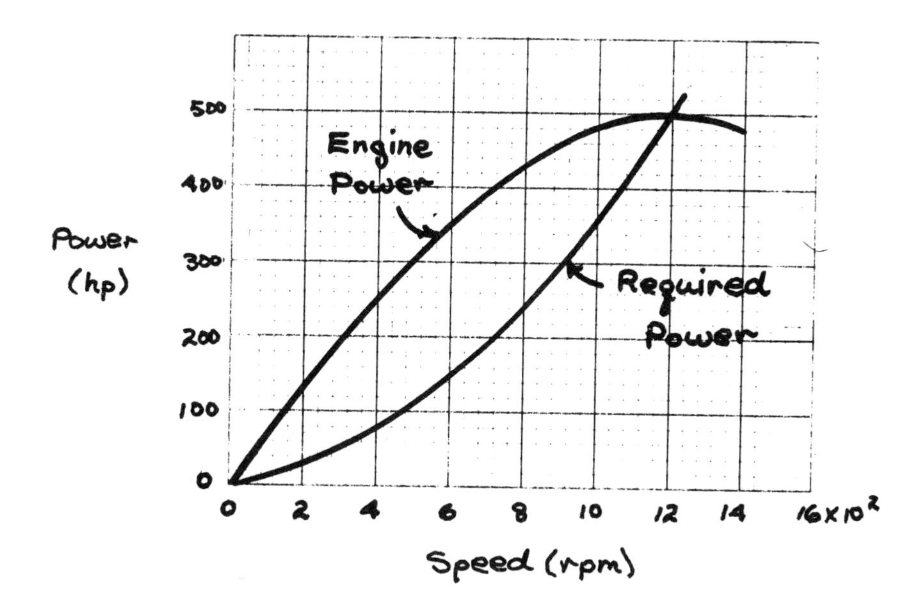

Solution

$$\text{Max. Engine Power} \propto \rho_{air\ intake} = \frac{P}{RT}\Big)_{intake}$$

$$\text{Required Power} \propto \rho_{water}$$

—continued—

$$\therefore \quad \frac{P_{eng.,\,S.L.}}{P_{eng.,\,Sea\ L.}} = \frac{p_{atm,\,S.L.}}{p_{atm,\,sea\ L.}} \quad \text{plot new engine curve}$$

$$\frac{P_{prop.,\,S.L.}}{P_{prop.,\,Sea\ L.}} = \frac{\rho_{water,\,S.L.}}{\rho_{water,\,Sea\ L.}} \quad \text{plot new required curve}$$

new intersection shows engine cannot come up to rated speed.

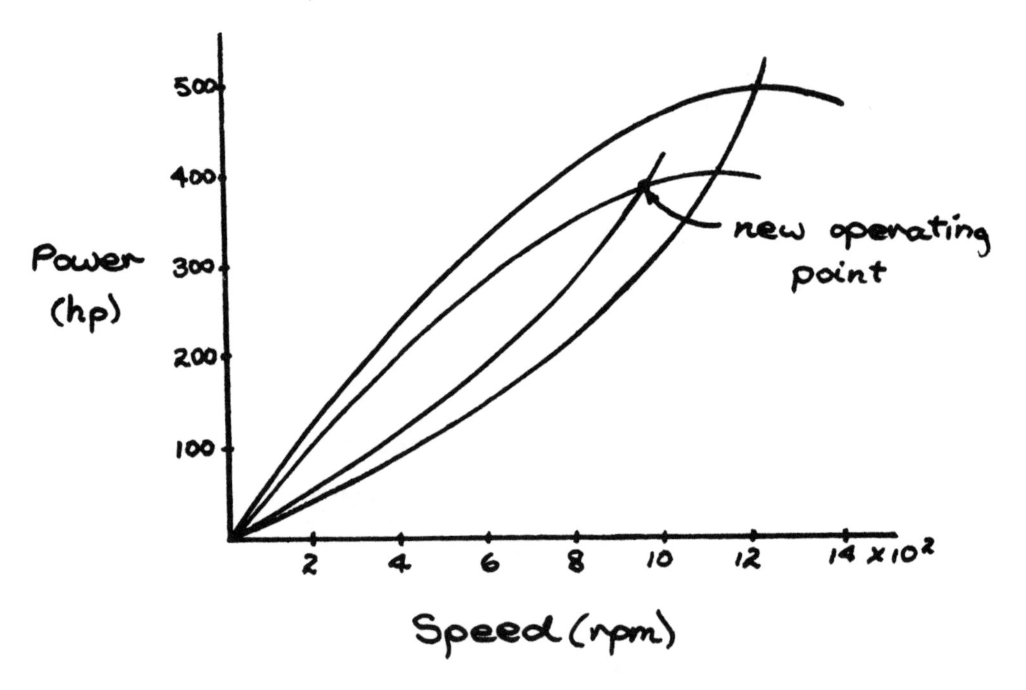

Possible modifications :

1) Reduce prop. dia or change prop.

2) Supercharge engine or change engine

POWER PLANTS 11

From the flow diagram below of a steam turbine generator unit, calculate the output in kw, the throttle steam flow and the unit heat rate in Btu per kwh with steam flow from the turbine to the condenser of 50,000 pounds per hour of 2.0 inches Hg absolute.

Mechanical efficiency of the unit 98.0% and generator efficiency 97.0%. Throttle conditions 750 psia and 760 FTT.

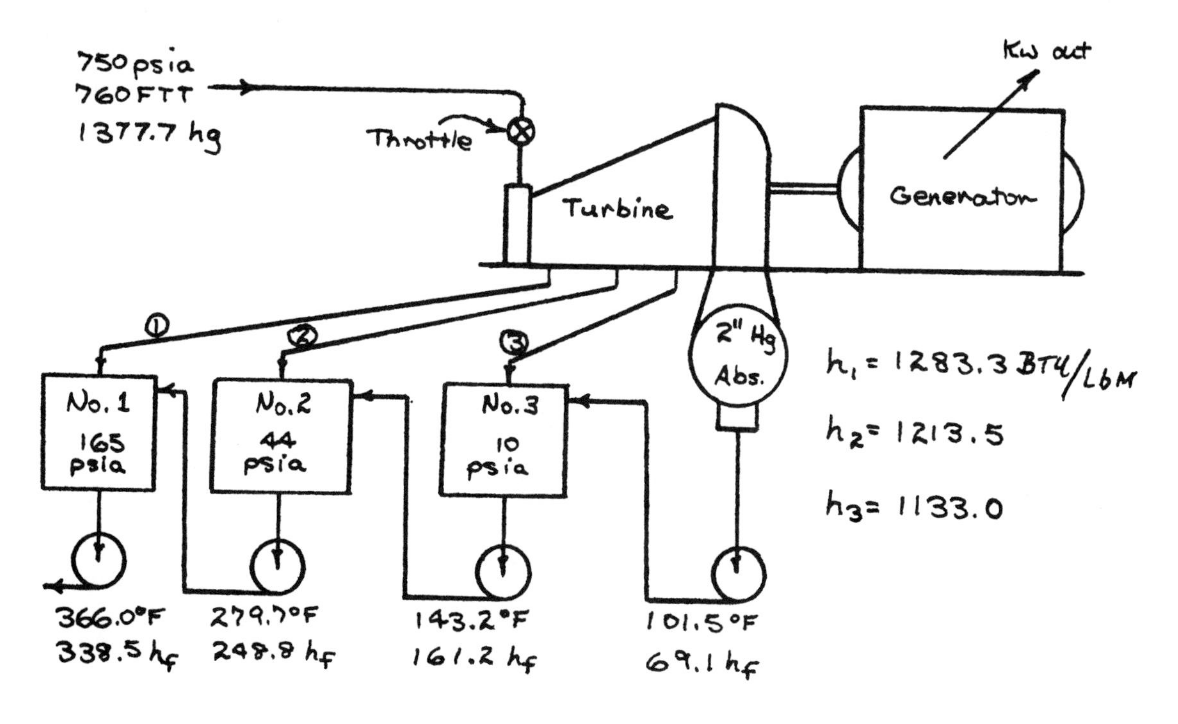

Solution

$$\text{Power output} = \eta_{mech}\,\eta_{gen.}\,\frac{\dot{m}_1(h_T-h_1)+\dot{m}_2(h_T-h_2)+\dot{m}_3(h_T-h_3)+\dot{m}_c(h_T-h_{exh})}{3413\ \text{Btu/kwh}}$$

* Note: Data for all enthalpies given except h_{exh}. Based on plot of intermediate turbine states on Mollier diag. suggests $h_{exh} \simeq 1050\ \frac{\text{Btu}}{\text{lbm}}$.

- continued -

To find $\dot{m}_{1,2,3}$ perform energy balences on feed water heaters.

$$\dot{m}_3 (h_3 - h_{out,3}) = \dot{m}_c (h_{o,3} - h_c)$$

$$\dot{m}_3 = 50{,}000 \frac{\text{lbm}}{\text{hr.}} \frac{(161.2 - 69.1)}{(1133.0 - 161.2)} = 4{,}750 \text{ lbm/hr.}$$

$$\dot{m}_2 = \dot{m}_3 + \dot{m}_c \frac{(h_{o,2} - h_{o,3})}{(h_2 - h_{o,2})} = 54{,}750 \frac{(248.8 - 161.2)}{(1213.5 - 248.8)}$$

$$= 4{,}970 \frac{\text{lbm}}{\text{hr.}}$$

$$\dot{m}_1 = \dot{m}_3 + \dot{m}_2 + \dot{m}_c \frac{(h_{o,1} - h_{o,2})}{(h_1 - h_{o,1})} = 59{,}720 \frac{(338.5 - 248.8)}{(1283.3 - 338.5)}$$

$$= 5{,}650 \frac{\text{lbm}}{\text{hr.}}$$

$$\dot{m}_1 (h_T - h_1) = 5{,}650 (1377.7 - 1283.3) = 0.534 \times 10^6 \text{ Btu/hr.}$$

$$\dot{m}_2 (h_T - h_2) = 4{,}970 (1377.7 - 1213.5) = 0.817 \times 10^6 \text{ Btu/hr.}$$

$$\dot{m}_3 (h_T - h_3) = 4{,}750 (1377.7 - 1133.0) = 1.162 \times 10^6 \text{ Btu/hr.}$$

$$\dot{m}_c (h_T - h_{exh}) = 50{,}000 (1377.7 - 1050.0) = \underline{16.38 \times 10^6 \text{ Btu/hr.}}$$

$$18.89 \times 10^6 \text{ Btu/hr.}$$

$$P_{out} = 0.98 \times 0.97 \frac{18.89 \times 10^6}{3.413 \times 10^3} = \underline{\underline{5261 \text{ Kw}}}$$

$$\dot{m}_t = \dot{m}_1 + \dot{m}_2 + \dot{m}_3 + \dot{m}_c = \underline{\underline{65{,}370 \frac{\text{lbm}}{\text{hr.}}}}$$

$$\text{heat rate} = \frac{\dot{m}_t (h_T - h_{o,1})}{P_{out}} = \frac{65{,}370}{5261} (1377.7 - 338.5)$$

$$= \underline{\underline{12{,}900 \frac{\text{Btu}}{\text{Kwh}}}}$$

$$(3.79 \text{ kw/kw})$$

POWER PLANTS 12

Many boiler explosions occur during the "light-off" period due to unorthodox operational procedures.

Program a safe procedure for igniting a single burner using natural gas as the fuel. Consider the burner equipped with adjustable register, pilot and igniter. Account for all permissive requirements prior to initial light-off.

(a) Program "light-off" sequence.

(b) List all factors which may cause fuel valve closure.

(c) Do you recommend "post purge"? Discuss this for normal shut-downs and after loss of flame.

Solution

a)
1. Check for presence of gas
2. Arm pilot light gas circuit
3. Operate igniter control
4. Check for automatic arming of main gas supply to burner.
5. Observe main burner ignition
6. adjust register for correct flame quality

b)
1. Loss of pilot flame
2. Loss of fuel pressure
3. flame out due to abnormal draft
4. Excessive flue or combustion zone temperature
5. Normal Temperature controller shutdown

c) Post purge is helpful in a safety and re-ignition sense because it eliminates combustion products, unburned fuel, and hot gases prior to restart thus assuring smooth ignition.

POWER PLANTS 13

A power plant supplies an auxiliary header with steam at 600 psig and 700°F. The total steam requirement for the auxiliaries is 25,000 lb/hr at 300 psig and 450°F.

REQUIRED: (a) Design a reducing and desuperheating station to achieve the above results. Water is available at 350 psig and 105°F.

(b) Compute the amount of spray water required and state your reasons for selecting the number and type of spray nozzles.

(c) Sketch the control and piping complex supporting your design.

Solution

a), c)

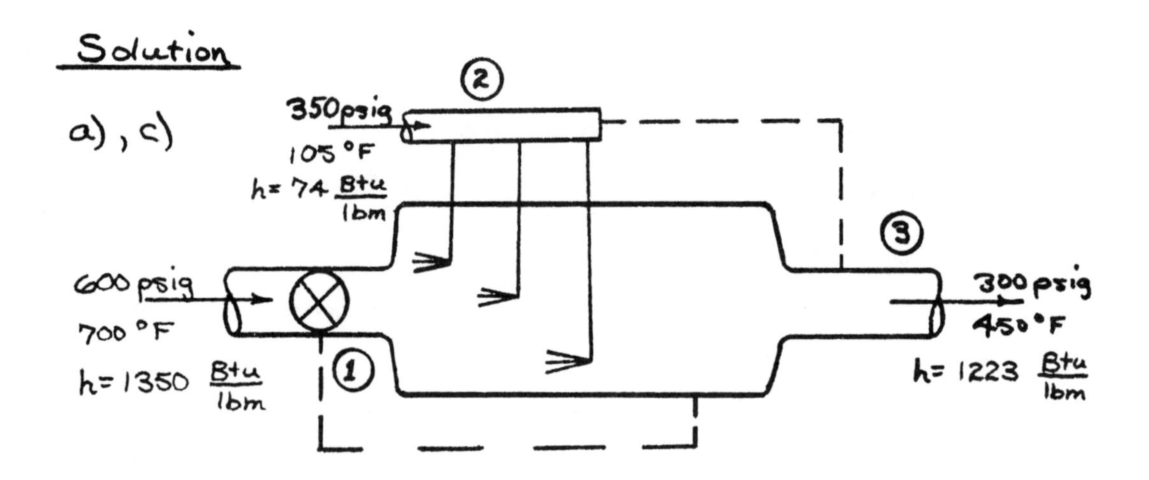

① main steam line throttle valve, holds down stream pressure @ 300 psig

② Temperature controlled nozzle actuating valve holds downstream temperature by supplying tempering water to appropriate number of spray nozzles.

-continued-

b) From 1st Law (assume ΔKE. negligible, adiabatic)

$$\dot{m}_w(h_3-h_2) = \dot{m}_s(h_1-h_3)$$

$$\dot{m}_s = 25{,}000 \text{ lb/hr} - \dot{m}_w$$

$$\therefore \dot{m}_w(h_3-h_2) + \dot{m}_w(h_1-h_3) = 25{,}000(h_1-h_3)$$

$$\dot{m}_w = 25{,}000 \frac{(h_1-h_3)}{(h_1-h_2)} = \underline{\underline{2490 \text{ lbm/hr}}}$$

$$(1{,}129 \text{ Kg/hr})$$

Since flow rate of water is only about 5 gpm, four to six orifice type nozzles should be able to handle the requirement.

VEHICLE EMISSIONS

SPECIAL PROBLEM 1

The NO_X is found to be 550 ppm equivalent of NO_2 in a vehicle exhaust sample using California Standard Smog Test. Determine the gr/mi if the exhaust is 700 ft^3/mi at standard pressure and temperature.

$$\text{gr/mi} = \frac{\text{ppm}}{10^6} \times \left(\rho_{NO_X}\right)_{P_{std},\, T_{std}} \nu$$

Assume $NO_X = NO_2$

$$\rho_{NO_2} = \frac{pM}{R_o T} = 54.16 \text{ gr/ft}^3$$

$$\text{gr/mi} = \frac{500}{10^6} \times 54.16 \times 700 = \underline{\underline{19.8}}$$

Note: CO(gr/mi) is determined in the same way.

SPECIAL PROBLEM 2

The hydrocarbons are found to be 1000 ppm equivalent of hexane in a vehicle exhaust sample using California Standard Smog Test. Determine the gr/mi if the exhaust volume is 600 ft^3/mi at standard pressure and temperature.

$$\rho_{HC} = 16.33 \text{ gr/ft}^3 \text{ assumes C:H} = 1:1.85$$

$$\text{gr/mi}_{\text{carbon equivalent}} = 1.8^* \times \frac{\text{ppm}}{10^6} \times \rho_{HC} \times c_{eq} \times \nu$$

$$= 1.8 \times \frac{10^3}{10^6} \times 16.33 \times 6 \times 600 = \underline{\underline{106}}$$

---

*1.8 is correction factor for NDIR instruments - not necessary for FID instruments.

11 Heating, Ventilating and Air Conditioning

RICHARD K. PEFLEY

Ventilation of an enclosed space is usually for the purposes of controlling the air temperature, humidity, and/or chemical composition. When humans occupy the space (room), the ventilation is necessary to provide adequate breathing oxygen and to transport the waste heat and moisture produced by the occupants. Normally, the waste heat and moisture far exceed the breathing oxygen ventilation requirement.

The heat and moisture release rates by the occupants are functions of their activity level, age, size, and sex. The fraction of energy associated with heat transferred to the air by convection and surrounding surfaces by radiation is called sensible heat while that associated with moisture evaporation is called latent heat. These fractions vary, depending on activity level and environmental state. In addition to the occupant heat and moisture load there may be an equipment and lighting load ventilation requirement.

It is normally assumed that the room air is homogeneous and is the same as the room air exhaust state.

Basic Refrigeration Cycles:

Carnot

$$C.O.P._{hp} = \frac{Q_r}{W_{net}} = \frac{1}{1 - \frac{T_c}{T_h}}$$

$$C.O.P._{ref} = \frac{Q_{abs}}{W_{net}} = C.O.P._{hp} - 1$$

T
T_h
T_c
s

Reference - ASHRAE Handbook of Fundamentals.

Rankine-Vapor

$$C.O.P._{hp} = \frac{Q_r}{W_{net}} = \frac{h_2 - h_3}{h_2 - h_1}$$

$$C.O.P._{ref} = C.O.P._{hp} - 1$$

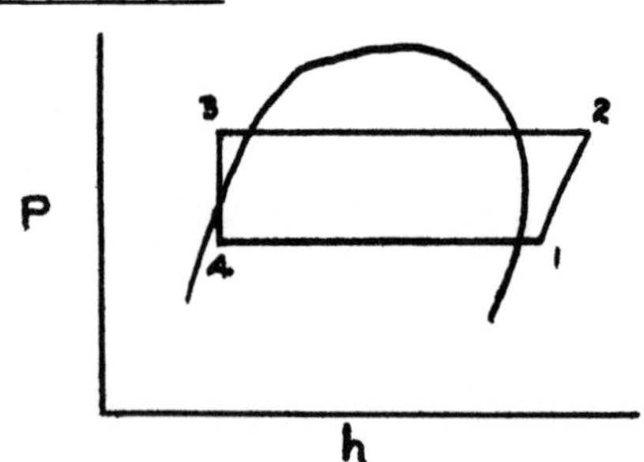

Gas Compression-Expansion Refrigeration Cycle:

$$C.O.P._{hp} = \frac{Q_r}{W_{net}} = \frac{1}{1 - \dfrac{1}{r_p^{\frac{k-1}{k}}}}$$

$$C.O.P._{ref} = C.O.P._{hp} - 1$$

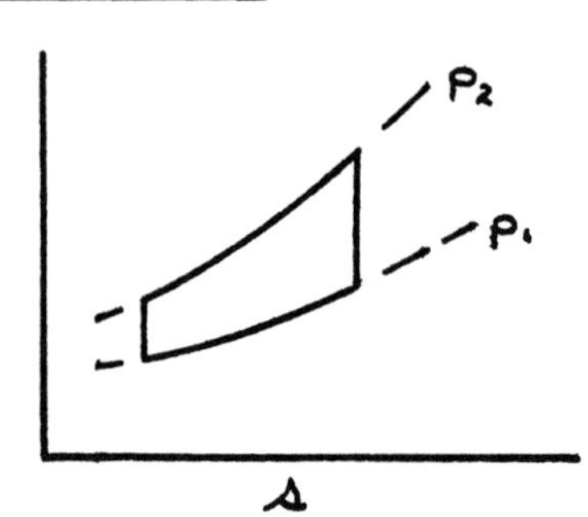

Human Factors and Constants:

Sedentary heat output rate	- 400 Btu/hr-person
Normal sedentary fresh air requirement	- 10 cfm/person
Minimum fresh air breathing requirement	- 3 cfm/person
Nominal comfort state ($q_s/q_\ell \simeq 3$)	- 75°F_{db} 60°F_{wb}
1 ton refrigeration = 12000 Btu/hr	

Terminology:

- c, c_p, c_v - specific heats
- $C.O.P._{hp}$ - coefficient of performance of a heat pump
- $C.O.P._{ref}$ - coefficient of performance of a refrigerator
- h_a, H - enthalpy of air water mixture per lbm dry air, enthalpy or head
- $m, \dot{m}$ - mass, mass flow rate
- P - pressure
- q_s - sensible heat rate
- q_ℓ - latent heat rate
- R.H. - relative humidity
- T_{db} - dry bulb temperature
- T_{dp} - dew point temperature
- T_e - effective temperature
- T_{wb} - wet bulb temperature
- v, V - specific, total volume
- w - specific humidity

Psychrometric Chart

100% RH
RH
w
h_a
T_{wb}
T_{dp} T_{db}

Subscripts:

wv - water vapor
O.A. - outside air
R.A. - room, exhaust air
M.A. - adiabatically mixed air
S.A. - supply air
C.A. - conditioned air
e - exit, exhaust
i - initial, in
s - steam
a,da - air, dry air

HEATING, VENTILATING AND AIR CONDITIONING 1

A steam line at 100 psia carrying dry saturated steam develops a leak as it passes through a sealed room, 20 ft by 20 ft by 12 ft. The air in the room is initially at 70°F, 50% relative humidity, 14.7 psia. The steam leaks into the room at the rate of 10 lbs per hour, diffuses thoroughly.

Assuming no heat exchange between the room and its surroundings, what is the relative humidity in the room at the end of one hour?

Solution

Treating room as a control volume and applying 1st Law:

$$m_s h_s = m_s u_f + m_a u_f - m_a u_i$$

By rearrangement:

$$m_s(h_s - u_f) = m_a(u_f - u_i) \quad \text{or} \quad m_s[h_s - h_f + (pv)_f] = m_a(u_f - u_i)$$

Assume steam and air behave as perfect gases

$$m_s[c_p(T_s - T_f) + RT_f] = m_a c_v (T_f - T_i)$$

Let: $c_{p_s} = 0.45 \frac{\text{Btu}}{\text{lbm °F}}$, $R_s = 0.11 \frac{\text{Btu}}{\text{lbm °F}}$, $c_{v_a} = 0.17 \frac{\text{Btu}}{\text{lbm °F}}$

$$m_a = \frac{V}{v} = \frac{4800 \text{ ft}^3}{13.5 \text{ ft}^3/\text{lbm}_{da}} = 356 \text{ lbm}_{da}$$

Substituting values:

$$10[0.45(787.8 - T_f) + 0.11\, T_f] = 356 \times 0.17(T_f - 530)$$

$$T_f = 556 \text{ R or } 96 \text{ F}$$

Initially, $m_{wv} = \omega \frac{V}{v} = 0.0078 \times 356 = 2.78$ lbm

-continued-

Finally $m_{wv} = 10 + 2.78 = 12.78$ lbm

Then, $p_{wv} = \frac{mRT}{V} = \frac{12.78 \times 86 \times 556}{4800 \times 144} = 0.885$ psia

$RH = \left.\frac{p_{wv}}{p_{wv_{sat}}}\right]_T = \frac{0.885}{0.840}$ (cannot be)

∴ air will be saturated and some water will be in liquid form. T_{room} will be somewhat higher, because vapor will have condensed.

HEATING, VENTILATING AND AIR CONDITIONING 2

A cooling tower is to supply 3500 kw of cooling under the following conditions. Air enters at 100°F DB and 70°F WB and leaves at 83°F DB 95% RH. Recirculating water enters at 105°F and leaves at 80°F. Make up water temperature is 65°F. Assume a reasonable figure for blowdown and windage losses.

Determine:

(a) The total air horsepower if the expected draft loss through the described tower is 0.40" H_2O excluding the fan, and the net fan disc area is 140 ft^2.

(b) The rate of gallons of make up water required under the above conditions.

Solution

Assume steady state cooling tower.

$\Sigma \dot{m}e)_{in} = 0$

83 F_{db}, 95% RH
h = 45.8 Btu/lb
ω = 0.0236
105 F
load
100 F_{db}
70 F_{wb}
h = 34.0
ω = 0.0089
$\dot{m}_m$
65 F
80 F
80 F $\dot{m}_d$

-continued-

$$\dot{m}_{da}(h_i - h_e) + \dot{m}_c(h_i - h_e) + \dot{m}_m h_i - \dot{m}_d h_e = 0$$

From Load,

$$\dot{m}_c = \frac{3500 \times 3413}{60 \times 1 \times 25} = 7960 \text{ lbm/min}$$

Since cooling comes essentially from evaporation, the make-up water can be estimated:

$$h_{fg}\,\dot{m}_m = \text{load} \quad \dot{m}_m = \frac{3500 \times 3413}{60 \times 1000} = 200 \text{ lbm/min}$$

Allowing 1/2% of circulation rate for blow down,

$$\dot{m}_d = 40 \text{ lbm/min}$$

Then,

$$\dot{m}_{da}(h_e - h_i) = \dot{m}_c c(T_i - T_e) + \dot{m}_m c(T_m - 32F) - \dot{m}_d c(T_d - 32F)$$

$$\dot{m}_{da}(45.8 - 34.0) = \text{load} + 200 \times 1(65-32) - 40 \times 1(80-32)$$

$$\dot{m}_{da} = 17{,}270 \text{ lbm/min} \quad \text{cfm} = 17{,}270 \times 14.2 = 2.45 \times 10^5 \text{ cfm}$$

$$V_e)_{\text{through fan}} = \frac{\text{cfm}}{60\,A} = 29.2 \text{ ft/sec}$$

$$\text{Dynamic pressure} = \frac{V^2}{2 g_c v} = \frac{851}{64.4 \times 14.2} = 0.93 \text{ psf}$$

$$\text{Static pressure} = \frac{0.4}{12} \times 62.4 = 2.08 \text{ psf}$$

$$\text{Ideal fan power} = Q \gamma \Delta H = \frac{245{,}000 \times 3.01}{33{,}000} = \underline{\underline{22.3 \text{ hp}}}$$

$$\dot{m}_m = \dot{m}_{da}(\omega_o - \omega_i) + \dot{m}_d = 17{,}270(0.0236 - 0.0089) + 40$$

$$= \underline{\underline{293 \text{ lbm/min}}}$$

* Note: original estimate is low but results will not change significantly

HEATING, VENTILATING AND AIR CONDITIONING 3

A room which is to be maintained at 76°F DB and 40% RH has an exterior wall consisting of exterior stucco on 1/2" gypsum sheathing with metal lath and plaster on the interior, all supported by two by four studding 16 inches on center. There are two 3'-0" by 4'-6" single glazed windows. Determine the external temperature at which condensation will begin to appear on the interior.

Solution

Since glass offers least resistance to heat transfer, its surface will be first to condense moisture.

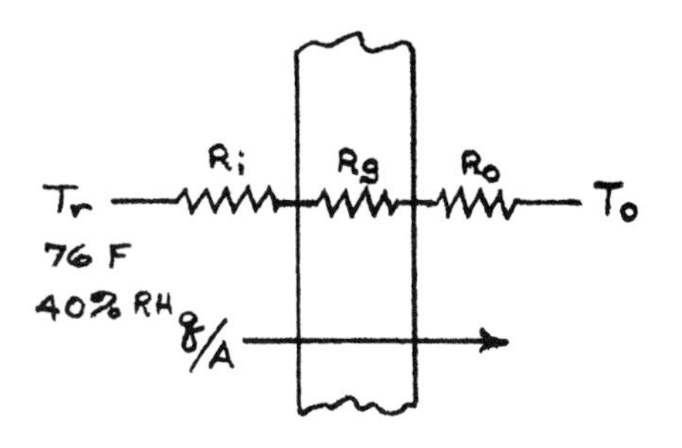

Room $T_{dp} = 49°F$

When inner surface of glass reaches 49°F, condensing will occur.

Assume: $h_i = 1.0 \frac{Btu}{hr\ ft^2\ °F}$ $\qquad R_i = \frac{1}{h_i} = 1.0 \frac{hr\ ft^2\ °F}{Btu}$

$h_o = 2.5 \frac{Btu}{hr\ ft^2\ °F}$ $\qquad R_o = \frac{1}{h_o} = 0.4 \frac{hr\ ft^2\ °F}{Btu}$

$k_{glass} = 0.40 \frac{Btu}{hr\ ft\ °F}$ $\qquad R_g = \frac{L}{k} = \frac{0.25}{12 \times .40} = 0.052 \frac{hr\ ft^2\ F}{Btu}$

glass thickness = 1/4"

no radiation, steady state

$$q/A = \frac{T_r - T_o}{R_i + R_g + R_o} = \frac{T_r - T_{dp}}{R_i}$$

$$\frac{T_r - T_o}{1.45} = \frac{T_r - T_{dp}}{1.0}$$

$$T_o = 1.45\ T_{dp} - 0.45\ T_r = 71 - 34.2 = \underline{\underline{36.8\ F}}$$

HEATING, VENTILATING AND AIR CONDITIONING 4

A vacuum refrigeration system consists of a large insulated flash chamber kept at low pressure by a steam ejector which pumps vapor to a condenser. Condensate is removed by a condensate pump, and an air ejector discharges air from the condenser to an air vent. Warm return water enters the flash chamber at 55°F; chilled water comes out of the flash chamber at 40°F. Vapor leaving the flash chamber has a quality of 0.97, and the temperature in the condenser is 90°F.

For 100 tons of refrigeration:

(a) How much chilled water at 40°F does this system provide in gallons per minute?

(b) How much make up water is needed in pounds per minute?

(c) How much vapor must the steam ejector remove from the flash chamber in cubic feet per minute?

Solution

a) $\dot{m}_c C (T_r - T_c) = \text{Load}$

$$\dot{m}_c = \frac{100 \times 12000}{1(55-40)}$$

$$= 8 \times 10^4 \ \frac{\text{lbm}}{\text{hr}}$$

$$= 160 \text{ gpm}$$

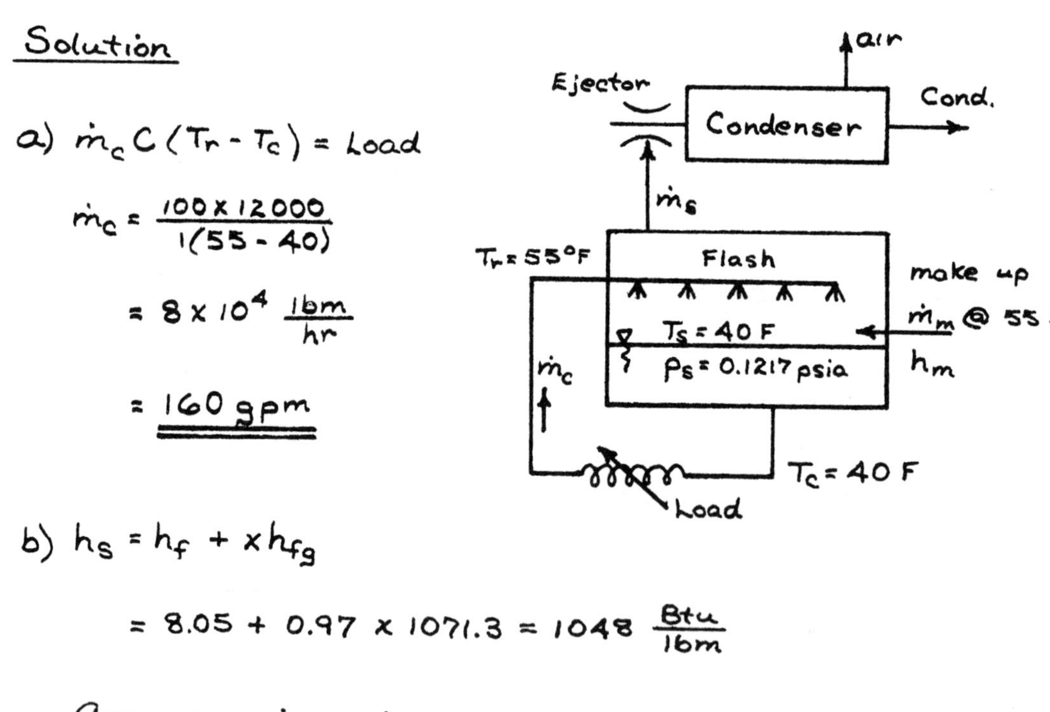

b) $h_s = h_f + x h_{fg}$

$$= 8.05 + 0.97 \times 1071.3 = 1048 \ \frac{\text{Btu}}{\text{lbm}}$$

Assume: $\dot{m}_m = \dot{m}_s$

$T_m = 55$ F

From 1st Law applied to flash chamber,

$$\dot{m}_s (h_s - h_m) = \dot{m}_c C (T_r - T_c) = \text{Load}$$

$$\dot{m}_s = \frac{1.2 \times 10^6}{(1048 - 23)} = 1.16 \times 10^3 \ \text{lbm/hr}$$

c) $v_s + v_f + x v_{fg} \simeq x v_{fg} = 0.97 \times 2444 = 2360 \text{ ft}^3/\text{lbm}$

$$Q_s = \frac{\dot{m}_s v_s}{60} = \frac{1.16 \times 10^3 \times 2.36 \times 10^3}{60} = \underline{\underline{4.55 \times 10^4 \text{ cfm}}}$$

HEATING, VENTILATING AND AIR CONDITIONING 5

An air conditioning system serving a hospital operating room requires 100% outside air as a safety precaution, in view of the hazard relating to the use of anesthetics. Outside design is 100°F DB and 79°F WB, while inside design is 80°F DB and 65°F WB. The air is supplied to the room at 65°F DB to minimize chilling drafts on the patient. The dehumidifying coil is supplied with the chilled water and controlled so as to provide a surface temperature of 50°F.

The operating room is a separate interior zone on the second floor of a three-story completely air conditioned building. Ten persons are present when the room is in use. There are ten kilowatts of lighting, and surgical and medical equipment add 12,600 Btuh latent, and 23,900 Btuh sensible load.

(a) What CFM of air should be supplied the room?

(b) What wet bulb temperature of the supply air as it leaves the dehumidifier would maintain the desired room condition?

(c) What is the coil by-pass factor?

(d) What should be the wet bulb temperature of the supply air to the room?

(e) What is the cooling load on the coil, in tons of refrigeration?

(f) Under design conditions, how many gallons of water must be supplied if the temperature difference between water entering and leaving is 10°F for chilled water and 20°F for hot?

Note: Base all calculations on standard air. An accurate graphic solution will be acceptable.

<u>Solution</u>

Assume:

1) human load @ 400 Btuh/person, $q_\ell = q_s$ due to extra clothing.

– continued –

2) Air leaves dehumidifier saturated at apparatus temp.

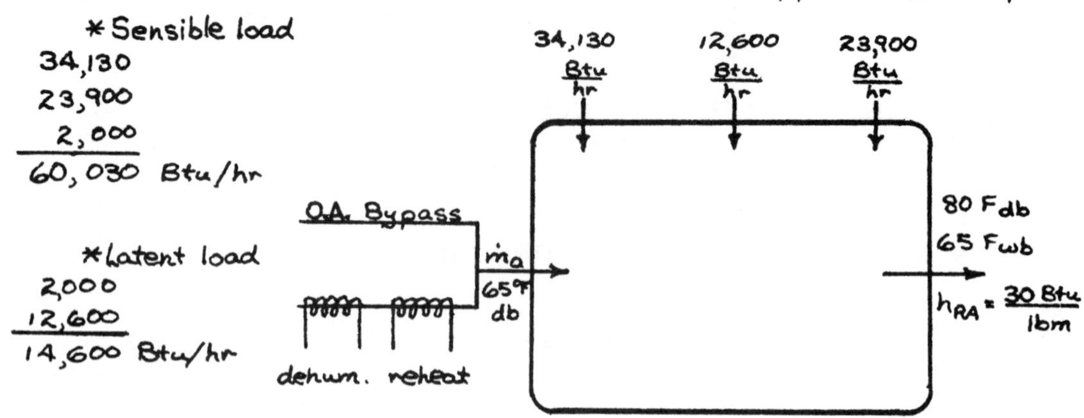

$$\frac{Q_{sensible}}{Q_{total}} = \frac{60{,}030}{74{,}630} = 0.805$$

Using psychrometric protractor, exit state and inlet dry bulb temperature locates inlet state as $65\,°F_{db}$ and 69% RH, $H_i = 25.5\ \frac{Btu}{lbm}$.

a) $Q/v\,(h_{RA} - h_{SA}) = \text{total load}$ $\quad Q = \dfrac{74{,}630 \times 13.4}{60\,(30 - 25.5)} = \underline{\underline{3710\ cfm}}$

b) Assuming air is saturated $\quad T_{wb} = \underline{\underline{50\,°F}}$

c) $(\dot{m}\,\Delta h)_{OA} = (\dot{m}\,\Delta h)_{coil}$ $\quad \dfrac{\dot{m}_{coil}}{\dot{m}_{OA}} = \dfrac{42.6 - 25.5}{25.5 - 22.6} = \underline{\underline{5.90}}$

∴ 14.5% of supply air bypasses humidifier cell.

d) $\underline{\underline{58.5\,°F}}$

e) Ref. load $= \dot{m}_{coil}\,(h_{OA} - h_{SA}) = \dfrac{3710 \times 0.86}{13.4}(42.6 - 20.4)\left(\dfrac{60}{12000}\right)$

$= \underline{\underline{26.5\ tons}}$

f) $\dot{m}_{water\ hum.}\,C\,(\Delta T) = \text{Ref. load}$ $\quad \dot{m}_{water\ hum.} = \dfrac{26.5 \times 12000}{1 \times 10 \times 60} = 530\ \dfrac{lbm}{min}$

$= \underline{\underline{63.0\ gpm}}$

$\dot{m}_{water\ reheat}\,C\,(\Delta T) = \dot{m}_{coil}\,C_p\,\Delta T$ $\quad \dot{m}_{water\ reheat} = \dfrac{3710 \times 0.86 \times 0.24\,(59 - 50)}{13.4 \times 20 \times 8.4}$

$= \underline{\underline{3.1\ gpm}}$

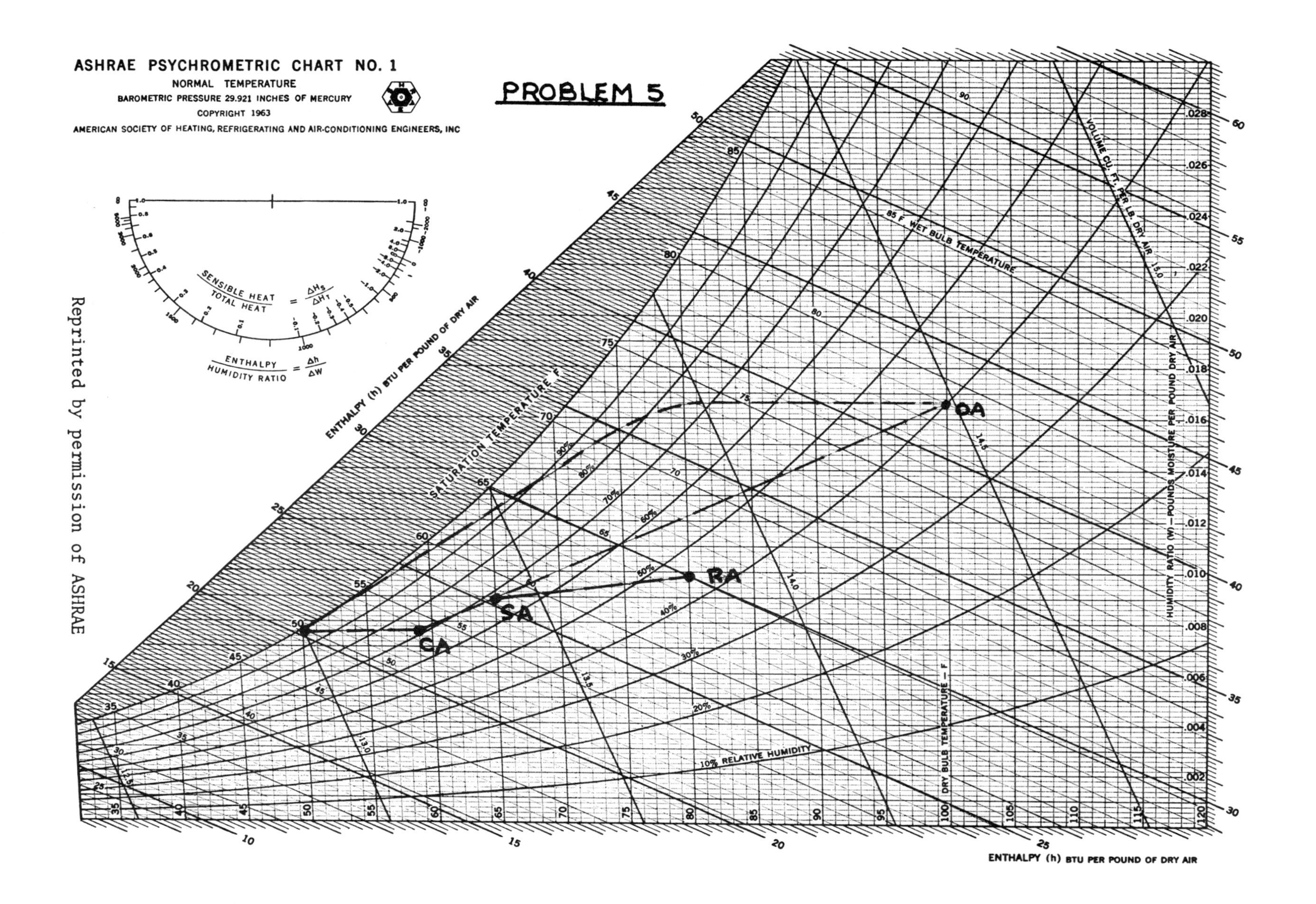

Reprinted by permission of ASHRAE

HEATING, VENTILATING AND AIR CONDITIONING 6

A cubic yard of standard air with dew point of 50°F is compressed isothermally until its volume is one cubic foot.

Find the dew point of the compressed air and the weight of water vapor, if any, that is condensed.

Solution

Assume T_{db} std. air is 70 F

Therefore,

have 27 ft^3 air @ 70 F_{db} 50 F_{dp} (50% RH) compressed to 1 ft^3 @ T=C.

Now, $p_{wv} = RH\, p_{sat} = 0.50 \times 0.363$ psia $= 0.182$ psia

$$m_{wv} = \frac{V}{v}\omega = \frac{27}{13.5} \times 0.0078 = 0.0156 \text{ lbm wv}$$

If there were no condensation $p_{wv} = 27 \times 0.182$ psia on compression, but p_{wv} cannot exceed 0.363 psia.

Therefore,

condensation will occur. From steam tables $m_{wv} = \frac{1}{868}$

$$m_{wv} = 0.00115 \frac{\text{lbm}}{ft^3}$$

Condensate $= 0.0156 - 0.0012 = 0.0144$ lbm

HEATING, VENTILATING AND AIR CONDITIONING 7

A 10,000 square foot church auditorium seating 1000 people is to be air conditioned. Outside design is 95°F DB and 76°F WB, inside design is 76°F DB and 64°F WB. Transmission load is 400,000 Btu/hr; lighting averages 3 watts per square foot; supply air at 55°F is to be 25% outside air.

Calculate:

(a) Sensible, latent and total heat loads.

(b) Tons of refrigeration and supply cfm.

(c) Coil entering and leaving conditions.

(d) Plot appropriate work on psychrometric chart. (Ask the proctor for a copy if you wish to work this problem.)

Solution

assume for humans, $q_s/q_e = 3$

a)

Sensible Load	Transmission load		400,000 Btuh
	Lighting	$3 \times 3.413 \times 10^4$	102,000 Btuh
	Human	1000×300	300,000 Btuh
			802,000 Btuh
Latent Load	Human	1000×100	100,000 Btuh
		Total	902,000 Btuh

$q_s/q_t = 0.89$

b) Assume outside air & return air mixed ahead of coil. Then air to coil $T_{db} = 81$ F $T_{wb} = 67^+$ F (see psyc. chart) Room process line known from q_s/q_t.

– continued –

∴ Since T_{db} supply air = 55 °F, it appears that supply air is approximately saturated.

$$\dot{m}_{SA}(h_{RA} - h_{SA}) = \text{Load}$$

$$\dot{m}_{SA} = \frac{902{,}000}{29.2 - 23.4} = 156{,}000 \text{ lbda/hr}$$

∴ $\dot{m}_{OA}$ = 39,000 lbda/hr or <u>9300 cfm</u>

$\dot{m}_{RA}$ = 117,000 lbda/hr or <u>26,700 cfm</u>

$$\text{Ref. Load} = \dot{m}_{SA}(h_{Ret.A} - h_{SA}) = \frac{1.56 \times 10^5 (31.8 - 23.4)}{12000}$$

$$= \underline{109 \text{ tons}}$$

c) Ret. A T_{db} = 81 F T_{wb} = 67 F

SA. approximately sat. @ 55 F

d) see chart

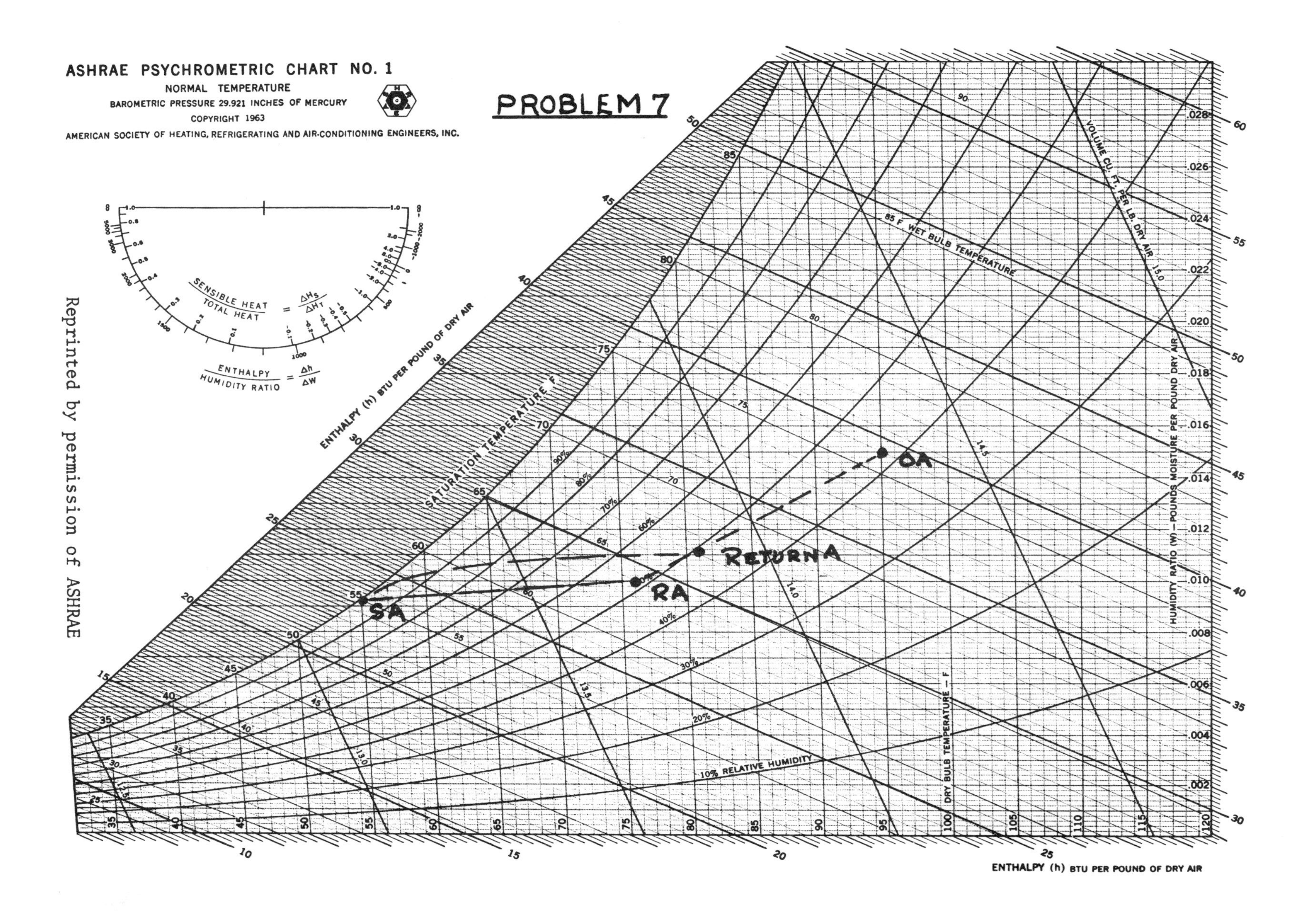

Reprinted by permission of ASHRAE

HEATING, VENTILATING AND AIR CONDITIONING 8

Liquid Refrigerant 12 must be lifted from a condenser operating at 100°F to a thermostatic expansion valve through a vertical height of 40 ft. The piping friction loss including that through valves and fittings is 7.5 psi. How much liquid subcooling is necessary to prevent the liquid from flashing into vapor prior to entering the expansion valve?

Solution

For refrigerant 12 @ 100°F:

$$P_{sat} = 131.9 \text{ psia} \qquad \rho = 78.8 \text{ ft}^3/\text{lbm}$$

$$\Delta P_{Total} = \Delta P_{friction} + \Delta P_{elevation}$$

$$= 7.5 + \frac{40 \times 78.8}{144}$$

$$= 7.5 + 21.9 = 29.4 \text{ psi}$$

$$\therefore P_{valve} = 131.9 - 29.4 = 102.5 \text{ psia}$$

$$T_{sat} = 82.5\ ^\circ\text{F} \ @ \ 102.5 \text{ psia}$$

Subcooling by 17.5 F should result in saturation state at throttling value.

HEATING, VENTILATING AND AIR CONDITIONING 9

A room is to be held at 80°F DB and a maximum of 50% RH when the outside temperature is 96°F DB and 77°F WB. The cubical contents of the room equals 120,000 cubic feet. The internal head load including transmission, occupants, lights, etc., is 286,000 Btu sensible heat and 15,000 Btu latent heat per hour. It is desired to provide one and one half air changes per hour of fresh air; 18,000 cfm will be circulated through the cooling coil under maximum load conditions. The coil will have face and by-pass dampers. The cooling coil will be supplied with chilled water entering at 46°F and leaving at 54°F; assume the characteristics of the coil such that the coil temperature is the average between the entering water and leaving wet bulb temperature.

Determine:

(a) The DB and WB entering the coil.

(b) The required DB and WB leaving the coil.

Solution

Initially assume outside air and return air are premixed in the proportions shown in the figure. Then check to see if resulting air supply state is a possible state from coil. If not, adjust accordingly.

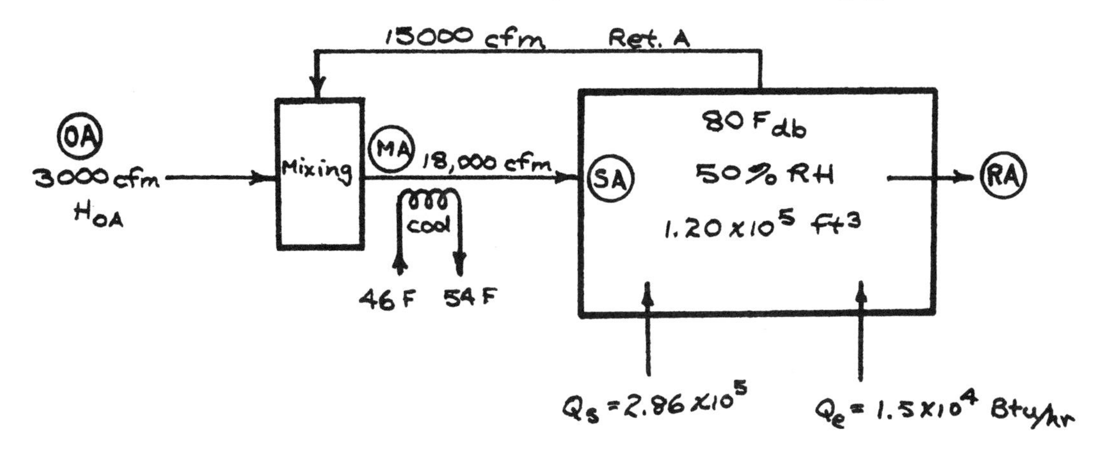

From given information:

$$\frac{q_s}{q_t} = \frac{2.86 \times 10^5}{3.01 \times 10^5} = 0.95$$

—continued—

$$q_t = \frac{cfm}{v}(h_{RA} - h_{SA}) \quad \therefore h_{SA} = h_{RA} - \frac{q_t v}{cfm}$$

$$h_{SA} = 31.4 - \frac{3.01 \times 10^5 \times 13.7}{1.8 \times 10^4 \times 60} = 31.4 - 3.82 = 27.6 \ \frac{Btu}{lbm}$$

$\therefore$ for $q_s/q_t = 0.95$ & $h_i = 27.6$ Btu,

$T_{SA\,db} = 66.5$ F $\quad T_{SA\,wb} = 61.5$ F

Since $\frac{OA}{Ret.A} = \frac{3000}{1500} = \frac{1}{5}$, use psychrometric chart

to find $T_{MA\,db} = 83.0$ F $\quad T_{MA\,wb} = 68.6$ F

For a chill coil temperature of 50°F, the process line will be as shown on psychrometric chart. For supply air to originate on this line means that air supply rate must be reduced by about 20%. Therefore, reduce return air to about 12,000 cfm and repeat solution.

$$\frac{q_s}{q_t} = 0.95$$

$$h_{SA} = 31.4 - \frac{3.01 \times 10^5 \times 13.7}{1.5 \times 10^4 \times 60} = 31.4 - 4.58 = 26.8 \ \frac{Btu}{lbm}$$

b) $T'_{SA\,db} = \underline{\underline{61.5\text{ F}}}$ $\quad T'_{SA\,wb} = \underline{\underline{60.0\text{ F}}}$

a) $\frac{OA}{Ret.\ A} = \frac{3000}{12000} = \frac{1}{4}$ $\quad T'_{MA\,db} = \underline{\underline{84.5\text{ F}}}$ $\quad T'_{MA\,wb} = \underline{\underline{69.5°\text{F}}}$

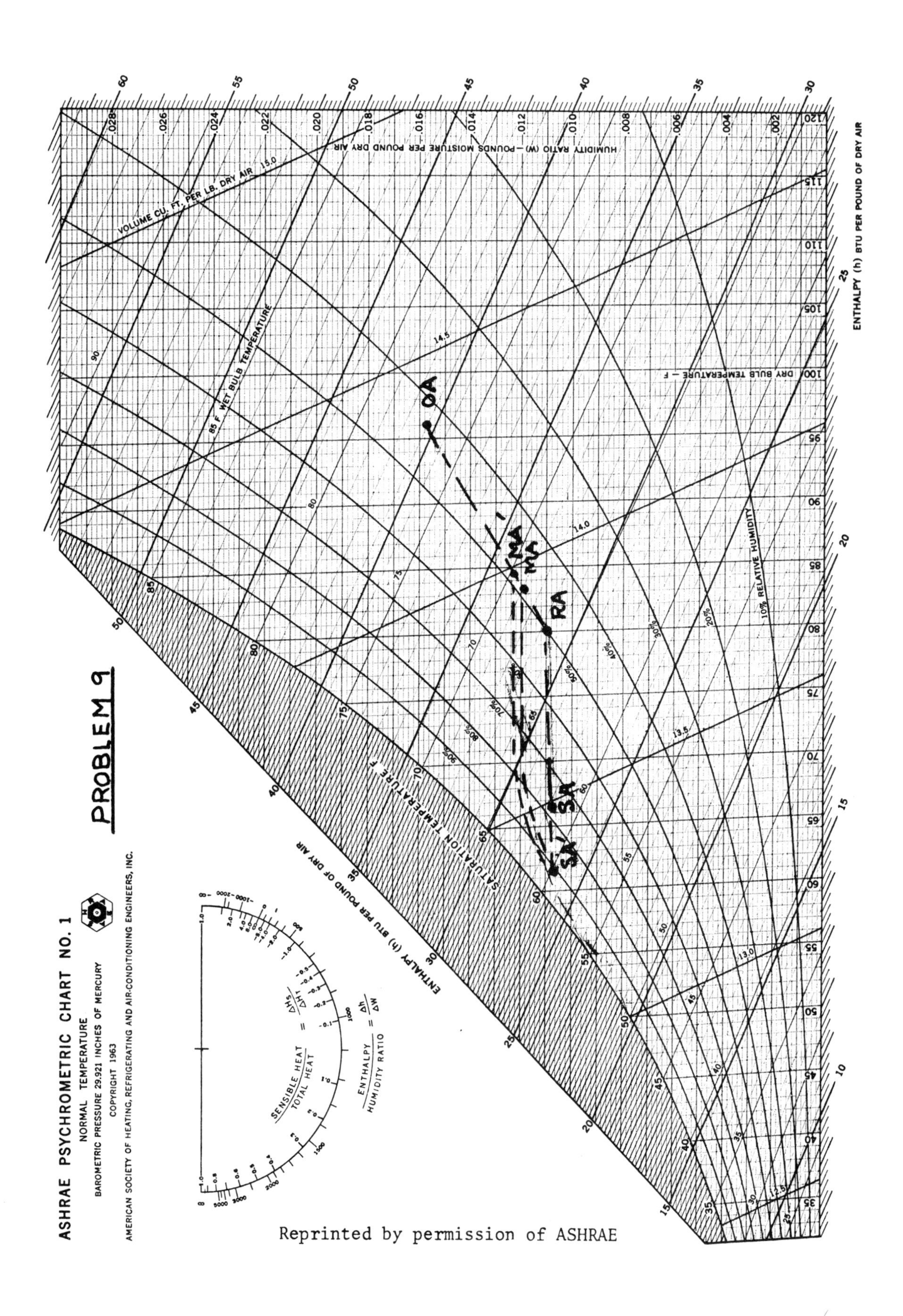

Reprinted by permission of ASHRAE

HEATING, VENTILATING AND AIR CONDITIONING 10

A small building is to be electrically heated. The building is located in Sacramento, California, where the following design data applies:

Heating season	- October 16 to May 14
Heating days	- 211
Avg. mean temp.	- 52.7°F
Degree days	- 2600
Outside design temp.	- 30°F
Indoor design temp.	- 70°F
Outside wind velocity	- 15 mph

(a) Compute the "U" factor (overall coefficient of heat transfer) for the typical wall section shown in Figure 1 using appropriate "R" values (thermal resistance) from the tables given. Units: Btu/hr/sq ft/degree temp. diff.

(b) Express the "U" factor of Part (a) in watts/sq ft/degree temp. diff.

(c) Compute the total heat loss for the room shown in Figure 2 using the "U" factors given. Answer must be expressed in watts.

(d) Using the NEMA formula for annual kilowatt hour consumption, compute the annual cost of heating. Energy cost: 1.5¢/kwhr. Assume an experience factor "C" of 14 for the NEMA formula.

THERMAL RESISTANCE TABLE CODE
(see next page for table)

A. Air Spaces (bounded by non-reflective materials, effective emissivity = 0.82

B. Air Surfaces

C. Insulation

D. Building Paper and Vapor Barriers

E. Wood

F. Plaster and Plasterboard

G. Masonry Materials

A.

Position of air space	Direction of heat flow	Thickness of air space (in.)	Resistance (R)
Horizontal	Up	¾ to 4	0.85
	Down	¾	1.02
		1½	1.15
		4	1.23
		8	1.25
Sloping (45°)	Up	¾ to 4	0.90
Vertical	Horizontal	¾ to 4	0.86

*Bounded by non-reflective materials, effective emissivity = 0.82

B.

Wind speed (mph)	Position of surface	Direction of heat flow	Resistance (R)
No wind; still air	Horizontal	Up	0.61
		Down	0.92
	Sloping (45°)	Up	0.62
	Vertical	Horizontal	0.68
7½	Any position	Any direction	0.25
15	Any position	Any direction	0.17

C.

Type	Material	Resistance (R) per inch thickness
Batts or blankets	Cotton fibre	3.85
	Mineral wool	3.70
	Wood fibre, multilayer	3.70
Loose fill	Macerated paper or pulp products	3.57
	Mineral wool	3.33
	Sawdust or shavings	2.22
	Expanded vermiculite	2.08
	Wood fibre (redwood, hemlock, fir)	3.33
Insulating board or slab	Glass fibre	4.00
	Cellular glass	2.50
	Corkboard (without binder)	3.70
	Hog hair (with asphalt binder)	3.00
	Foamed plastic	3.45
	Shredded wood (cemented)	1.82

D.

Material	Resistance (R)
Vapor-permeable felt	0.06
Vapor seal, 2 layers of mopped 15-lb felt	0.12
Vapor seal, plastic film	Negligible

E.

Material		Thickness of heat path (inches)	Resistance (R)
Wood framing (soft wood)	1 x 2	25/32	0.98
	2 x 2	1 5/8	2.03
	2 x 3	2 5/8	3.28
	2 x 4	3 5/8	4.53
	2 x 6	5 5/8	7.03
	2 x 8	7 5/8	9.53
	2 x 10	9 5/8	12.03
	2 x 12	11 5/8	14.53
Plywood		1/4	0.31
		3/8	0.47
		1/2	0.63
		5/8	0.78
		3/4	0.94
Hardboard (fibreboard)		1/4	0.18
Soft wood (fir, pine, etc.)		25/32	0.98
		1	1.25
Hard wood (maple, oak, etc.)		25/32	0.71
		1	0.91

F.

Material	Thickness (inches)	Resistance (R)
Cement plaster, sand aggregate	1/2	0.10
	3/4	0.15
Gypsum plaster, lightweight aggregate	1/2	0.32
	5/8	0.39
	3/4	0.47
Gypsum board (sheetrock; plasterboard)	3/8	0.32
	1/2	0.45

G.

Material		Density (lb/cu ft)	Resistance (R) per inch thickness
Concrete	Sand and gravel or stone aggregate	140	0.08
	Gypsum fibre	51	0.60
	Lightweight aggregate	100	0.28
		80	0.40
		60	0.59
		40	0.86
		20	1.43
Cement mortar		116	0.20
Stucco		116	0.20
Brick	Common	120	0.20
	Face	130	0.11
Stone, lime, and sand			0.08

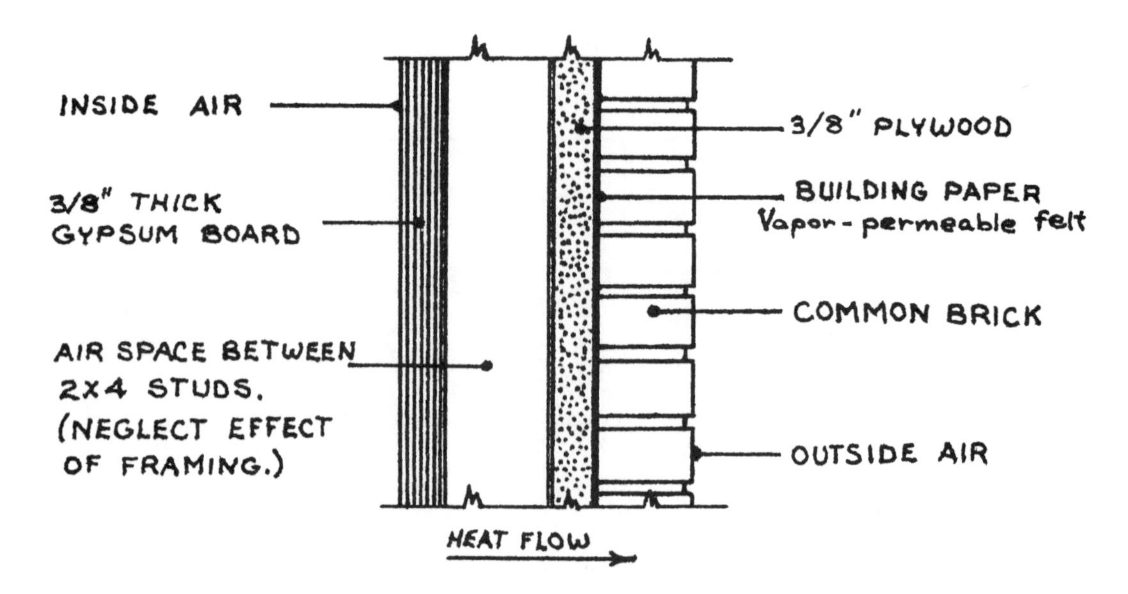

FIGURE 1 TYPICAL WALL SECTION N.T.S.

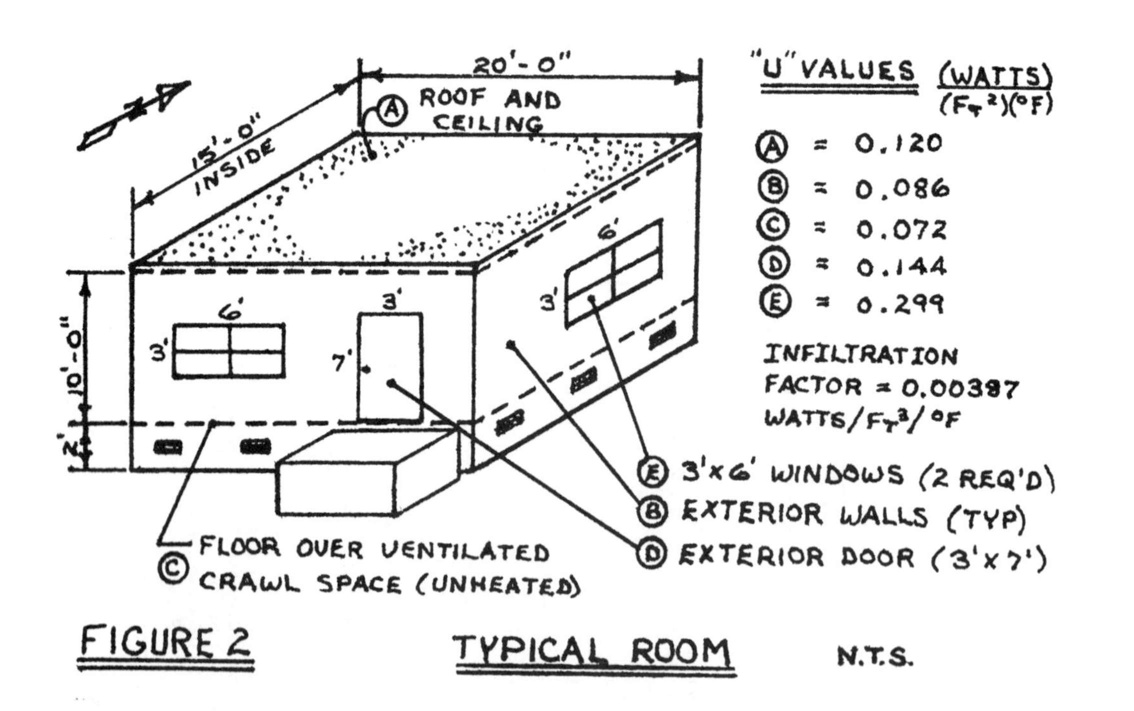

FIGURE 2 TYPICAL ROOM N.T.S.

SPACE HEATING PROBLEM

Solution

a) $R_{c,i} = 0.68$

$R_{gyp} = 0.32$

$R_{air} = 0.86$

$R_{ply} = 0.47$

$R_p = 0.06$

$R_b = 0.60$

$R_{c,o} = 0.17$

$\Sigma R = 3.16$ $\quad U = \frac{1}{\Sigma R} = 0.32 \frac{Btu}{hr\ ft^2\ °F}$

b) $0.32/3.41 = 0.094 \frac{watts}{ft^2\ °F}$

c) Total wall area $= 700\ ft^2$ $\quad A_{windows} = 36\ ft^2$

$A_{door} = 21\ ft^2$ $\quad A_{wall,net} = 643\ ft^2$

Infiltration $= 1.5\ Vol._{room} = 1.5 \times 3000 = 4500\ ft^3/hr$

$q = \Sigma UA\Delta T$

$q_A = 0.120 \times 300\Delta T = 36.0\ \Delta T$

$q_B = 0.086 \times 643\Delta T = 55.3\ \Delta T$

$q_C = 0.072 \times 300\Delta T = 21.6\ \Delta T$

$q_D = 0.144 \times 21\ \Delta T = 3.0\ \Delta T$

$q_E = 0.299 \times 36\Delta T = 10.8\ \Delta T$

$q_{inf.} = 4.5 \times 10^3 \times 3.87 \times 10^{-3} = 17.4\ \Delta T$

$144.1\ \Delta T$

$q = 144.1(70-30) = 5760\ watts$

d) NEMA formula $\quad Kwh = \frac{\text{heat load (kw)} \times \text{degree day} \times C}{\text{Temp. Diff.}}$

$$Kwh = \frac{5.76 \times 2600 \times 14}{40} = 5{,}250$$

$$Cost = 5{,}250 \times .015 = \$78.70$$

HEATING, VENTILATING AND AIR CONDITIONING 11

The duct system for an office is shown below:

The velocity of the air entering at A is 1800 feet per minute. Total quantity of air entering A is 6000 cubic feet per minute. The static pressure regain is 50%.

(Note: Use the equivalent lengths (includes elbows and fittings) listed in the table for the calculations.)

Static pressure regain is the velocity pressure at the fan outlet minus the velocity pressure at the end of the system.

REQUIRED: (Using the equal friction method)

(a) From the information given in the table below, determine the friction loss in inches water gage per 100 feet of equivalent length.

(b) Determine the following for each section in the table below:

(c) Calculate the static pressure in inches water gage for points A, B, C, D, E. F, G in the duct layout using the equivalent lengths.

(d) What is the total static pressure in inches water gage required by the duct system when the pressure static regain is considered?

Section	Air Qty. cfm	Equivalent length/ft	Duct Size inches/dia.	Velocity ft/min
A-B	6000	78	______	1800
B-C	______	25	______	______
C-D	______	53	______	______
D-E	______	20	______	______
E-F	______	20	______	______
F-G	______	20	______	______
G-H	______	20	______	______

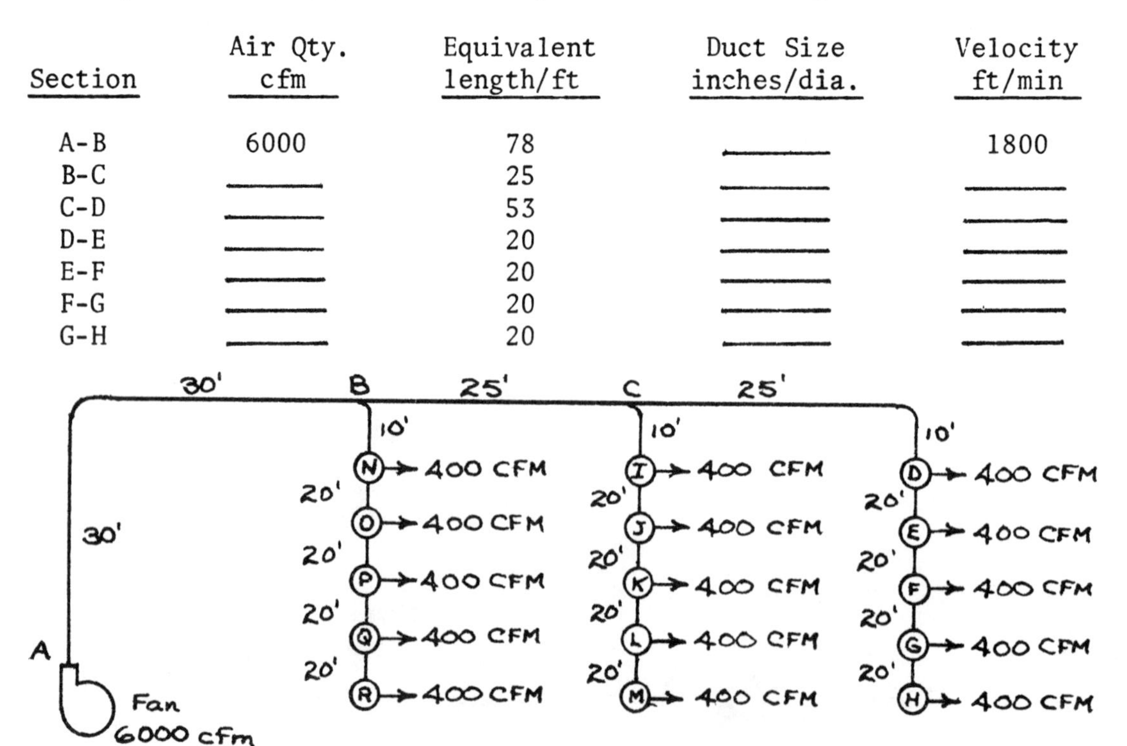

Solution

a) From ASHRAE Handbook "Fundamentals"

$\Delta p = 0.16''\ H_2O/100\ ft$

or $A = Q/v = \frac{6000}{1800}$ $d = 2.05$ ft est. $f = 0.02$

$$\Delta p = f \frac{l}{d} \frac{v^2}{2g} \frac{\rho_{air}}{\rho_{water}} = 0.02 \times \frac{100}{2.05} \times \frac{(30)^2}{64.4} \times \frac{.075}{62.3} \times 12$$

$$= 0.14''\ H_2O$$

b) Equal friction -- constant $\Delta p_{static}/l$

	Q(cfm)	V(ft/sec.)*	d(in.)*	Eq. l(ft)	Δp ("H_2O)
A-B	6000	1800	25	78	0.124
B-C	4000	1610	23.5	25	0.046
C-D	2000	1360	16.5	53	0.085
D-E	1600	1310	15	20	0.032
E-F	1200	1210	13.5	20	0.032
F-G	800	1100	11.5	20	0.032
G-H	400	910	10	20	0.032

$\Delta p_{total} = 0.377''\ H_2O$

* from Handbook

c) $p_A = 0.377 + p_{ref.}$ **

$p_B = 0.253 + p_{ref}$

$p_C = 0.213 + p_{ref}$

$p_D = 0.128 + p_{ref.}$

$p_E = 0.096 + p_{ref.}$

$p_F = 0.064 + p_{ref.}$

$p_G = 0.032 + p_{ref.}$

$p_H = 0 + p_{ref.}$

** Static pressure is not given at any point so pref. is not known.

-continued-

$$d)\quad H_{vel,A} = \frac{V_A^2}{2g} \times \frac{\rho_{air}}{\rho_{water}} = \frac{(30)^2}{64.4} \times \frac{.075}{62.3} \times 12 = 0.202''$$

$$H_{vel,H} = 1/4 \times H_{vel,A} \qquad = 0.050''$$

$$\Delta P_{static\ recovery} = (0.202 - 0.050)\,0.50 = 0.076''\ H_2O$$

$$\therefore \Delta P_{static,net} = 0.377 - .076 = \underline{\underline{0.301''\ H_2O}}$$

A

Engineering Economics

DONALD G. NEWNAN

This is a review of the field known variously as *engineering economics, engineering economy* or *engineering economic analysis.* Since engineering economics is straightforward and logical, even people who have not had a formal course should be able to gain sufficient knowledge from this chapter to successfully solve most engineering economics problems.

There are 35 example problems throughout the review. These examples are an integral part of the review and should be examined as you come to them.

The field of engineering economics uses mathematical and economic techniques to systematically analyze situations which pose alternative courses of action. The initial step in engineering economics problems is to resolve a situation, or each alternative in a given situation, into its favorable and unfavorable consequences or factors. These are then measured in some common unit—usually money. Factors which cannot readily be equated to money are called intangible or irreducible factors. Such factors are considered in conjunction with the monetary analysis when making the final decision on proposed courses of action.

Cash Flow

A cash flow table shows the "money consequences" of a situation and its timing. For example, a simple problem might be to list the year-by-year consequences of purchasing and owning a used car:

Year	Cash Flow	
Beginning of first Year 0	–$4500	Car purchased "now" for $4500 cash. The minus sign indicates a disbursement.
End of Year 1	–350	Maintenance costs are $350 per year.
End of Year 2	–350	
End of Year 3	–350	
End of Year 4	350 +2000	The car is sold at the end of the 4th year for $2000. The plus sign represents a receipt of money.

This same cash flow may be represented graphically, as shown in Fig. A-1. The upward arrow represents a receipt of money, and the downward arrows represent disbursements. The horizontal axis represents the passage of time.

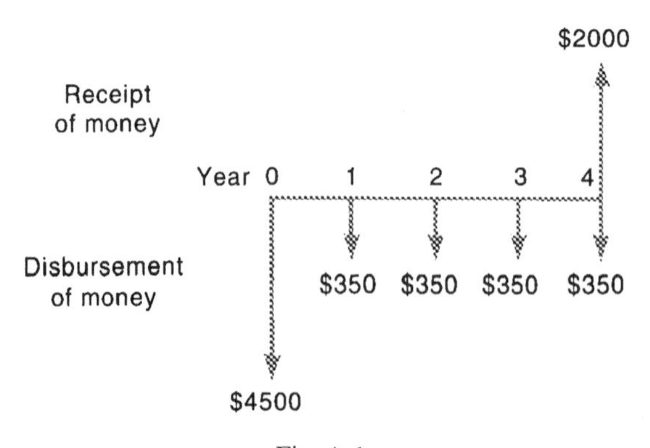

Fig. A-1

Example 1

In January, 1993 a firm purchased a used typewriter for $500. Repairs cost nothing in 1993 or 1994. Repairs are $85 in 1995, $130 in 1996, and $140 in 1997. The machine is sold in 1997 for $300. Compute the cash flow table.

Solution

Unless otherwise stated, the customary assumption is a beginning-of-year purchase, followed by end-of-year receipts or disbursements, and an end-of-year resale or salvage value. Thus the typewriter repairs and the typewriter sale are assumed to occur at the end of the year. Letting a minus sign represent a disbursement of money, and a plus sign a receipt of money, we are able to set up the cash flow table:

Year	Cash Flow
Beginning of 1993	–$500
End of 1993	0
End of 1994	0
End of 1995	–85
End of 1996	–130
End of 1997	+160

Notice that at the end of 1997 the cash flow table shows +160 which is the net sum of –140 and +300. If we define Year 0 as the beginning of 1993, the cash flow table becomes

Year	Cash Flow
0	–$500
1	0
2	0
3	–85
4	–130
5	+160

From this cash flow table, the definitions of Year 0 and Year 1 become clear. Year 0 is defined as the *beginning* of Year 1. Year 1 is the *end* of Year 1, and so forth.

Time Value of Money

When the money consequences of an alternative occur in a short period of time—say, less than one year—we might simply add up the various sums of money and obtain the net result. But we cannot treat money this same way over longer periods of time. This is because money today does not have the same value as money at some future time.

Consider this question: Which would you prefer, $100 today or the assurance of receiving $100 a year from now? Clearly, you would prefer the $100 today. If you had the money today, rather than a year from now, you could use it for the year. And if you had no use for it, you could lend it to someone who would pay interest for the privilege of using your money for the year.

Simple Interest

Simple interest is interest that is computed on the original sum. Thus if one were to lend a present sum P to someone at a simple annual interest rate i, the future amount F due at the end of n years would be

$$F = P + Pin$$

Example 2

How much will you receive back from a $500 loan to a friend for three years at 10% simple annual interest?

Solution

$$F = P + Pin = 500 + 500 \times 0.10 \times 3 = \$650$$

In Example 2 one observes that the amount owed, based on 10% simple interest at the end of one year, is $500 + 500 \times 0.10 \times 1 = \$550$. But at simple interest there is no interest charged on the $50 interest even though it is not paid until the end of the third year. Thus simple interest is not realistic and is seldom used. *Compound interest* charges interest on the principal owed plus the interest earned to date. This produces a charge of interest on interest, or compound interest. Engineering economics uses compound interest computations.

Equivalence

In the preceding section we saw that money at different points in time (for example, $100 today or $100 one year hence) may be equal in the sense that they both are $100, but $100 a year hence is *not* an acceptable substitute for $100 today. When we have acceptable substitutes, we say they are *equivalent* to each other. Thus at 8% interest, $108 a year hence is equivalent to $100 today.

Example 3

At a 10% per year (compound) interest rate, $500 now is *equivalent* to how much three years hence?

Solution

$500 now will increase by 10% in each of the three years.

$$\begin{aligned} \text{Now} &= \$500.00 \\ \text{End of 1st year} = 500 + 10\%(500) &= 550.00 \\ \text{End of 2nd year} = 550 + 10\%(550) &= 605.00 \\ \text{End of 3rd year} = 605 + 10\%(605) &= 665.50 \end{aligned}$$

Thus $500 now is *equivalent* to $665.50 at the end of three years. Note that interest is charged each year on the original $500 plus the unpaid interest. This compound interest computation gives an answer that is $15.50 higher than the simple interest computation in Example 2.

Equivalence is an essential factor in engineering economics. Suppose we wish to select the better of two alternatives. First, we must compute their cash flows. For example

	Alternative	
Year	A	B
0	–$2000	–$2800
1	+800	+1100
2	+800	+1100
3	+800	+1100

The larger investment in Alternative B results in larger subsequent benefits, but we have no direct way of knowing whether it is better than Alternative A. So we do not know which to select. To make a decision, we must resolve the alternatives into *equivalent* sums so they may be compared accurately.

Compound Interest

To facilitate equivalence computations, a series of compound interest factors will be derived here, and their use will be illustrated in examples.

Symbols and Functional Notation

i = Effective interest rate per interest period. In equations, the interest rate is stated as a decimal (that is, 8% interest is 0.08).

n = Number of interest periods. Usually the interest period is one year, but it could be something else.

P = A present sum of money.

F = A future sum of money. The future sum F is an amount n interest periods from the present that is equivalent to P at interest rate i.

A = An end-of-period cash receipt or disbursement in a uniform series continuing for n periods, the entire series is equivalent to P or F at interest rate i.

G = Uniform period-by-period increase in cash flows; the uniform gradient.

r = Nominal annual interest rate.

From Table A-1 we can see that the functional notation scheme is based on writing (To Find/ Given, i, n). Thus, if we wished to find the future sum F, given a uniform series of receipts A, the proper compound interest factor to use would be $(F/A,i,n)$.

Table A-1. Periodic Compounding: Functional Notation and Formulas

Factor	Given	To Find	Functional Notation	Formula
• ***Single Payment***				
Compound Amount Factor	P	F	$(F/P, i\%, n)$	$F = P(1+i)^n$
Present Worth Factor	F	P	$(P/F, i\%, n)$	$P = F(1+i)^{-n}$
• ***Uniform Payment Series***				
Sinking Fund Factor	F	A	$(A/F, i\%, n)$	$A = F\left[\frac{i}{(1+i)^n - 1}\right]$
Capital Recovery Factor	P	A	$(A/P, i\%, n)$	$A = P\left[\frac{i(1+i)^n}{(1+i)^n - 1}\right]$
Compound Amount Factor	A	F	$(F/A, i\%, n)$	$F = A\left[\frac{(1+i)^n - 1}{i}\right]$
Present Worth Factor	A	P	$(P/A, i\%, n)$	$P = A\left[\frac{(1+i)^n - 1}{i(1+i)^n}\right]$
• ***Uniform Gradient***				
Gradient Present Worth	G	P	$(P/G, i\%, n)$	$P = G\left[\frac{(1+i)^n - 1}{i^2(1+i)^n} - \frac{n}{i(1+i)^n}\right]$
Gradient Future Worth	G	F	$(F/G, i\%, n)$	$F = G\left[\frac{(1+i)^n - 1}{i^2} - \frac{n}{i}\right]$
Gradient Uniform Series	G	A	$(A/G, i\%, n)$	$A = G\left[\frac{1}{i} - \frac{n}{(1+i)^n - 1}\right]$

Single Payment Formulas

Suppose a present sum of money P is invested for one year at interest rate i. At the end of the year, the initial investment P is received together with interest equal to Pi or a total amount $P + Pi$. Factoring P, the sum at the end of one year is $P(1 + i)$. If the investment is allowed to remain for subsequent years, the progression is as follows:

Amount at Beginning of Period	+	Interest for the Period	=	Amount at End of the Period
1st year P	+	Pi	=	$P(1 + i)$
2nd year $P(1 + i)$	+	$Pi(1 + i)$	=	$P(1 + i)^2$
3rd year $P(1 + i)^2$	+	$Pi(1 + i)^2$	=	$P(1 + i)^3$
nth year $P(1 + i)^{n-1}$	+	$Pi(1 + i)^{n-1}$	=	$P(1 + i)^n$

The present sum P increases in n periods to $P(1 + i)^n$. This gives a relation between a present sum P and its equivalent future sum F:

$$\text{Future Sum} = (\text{Present Sum})(1 + i)^n$$

$$F = P(1 + i)^n$$

This is the **Single Payment Compound Amount formula**. In functional notation it is written:

$$F = P(F/P,i,n)$$

The relationship may be rewritten as

$$\text{Present Sum} = (\text{Future Sum})(1 + i)^{-n}$$

$$P = F(1 + i)^{-n}$$

This is the **Single Payment Present Worth formula**. It is written:

$$P = F(P/F,i,n)$$

Example 4

At a 10% per year interest rate, \$500 now is *equivalent* to how much three years hence?

Solution

This problem was solved in Example 3. Now it can be solved using a single payment formula. P = \$500, n = 3 years, i = 10%, and F = unknown:

$$F = P(1 + i)^n = 500(1 + 0.10)^3 = \$665.50$$

This problem also may be solved using a compound interest table:

$$F = P(F/P,i,n) = 500(F/P,10\%,3)$$

From the 10% compound interest table, read $(F/P,10\%,3) = 1.331$.

$$F = 500(F/P,10\%,3) = 500(1.331) = \$665.50$$

Example 5

To raise money for a new business, a man asks you to lend him some money. He offers to pay you \$3000 at the end of four years. How much should you give him now if you want 12% interest per year?

Solution

P = unknown, F = \$3000, n = 4 years, and i = 12%:

$$P = F(1 + i)^{-n} = 3000(1 + 0.12)^{-4} = \$1906.55$$

Alternate computation using a compound interest table:

$$P = F(P/F,i,n) = 3000(P/F,12\%,4) = 3000(0.6355) = \$1906.50$$

Note that the solution based on the compound interest table is slightly different from the exact solution using a hand-held calculator. In engineering economics the compound interest tables are always considered to be sufficiently accurate.

Uniform Payment Series Formulas

Consider the situation shown in Fig. A-2. Using the single payment compound amount factor, we can write an equation for F in terms of A:

$$F = A + A(1 + i) + A(1 + i)^2 \qquad \text{(i)}$$

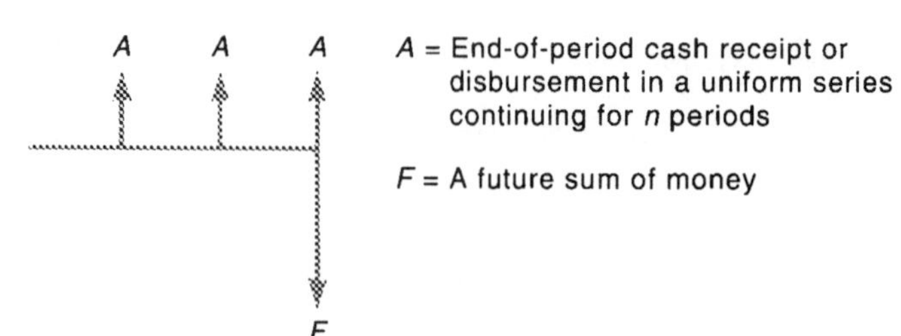

Fig. A-2

In this situation, with $n = 3$, Equation (i) may be written in a more general form:

$$F = A + A(1 + i) + A(1 + i)^{n-1} \qquad \text{(ii)}$$

Multiply Eq. (ii) by $(1 + i)$: $\quad (1 + i)F = A(1 + i) + A(1 + i)^{n-1} + A(1 + i)^n \qquad$ (iii)

Subtract Eq. (ii): $\quad - \quad F = A + A(1 + i) + A(1 + i)^{n-1} \qquad$ (ii)

(iii) – (ii): $\quad iF = -A + A(1 + i)^n$

This produces the **Uniform Series Compound Amount formula:**

$$F = A\left(\frac{(1+i)^n - 1}{i}\right)$$

Solving this equation for A produces the **Uniform Series Sinking Fund formula:**

$$A = F\left(\frac{i}{(1+i)^n - 1}\right)$$

Since $F = P(1 + i)^n$, we can substitute this expression for F in the equation and obtain the **Uniform Series Capital Recovery formula:**

$$A = P\left(\frac{i(1+i)^n}{(1+i)^n - 1}\right)$$

Solving the equation for P produces the **Uniform Series Present Worth formula:**

$$P = A\left(\frac{(1+i)^n - 1}{i(1+i)^n}\right)$$

In functional notation, the uniform series factors are:

Compound Amount $\quad (F/A,i,n)$

Sinking Fund $\quad (A/F,i,n)$

Capital Recovery $\quad (A/P,i,n)$

Present Worth $\quad (P/A,i,n)$

Example 6

If \$100 is deposited at the end of each year in a savings account that pays 6% interest per year, how much will be in the account at the end of five years?

Solution

$A = \$100$, $F =$ unknown, $n = 5$ years, and $i = 6\%$:

$$F = A(F/A,i,n) = 100(F/A,6\%,5) = 100(5.637) = \$563.70$$

Example 7

A fund established to produce a desired amount at the end of a given period, by means of a series of payments throughout the period, is called a **sinking fund**. A sinking fund is to be established to accumulate money to replace a $10,000 machine. If the machine is to be replaced at the end of 12 years, how much should be deposited in the sinking fund each year? Assume the fund earns 10% annual interest.

Solution

$$\text{Annual sinking fund deposit } A = 10{,}000(A/F,10\%,12) = 10{,}000(0.0468) = \$468$$

Example 8

An individual is considering the purchase of a used automobile. The total price is $6200. With $1240 as a downpayment, and the balance paid in 48 equal monthly payments with interest at 1% per month, compute the monthly payment. The payments are due at the end of each month.

Solution

The amount to be repaid by the 48 monthly payments is the cost of the automobile *minus* the $1240 downpayment.

P = $4960, A = unknown, $n = 48$ monthly payments, and $i = 1\%$ per month:

$$A = P(A/P,1\%,48) = 4960(0.0263) = \$130.45$$

Example 9

A couple sells their home. In addition to cash, they take a mortgage on the house. The mortgage will be paid off by monthly payments of $450 for 50 months. The couple decides to sell the mortgage to a local bank. The bank will buy the mortgage, but it requires a 1% per month interest rate on their investment. How much will the bank pay for the mortgage?

Solution

A = $450, $n = 50$ months, $i = 1\%$ per month, and P = unknown:

$$P = A(P/A,i,n) = 450(P/A,1\%,50) = 450(39.196) = \$17{,}638.20$$

Uniform Gradient

At times one will encounter a situation where the cash flow series is not a constant amount A. Instead, it is an increasing series. The cash flow shown in Fig. A-3 may be resolved into two components (Fig. A-4). We can compute the value of P^* as equal to P' plus P. And, we already have the equation for P': $P' = A(P/A,i,n)$. The value for P in the right-hand diagram is:

$$P = G\left[\frac{(1+i)^n - 1}{i^2(1+i)^n} - \frac{n}{i(1+i)^n}\right]$$

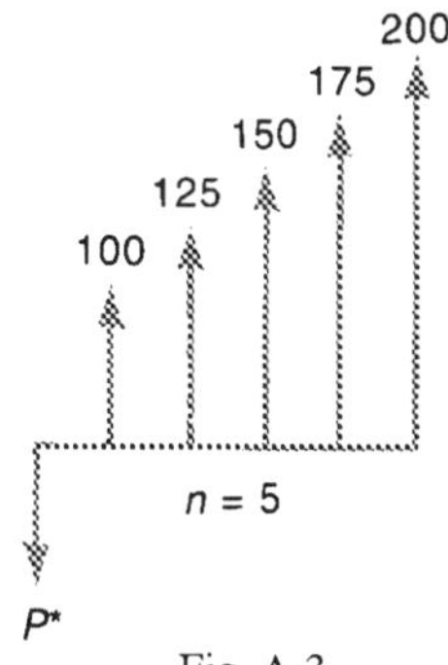

Fig. A-3

This is the **Uniform Gradient Present Worth formula.** In functional notation, the relationship is $P = G(P/G,i,n)$.

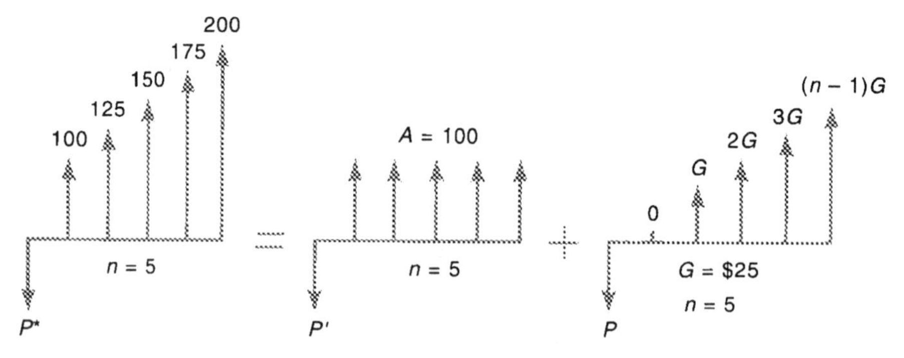

Fig. A-4

Example 10

The maintenance on a machine is expected to be $155 at the end of the first year, and it is expected to increase $35 each year for the following seven years (Fig. A-5). What sum of money should be set aside now to pay the maintenance for the eight-year period? Assume 6% interest.

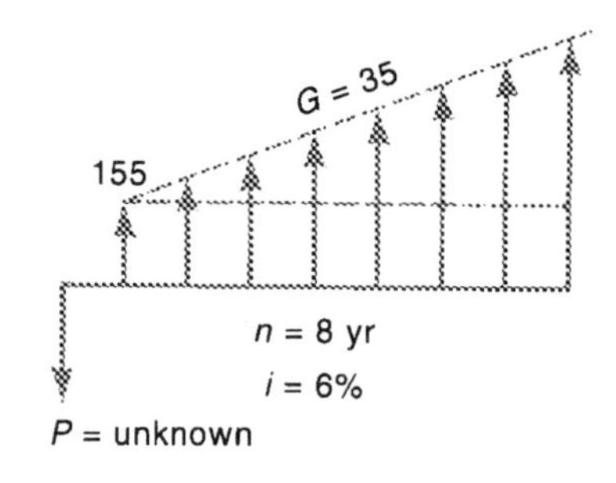

Fig. A-5

Solution

$$P = 155(P/A,6\%,8) + 35(P/G,6\%,8)$$
$$= 155(6.210) + 35(19.841) = \$1656.99$$

In the gradient series, if—instead of the present sum P—an equivalent uniform series A is desired, the problem might appear as shown in Fig A-6. The relationship between A' and G in the right-hand diagram is:

$$A' = G\left[\frac{1}{i} - \frac{n}{(1+i)^n - 1}\right]$$

In functional notation, the Uniform Gradient (to) Uniform Series factor is: $A' = G(A/G,i,n)$.

The Uniform Gradient Uniform Series factor may be read from the compound interest tables directly, or computed as:

$$(A/G,i,n) = \frac{1 - n(A/F,i,n)}{i}$$

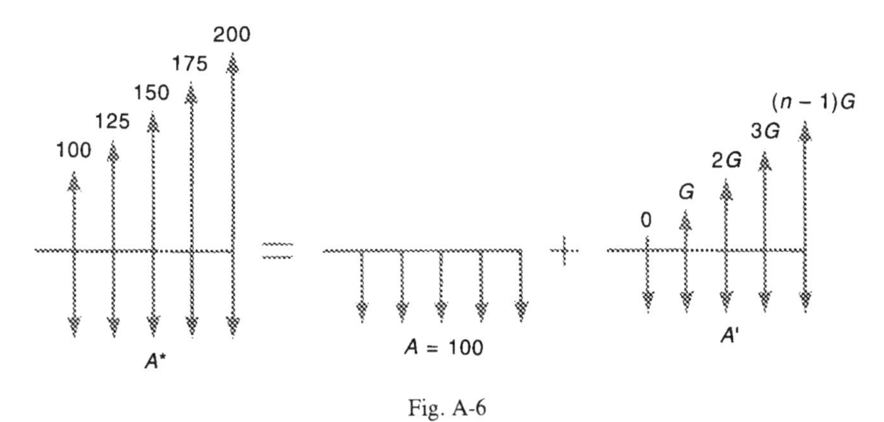

Fig. A-6

Note carefully the diagrams for the uniform gradient factors. The first term in the uniform gradient is zero and the last term is $(n - 1)G$. But we use n in the equations and function notation. The derivations (not shown here) were done on this basis, and the uniform gradient compound interest tables are computed this way.

Example 11

For the situation in Example 10, we wish now to know the uniform annual maintenance cost. Compute an equivalent A for the maintenance costs.

Solution

Refer to Fig. A-7. The equivalent uniform annual maintenance cost is

$$A = 155 + 35(A/G,6\%,8) = 155 + 35(3.195) = \$266.83$$

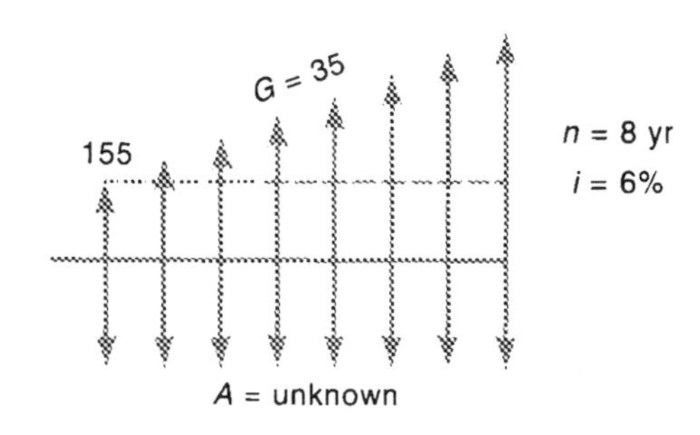

Fig. A-7

Standard compound interest tables give values for eight interest factors: two single payment, four uniform-payment series, and two uniform gradients. The tables do *not* give the Uniform Gradient Future Worth factor, $(F/G,i,n)$. If it is needed, it may be computed from two tabulated factors:

$$(F/G,i,n) = (P/G,i,n)(F/P,i,n)$$

For example, if $i = 10\%$, and $n = 12$ years, then $(F/G,10\%,12) = (P/G,10\%,12)(F/P,10\%,12) = (29.901)(3.138) = 93.83$

A second method of computing the Uniform Gradient Future Worth factor is:

$$(F/G,i,n) = \frac{(F/A,i,n)-n}{i}$$

Using this equation, for $i = 10\%$, and $n = 12$ years, $(F/G,10\%,12) = [(F/A,10\%,12) - 12]/0.10$ $= (21.384 - 12)/0.10 = 93.84$

Continuous Compounding

Table A-2. Continuous Compounding: Functional Notation and Formulas

Factor	Given	To Find	Functional Notation	Formula
• ***Single Payment***				
Compound Amount Factor	P	F	$(F/P, r\%, n)$	$F = P[e^{rn}]$
Present Worth Factor	F	P	$(P/F, r\%, n)$	$P = F[e^{-rn}]$
• ***Uniform Payment Series***				
Sinking Fund Factor	F	A	$(A/F, r\%, n)$	$A = F\left[\frac{e^r - 1}{e^{rn} - 1}\right]$
Capital Recovery Factor	P	A	$(A/P, r\%, n)$	$A = P\left[\frac{e^r - 1}{1 - e^{-rn}}\right]$
Compound Amount Factor	A	F	$(F/A, r\%, n)$	$F = A\left[\frac{e^{rn} - 1}{e^r - 1}\right]$
Present Worth Factor	A	P	$(P/A, r\%, n)$	$P = A\left[\frac{1 - e^{-rn}}{e^r - 1}\right]$

r = nominal annual interest rate; n = number of years

Example 12

Five hundred dollars is deposited each year into a savings bank account that pays 5% nominal interest, compounded continuously. How much will be in the account at the end of five years?

Solution

$A = \$500, \quad r = 0.05, \quad n = 5$ years

$$F = A(F/A,r\%,n) = A\left[\frac{e^{rn}-1}{e^r-1}\right] = 500\left[\frac{e^{0.05(5)}-1}{e^{0.05}-1}\right] = \$2769.84$$

Nominal and Effective Interest

Nominal interest is the annual interest rate without considering the effect of any compounding. **Effective interest** is the annual interest rate taking into account the effect of any compounding during the year.

Non-Annual Compounding

Frequently an interest rate is described as an annual rate, even though the interest period may be something other than one year. A bank may pay 1% interest on the amount in a savings account every three months. The *nominal* interest rate in this situation is 4 ¥ 1% = 4%. But if you deposited $1000 in such an account, would you have 104%(1000) = $1040 in the account at the end of one year? The answer is no, you would have more. The amount in the account would increase as follows:

Amount in Account

At beginning of year = $1000.00

End of 3 months: 1000.00 + 1%(1000.00) = 1010.00

End of 6 months: 1010.00 + 1%(1010.00) = 1020.10

End of 9 months: 1020.10 + 1%(1020.10) = 1030.30

End of one year: 1030.30 + 1%(1030.30) = 1040.60

At the end of one year, the interest of $40.60, divided by the original $1000, gives a rate of 4.06%. This is the *effective* interest rate.

Effective interest rate per year, $i_{eff} = (1 + r/m)^m - 1$

where r = Nominal annual interest rate

m = Number of compound periods per year

r/m = Effective interest rate per period

Example 13

A bank charges $1\frac{1}{2}\%$ interest per month on the unpaid balance for purchases made on its credit card. What nominal interest rate is it charging? What effective interest rate?

Solution

The nominal interest rate is simply the annual interest ignoring compounding, or $12(1\frac{1}{2}\%) = 18\%$.

Effective interest rate = $(1 + 0.015)^{12} - 1 = 0.1956 = 19.56\%$

Continuous Compounding

When m, the number of compound periods per year, becomes very large and approaches infinity, the duration of the interest period decreases from Δt to dt. For this condition of *continuous compounding*, the effective interest rate per year is

$$i_{eff} = e^r - 1$$

where r = nominal annual interest rate.

Example 14

If the bank in Example 13 changes its policy and charges $1\frac{1}{2}\%$ per month, compounded continuously, what nominal and what effective interest rate is it charging?

Solution

Nominal annual interest rate, $r = 12 \times 1\frac{1}{2}\% = 18\%$

Effective interest rate per year, $i_{eff} = e^{0.18} - 1 = 0.1972 = 19.72\%$

Solving Engineering Economics Problems

The techniques presented so far illustrate how to convert single amounts of money, and uniform or gradient series of money, into some equivalent sum at another point in time. These compound interest computations are an essential part of engineering economics problems.

The typical situation is that we have a number of alternatives; the question is, which alternative should we select? The customary method of solution is to express each alternative in some common form and then choose the best, taking both the monetary and intangible factors into account. In most computations an interest rate must be used. It is often called the Minimum Attractive Rate of Return (MARR) to indicate that this is the smallest interest rate, or rate of return, at which one is willing to invest money.

Criteria

Engineering economics problems inevitably fall into one of three categories:

1. Fixed input. The amount of money or other input resources is fixed.
   Example: A project engineer has a budget of $450,000 to overhaul a plant.
2. Fixed output. There is a fixed task, or other output to be accomplished.
   Example: A mechanical contractor has been awarded a fixed price contract to air-condition a building.
3. Neither input nor output fixed. This is the general situation where neither the amount of money (or other inputs), nor the amount of benefits (or other outputs) are fixed. *Example*: A consulting engineering firm has more work available than it can handle. It is considering paying the staff to work evenings to increase the amount of design work it can perform.

There are five major methods of comparing alternatives: present worth; future worth; annual cost; rate of return; and benefit-cost analysis. These are presented in the sections that follow.

Present Worth

Present-Worth analysis converts all of the money consequences of an alternative into an equivalent present sum. The criteria are listed here:

Category	Present-Worth Criterion
Fixed Input.	Maximize the Present Worth of benefits or other outputs.
Fixed Output.	Minimize the Present Worth of costs or other inputs
Neither Input nor Output fixed.	Maximize Present Worth of benefits minus Present Worth of costs, or Maximize Net Present Worth.

Appropriate Problems

Present-Worth analysis is most frequently used to determine the present value of future money receipts and disbursements. We might want to know, for example, the present worth of an income producing property, like an oil well. This should provide an estimate of the price at which the property could be bought or sold.

An important restriction in the use of present worth calculations is that there must be a common analysis period when comparing alternatives. It would be incorrect, for example, to compare the present worth (PW) of cost of Pump *A*, expected to last 6 years, with the PW of cost of Pump *B*, expected to last 12 years (Fig. A-8). In situations like this, the solution is either to use some other analysis technique (generally the annual cost method is suitable in these situations) or to restructure the problem so there is a common analysis period.

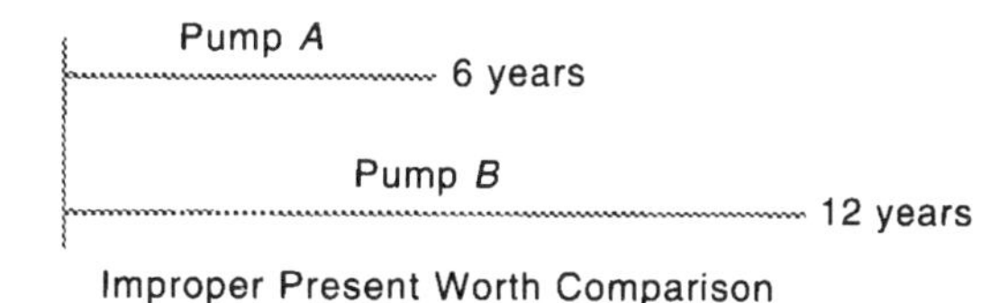

Fig. A-8

In the example above, a customary assumption would be that a pump is needed for 12 years and that Pump *A* will be replaced by an identical Pump *A* at the end of 6 years. This gives a 12-year common analysis period (Fig. A-9). This approach is easy to use when the different lives of the alternatives have a practical least common multiple life. When this is not true (for example, life of *J* equals 7 years and the life of *K* equals 11 years), some assumptions must be made to select a suitable common analysis period, or the present worth method should not be used.

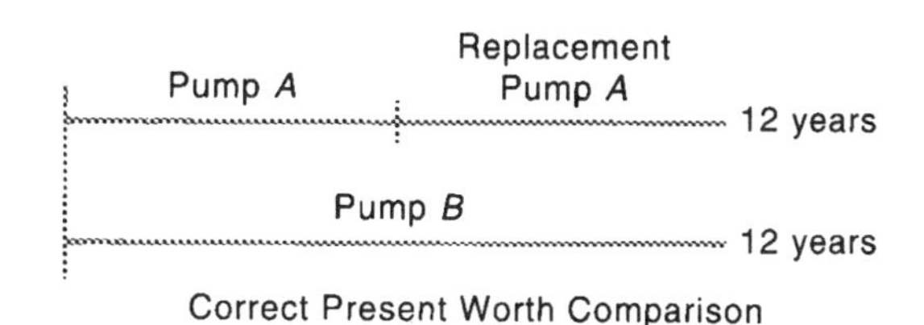

Fig. A-9

Example 15

Machine *X* has an initial cost of $10,000, an annual maintenance of $500 per year, and no salvage value at the end of its four-year useful life. Machine *Y* costs $20,000, and the first year there is no maintenance cost. Maintenance is $100 the second year, and it increases $100 per year thereafter. The machine has an anticipated $5000 salvage value at the end of its 12-year useful life. If the minimum attractive rate of return (MARR) is 8%, which machine should be selected?

Solution

The analysis period is not stated in the problem. Therefore, we select the least common multiple of the lives, or 12 years, as the analysis period.

Present worth of cost of 12 years of Machine *X*

$$= 10{,}000 + 10{,}000(P/F,8\%,4) + 10{,}000(P/F,8\%,8) + 500(P/A,8\%,12)$$
$$= 10{,}000 + 10{,}000(0.7350) + 10{,}000(0.5403) + 500(7.536) = \$26{,}521$$

Present worth of cost of 12 years of Machine *Y*

$$= 20{,}000 + 100(P/G,8\%,12) - 5000(P/F,8\%,12)$$
$$= 20{,}000 + 100(34.634) - 5000(0.3971) = \$21{,}478$$

Choose Machine *Y* with its smaller PW of cost.

Example 16

Two alternatives have the following cash flows:

	Alternative	
Year	A	B
0	–$2000	–$2800
1	+800	+1100
2	+800	+1100
3	+800	+1100

At a 4% interest rate, which alternative should be selected?

Solution

The Net Present Worth of each alternative is computed:

Net Present Worth (NPW) = PW of benefits – PW of cost

$NPW_A = 800(P/A,4\%,3) - 2000 = 800(2.775) - 2000 = \$220.00$

$NPW_B = 1100(P/A,4\%,3) - 2800 = 1100(2.775) - 2800 = \$252.50$

To maximize NPW, choose Alternative *B*.

Infinite Life and Capitalized Cost

In the special situation where the analysis period is infinite ($n = \infty$), an analysis of the present worth of cost is called **capitalized cost**. There are a few public projects where the analysis period is infinity. Other examples are permanent endowments and cemetery perpetual care.

When n equals infinity, a present sum P will accrue interest of Pi for every future interest period. For the principal sum P to continue undiminished (an essential requirement for n equal to infinity), the end-of-period sum A that can be disbursed is Pi (Fig. A-10). When $n = \infty$, the fundamental relationship is

$$A = Pi$$

Some form of this equation is used whenever there is a problem with an infinite analysis period.

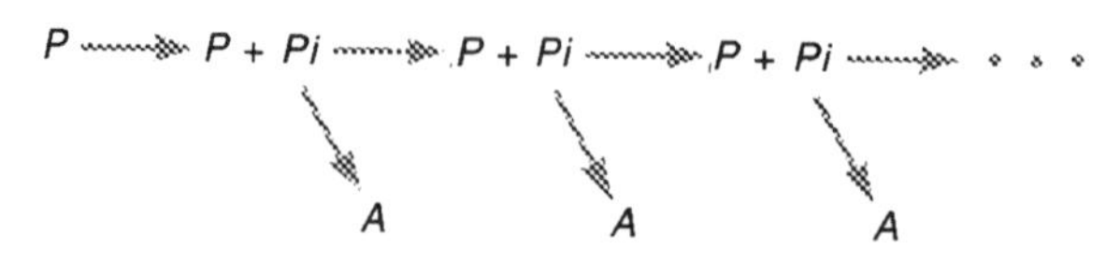

Fig. A-10

Example 17

In his will, a man wishes to establish a perpetual trust to provide for the maintenance of a small local park. If the annual maintenance is $7500 per year and the trust account can earn 5% interest, how much money must be set aside in the trust?

Solution

When $n = \infty$, $A = Pi$ or $P = A/i$

The capitalized cost is $P = A/i = \$7500/0.05 = \$150{,}000$

Future Worth or Value

In present-worth analysis, the comparison is made in terms of the equivalent present costs and benefits. But the analysis need not be made in terms of the present—it can be made in terms of a past, present, or future time. Although the numerical calculations may look different, the decision is unaffected by the selected point in time. Often we do want to know what the future situation will be if we take some particular couse of action now. An analysis based on some future point in time is called **Future-Worth Analysis**.

Category	Future-Worth Criterion
Fixed Input	Maximize the Future Worth of benefits or other outputs.
Fixed Output	Minimize the Furture Worth of costs or other inputs.
Neither Input nor Output fixed	Maximize Future Worth of benefits minus Future Worth of costs, or Maximize Net Future Worth.

Example 18

Two alternatives have the following cash flows:

	Alternative	
Year	A	B
0	–\$2000	–\$2800
1	+800	+1100
2	+800	+1100
3	+800	+1100

At a 4% interest rate, which alternative should be selected?

Solution

In Example 16, this problem was solved by Present-Worth analysis at Year 0. Here it will be solved by Future-Worth analysis at the end of Year 3.

$$\text{Net Future Worth (NFW)} = \text{FW of benefits} - \text{FW of cost}$$

$$NFW_A = 800(F/A,4\%,3) - 2000(F/P,4\%,3) = 800(3.122) - 2000(1.125) = +\$247.60$$

$$NFW_B = 1100(F/A,4\%,3) - 2800(F/P,4\%,3) = 1100(3.122) - 2800(1.125) = +\$284.20$$

To maximize NFW, choose Alternative *B*.

Annual Cost

The annual cost method is more accurately described as the method of Equivalent Uniform Annual Cost (EUAC). Or, where the computation is of benefits, it is called the method of Equivalent Uniform Annual Benefits (EUAB).

Criteria

For each of the three possible categories of problems, there is an annual cost criterion for economic efficiency.

Category	Annual Cost Criterion
Fixed Input	Maximize the Equivalent Uniform Annual Benefits (EUAB).
Fixed Output	Minimize the Equivalent Uniform Annual Cost (EUAC).
Neither Input nor Output fixed	Maximize [EUAB-EUAC]

Application of Annual Cost Analysis

In the section on present worth, we pointed out that the present worth method requires a common analysis period for all alternatives. This restriction does not apply in all annual cost calculations, but it is important to understand the circumstances that justify comparing alternatives with different service lives.

Frequently, an analysis is to provide for a more-or-less continuing requirement. For example, one might need to pump water from a well on a continuing basis. Regardless of whether the pump has a useful service life of 6 years or 12 years, we would select the one whose annual cost is a minimum. And this still would be the case if the pump's useful lives were the more troublesome 7 and 11 years. Thus, if we can assume a continuing need for an item, an annual cost comparison among alternatives of differing service lives is valid. This is because the underlying assumption made in these situations is that the shorter-lived alternative can be replaced with an identical item with identical costs, when it has reached the end of its useful life. This means the EUAC of the initial alternative is equal to the EUAC for the continuing series of replacements.

On the other hand, if there is a specific requirement to pump water for 10 years, then each pump must be evaluated to see what costs will be incurred during the analysis period and what salvage value, if any, may be recovered at the end of the analysis period. The annual cost comparison needs to consider the actual circumstances of the situation.

Examination problems are often readily solved by the annual cost method. And the underlying "continuing requirement" is usually present, so an annual cost comparison of unequal-lived alternatives is an appropriate method of analysis.

Example 19

Consider the following alternatives:

	A	B
First cost	$5000	$10,000
Annual maintenance	500	200
End-of-useful-life salvage value	600	1000
Useful life	5 years	15 years

Based on an 8% interest rate, which alternative should be selected?

Solution

Assuming both alternatives perform the same task and there is a continuing requirement, the goal is to minimize EUAC.

Alternative A:

$$\begin{aligned} \text{EUAC} &= 5000(A/P,8\%,5) + 500 - 600(A/F,8\%,5) \\ &= 5000(0.2505) + 500 - 600(0.1705) = \$1650 \end{aligned}$$

Alternative B:

$$\text{EUAC} = 10{,}000(A/P,8\%,15) + 200 - 1000(A/F,8\%,15)$$
$$= 10{,}000(0.1168) + 200 - 1000(0.0368) = \$1331$$

To minimize EUAC, select Alternative *B*.

Rate of Return Analysis

A typical situation is a cash flow representing the costs and benefits. The rate of return may be defined as the interest rate where PW of cost = PW of benefits, EUAC = EUAB, or PW of cost – PW of benefits = 0.

Example 20

Compute the rate of return for the investment represented by the following cash flow table.

Year:	0	1	2	3	4	5
Cash Flow:	–\$595	+250	+200	+150	+100	+50

Solution

This declining uniform gradient series may be separated into two cash flows for which compound interest factors are available.

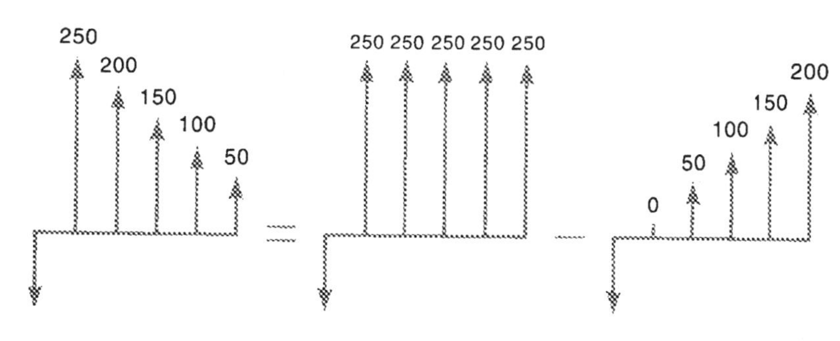

Fig. A-11

Note that the gradient series factors are based on an *increasing* gradient. Here the declining cash flow is solved by subtracting an increasing uniform gradient, as indicated in the figure.

PW of cost – PW of benefits = 0

$$595 - [250(P/A,i,5) - 50(P/G,i,5)] = 0$$

Try $i = 10\%$:

$$595 - [250(3.791) - 50(6.862)] = -9.65$$

Try $i = 12\%$

$$595 - [250(3.605) - 50(6.397)] = +13.60$$

The rate of return is between 10% and 12%. It may be computed more accurately by linear interpolation:

$$\text{Rate of return} = 10\% + (2\%)\left(\frac{9.65 - 0}{13.60 + 9.65}\right) = 10.83\%$$

Two Alternatives

Compute the incremental rate of return on the cash flow representing the difference between the two alternatives. Since we want to look at increments of *investment*, the cash flow for the difference between the alternatives is computed by taking the higher initial-cost alternative minus the lower initial-cost alternative. If the incremental rate of return is greater than or equal to the predetermined minimum attractive rate of return (MARR), choose the higher-cost alternative; otherwise, choose the lower-cost alternative.

Example 21

Two alternatives have the following cash flows:

	Alternative	
Year	A	B
0	–$2000	–$2800
1	+800	+1100
2	+800	+1100
3	+880	+1100

If 4% is considered the minimum attractive rate of return (MARR), which alternative should be selected?

Solution

These two alternatives were previously examined in Examples 16 and 18 by Present-Worth and Future-Worth analysis. This time, the alternatives will be resolved using a rate-of-return analysis.

Note that the problem statement specifies a 4% minimum attractive rate of return (MARR), whereas Examples 16 and 18 referred to a 4% interest rate. These are really two different ways of saying the same thing: the minimum acceptable time value of money is 4%.

First, tabulate the cash flow that represents the increment of investment between the alternatives. This is done by taking the higher initial-cost alternative minus the lower initial-cost alternative:

	Alternative		Difference between Alternatives
Year	A	B	B – A
0	–$2000	–$2800	–$800
1	+800	+1100	+300
2	+800	+1100	+300
3	+800	+1100	+300

Then compute the rate of return on the increment of investment represented by the difference between the alternatives:

$$\text{PW of cost} = \text{PW of benefits}$$

$$800 = 300(P/A,i,3)$$

$$(P/A,i,3) = 800/300 = 2.67$$

$$i = 6.1\%$$

Since the incremental rate of return exceeds the 4% MARR, the increment of investment is desirable. Choose the higher-cost Alternative *B*.

Before leaving this example, one should note something that relates to the rates of return on Alternative *A* and on Alternative *B*. These rates of return, if calculated, are:

	Rate of Return
Alternative *A*	9.7%
Alternative *B*	8.7%

The correct answer to this problem has been shown to be Alternative *B*, even though Alternative *A* has a higher rate of return. The higher-cost alternative may be thought of as the lower-cost alternative, plus the increment of investment between them. Looked at this way, the higher-cost Alternative *B* is equal to the desirable lower-cost Alternative *A* plus the differences between the alternatives.

The important conclusion is that computing the rate of return for each alternative does *not* provide the basis for choosing between alternatives. Instead, incremental analysis is required.

Example 22

Consider the following:

	Alternative	
Year	A	B
0	–$200.0	–$131.0
1	+77.6	+48.1
2	+77.6	+48.1
3	+77.6	+48.1

If the minimum attractive rate of return (MARR) is 10%, which alternative should be selected?

Solution

To examine the increment of investment between the alternatives, we will examine the higher initial-cost alternative minus the lower initial-cost alternative, or *A* – *B*.

	Alternative		Increment
Year	A	B	A – B
0	–$200.0	–$131.0	–$69.0
1	+77.6	+48.1	+29.5
2	+77.6	+48.1	+29.5
3	+77.6	+48.1	+29.5

Solve for the incremental rate of return:

$$\text{PW of cost} = \text{PW of benefits}$$

$$69.0 = 29.5(P/A,i,3)$$

$$(P/A,i,3) = 69.0/29.5 = 2.339$$

From compound interest tables, the incremental rate of return is between 12% and 18%. This is a desirable increment of investment, hence we select the higher initial-cost Alternative *A*.

Three or More Alternatives

When there are three or more mutually exclusive alternatives, proceed with the same logic presented for two alternatives. The components of incremental analysis are listed below:

Step 1. Compute the rate of return for each alternative. Reject any alternative where the rate of return is less than the desired MARR. (This step is not essential, but helps to immediately identify unacceptable alternatives.)

Step 2. Rank the remaining alternatives in their order of increasing initial cost.

Step 3. Examine the increment of investment between the two lowest-cost alternatives as described for the two-alternative problem. Select the better of the two alternatives and reject the other one.

Step 4. Take the preferred alternative from Step 3. Consider the next higher initial-cost alternative and proceed with another two-alternative comparison.

Step 5. Continue until all alternatives have been examined and the best of the multiple alternatives has been identified.

Example 23

Consider the following:

	Alternative	
Year	A	B
0	–$200.0	–$131.0
1	+77.6	+48.1
2	+77.6	+48.1
3	+77.6	+48.1

If the minimum attractive rate of return (MARR) is 10%, which alternative, if any, should be selected?

Solution

One should carefully note that this is a *three-alternative* problem where the alternatives are *A*, *B*, and *Do Nothing*. In this solution we will skip Step 1. Reorganize the problem by placing the alternatives in order of increasing initial cost:

	Alternative		
Year	Do Nothing	B	A
0	0	–$131.0	–$200.0
1	0	+48.1	+77.6
2	0	+48.1	+77.6
3	0	+48.1	+77.6

Examine the *B – Do Nothing* increment of investment:

Year	*B – Do Nothing*
0	–\$131.0 – 0 = –\$131.0
1	+48.1 – 0 = +48.1
2	+48.1 – 0 = +48.1
3	+48.1 – 0 = +48.1

Solve for the incremental rate of return.

$$\text{PW of cost} = \text{PW of benefits}$$

$$131.0 = 48.1(P/A,i,3)$$

$$(P/A,i,3) = 131.0/48.1 = 2.723$$

From compound interest tables, the incremental rate of return is about 5%. Since the incremental rate of return is less than 10%, the *B – Do Nothing* increment is not desirable. Reject Alternative *B*.

Next, consider the increment of investment between the two remaining alternatives.

Year	*A – Do Nothing*
0	–\$200.0 – 0 = –\$200.0
1	+77.6 – 0 = +77.6
2	+77.6 – 0 = +77.6
3	+77.6 – 0 = +77.6

Solve for the incremental rate of return.

$$\text{PW of cost} = \text{PW of benefits}$$

$$200.0 = 77.6(P/A,i,3)$$

$$(P/A,i,3) = 200.0/77.6 = 2.577$$

The incremental rate of return is 8%, less than the desired 10%. Reject the increment and select the remaining alternative: *Do nothing.*

If you have not already done so, you should go back to Example 22 and see how the slightly changed wording of the problem radically altered it. Example 22 required a choice between two undesirable alternatives. Example 23 adds the *Do-nothing* alternative which is superior to *A* and *B*.

Benefit-Cost Analysis

Generally, in public works and governmental economic analyses, the dominant method of analysis is the **Benefit-Cost ratio**. It is simply the ratio of benefits divided by costs, taking into account the time value of money.

$$\text{B/C} = \frac{\text{PW of benefits}}{\text{PW of cost}} = \frac{\text{Equivalent Uniform Annual Benefits}}{\text{Equivalent Uniform Annual Cost}}$$

For a given interest rate, a B/C ratio ≥ 1 reflects an acceptable project. The B/C analysis method is parallel to that of rate-of-return analysis. The same kind of incremental analysis is required.

Example 24

Solve Example 22 by Benefit-Cost analysis.

Solution

	Alternative		Increment
Year	A	B	A – B
0	–$200.0	–$131.0	–$69.0
1	+77.6	+48.1	+29.5
2	+77.6	+48.1	+29.5
3	+77.6	+48.1	+29.5

The benefit-cost ratio for the $A - B$ increment is

$$\text{B/C} = \frac{\text{PW of benefits}}{\text{PW of cost}} = \frac{29.5(P/A,10\%,3)}{69.0} = \frac{73.37}{69.0} = 1.06$$

Since the B/C ratio exceeds 1, the increment of investment is desirable. Select the higher cost Alternative A.

Breakeven Analysis

In business, "breakeven" is defined as the point where income just covers costs. In engineering economics, the breakeven point is defined as the point where two alternatives are equivalent.

Example 25

A city is considering a new $50,000 snowplow. The new machine will operate at a savings of $600 per day compared to the present equipment. Assume the minimum attractive rate of return (MARR) is 12%, and the machine's life is 10 years with zero resale value at that time. How many days per year must the machine be used to justify the investment?

Solution

This breakeven problem may be readily solved by annual cost computations. We will set the equivalent uniform annual cost (EUAC) of the snowplow equal to its annual benefit and solve for the required annual utilization. Let X = breakeven point = days of operation per year.

$$\text{EUAC} = \text{EUAB}$$
$$50{,}000(A/P,12\%,10) = 600X$$
$$X = 50{,}000(0.1770)/600 = 14.8 \text{ days/year}$$

Optimization

Optimization is the determination of the best or most favorable situation.

Minima-Maxima

In problems where the situation can be represented by a function, the customary approach is to set the first derivative of the function to zero and solve for the root(s) of this equation. If the second derivative is *positive,* the function is a minimum for the critical value; if it is *negative*, the function is a maximum.

Example 26

A consulting engineering firm estimates their net profit is given by the equation

$$P(x) = -0.03x^3 + 36x + 500 \quad x \geq 0$$

where x = number of employees and $P(x)$ = net profit. What is the optimum number of employees?

Solution

$P'(x) = -0.09x^2 + 36 = 0 \quad P''(x) = -0.18x$

$x^2 = 36/0.09 = 400$

$x = 20$ employees.

$P''(20) = -0.18(20) = -3.6$

Since $P''(20) < 0$, the net profit is maximized for 20 employees.

Economic Problem—Best Alternative

Since engineering economics problems seek to identify the best or more favorable situation, they are by definition optimization problems. Most use compound interest computations in their solution, but some do not. Consider the following example.

Example 27

A firm must decide which of three alternatives to adopt to expand its capacity. It wants a minimum annual profit of 20% of the initial cost of each increment of investment. Any money not invested in capacity expansion can be invested elsewhere for an annual yield of 20% of the initial cost.

Alternative	Initial cost	Annual profit	Profit rate
A	$100,000	$30,000	30%
B	300,000	66,000	22
C	500,000	80,000	16

Which alternative should be selected?

Solution

Since Alternative *C* fails to produce the 20% minimum annual profit, it is rejected. To decide between Alternatives *A* and *B*, examine the profit rate for the *B* – *A* increment.

Alt	Initial Cost	Annual Profit	Incremental Cost	Incremental Profit	Incremental Profit Rate
A	$100,000	$30,000			
			$200,000	$36,000	18%
B	300,000	66,000			

The *B* – *A* incremental profit rate is less than the minimum 20%, so Alternative *B* should be rejected. Thus the best investment of $300,000, for example, would be Alternative *A* (annual profit = \$30,000) plus \$200,000 invested elsewhere at 20% (annual profit = \$40,000). This combination would yield a \$70,000 annual profit, which is better than the Alternative *B* profit of \$66,000. Select *A*.

Economic Order Quantity

One special case of optimization occurs when an item is used continuously and is periodically purchased. Thus the inventory of the item fluctuates from zero (just prior to the receipt of the purchased quantity) to the purchased quantity (just after receipt). The simplest model for the Economic Order Quantity (EOQ) is

$$\text{EOQ} = \sqrt{\frac{2\text{BD}}{\text{E}}}$$

where B = Ordering cost, \$/order

D = Demand per period, units

E = Inventory holding cost, \$/unit/period

EOQ = Economic order quantity, units

Example 28

A company uses 8000 wheels per year in its manufacture of golf carts. The wheels cost \$15 each and are purchased from an outside supplier. The money invested in the inventory costs 10% per year, and the warehousing cost amounts to an additional 2% per year. It costs \$150 to process each purchase order. When an order is placed, how many wheels should be ordered?

Solution

$$\text{EOQ} = \sqrt{\frac{2 \times \$150 \times 8000}{(10\% + 2\%)\ (15.00)}} = 1155 \text{ wheels}$$

Valuation and Depreciation

Depreciation of capital equipment is an important component of many after-tax economic analyses. For this reason, one must understand the fundamentals of depreciation accounting.

Notation

BV = Book value

C = Cost of the property (Basis)

D_j = Depreciation in Year *j*

S_n = Salvage value in year *n*

Depreciation is the systematic allocation of the cost of a capital asset over its useful life. **Book value** is the original cost of an asset, minus the accumulated depreciation of the asset.

$$\text{BV} = \text{C} - \Sigma(\text{D}_j)$$

In computing a schedule of depreciation charges four items are considered.

1. Cost of the property, C (called the *basis* in tax law).

2. Type of property. Property is classified either as *tangible* (like machinery) or *intangible* (like a franchise or a copyright) and either *real property* (real estate) or *personal property* (everything not real property).
3. Depreciable life in years, n.
4. Salvage value of the property at the end of its depreciable (usable) life, S_n.

Straight Line Depreciation

Depreciation charge in any year,

$$D_j = \frac{C - S_n}{n}$$

An alternate computation is

$$\text{Depreciation charge in any year, } D_j = \frac{\text{C – depreciation taken to beginning of year } j - \text{S}_n}{\text{Remaining useful life beginning of year } j}$$

Sum-Of-Years-Digits Depreciation

$$\text{Depreciation charge in any year, } D_j = \frac{\text{Remaining depreciable life at beginning of year}}{\text{Sum-of-Years Digits for Total Useful Life}} \times (\text{C} - \text{S}_n)$$

$$D_j = \frac{n - j + 1}{\frac{n}{2}(n+1)}(\text{C} - \text{S}_n)$$

Declining-Balance Depreciation

$$\text{Double Declining Balance depreciation charge in any year, } D_j = \frac{2C}{m}\left(1 - \frac{2}{n}\right)^{j-1}$$

$$\text{Total depreciation at the end of } n \text{ years, } C = \left[1 - \left(1 - \frac{2}{n}\right)^n\right]$$

$$\text{Book value at the end of } j \text{ years, } BV_j = C\left(1 - \frac{2}{n}\right)^j$$

For 150% declining balance depreciation, replace the “2” in the three equations above with “1.5.”

Sinking-Fund Depreciation

$$\text{Depreciation charge in any year, } D_j = (C - S_n)(A/F, i\%, n)(F/P, i\%, j - 1)$$

Modified Accelerated Cost Recovery System Depreciation

The Modified-Accelerated-Cost-Recovery-System (MACRS) depreciation method generally applies to property placed in service after 1986. To compute the MACRS depreciation for an item one must know:

1. Cost (basis) of the item.
2. Property class. All tangible property is classified in one of six classes (3, 5, 7, 10, 15, and 20 years), which is the life over which it is depreciated (see Table A-3). Residential real estate and nonresidential real estate are in two separate real property classes of 27.5 years and 39 years, respectively.
3. Depreciation computation.
   - 3, 5, 7, and 10-year property classes use double-declining-balance depreciation with conversion to straight-line depreciation in the year that increases the deduction.
   - 15 and 20-year property classes use 150%-declining-balance depreciation with conversion to straight-line depreciation in the year that increases the deduction.
   - In MACRS, the salvage value is assumed to be zero.

Table A-3. MACRS Classes of Depreciable Property

Property class	*Personal property (all property except real estate)*
Three-Year Property	Special handling devices for food and beverage manufacture; Special tools for the manufacture of finished plastic products, fabricated metal products, and motor vehicles; Property with Asset Depreciation Range (ADR) midpoint life of 4 years or less.
Five-Year Property	Automobiles* and trucks; Aircraft (of non-air-transport companies); Equipment used in research and experimentation; Computers; Petroleum drilling equipment; Property with ADR midpoint life of more than 4 years and less than 10 years.
Seven-Year Property	All other property not assigned to another class; Office furniture, fixtures, and equipment; Property with ADR midpoint life of 10 years or more and less than 16 years.
Ten-Year Property	Assets used in petroleum refining and preparation of certain food products; Vessels and water transportation equipment; Property with ADR midpoint life of 16 years or more and less than 20 years
Fifteen-Year Property	Telephone distribution plants; Municipal sewage treatment plants; Property with ADR midpoint life of 20 years or more and less than 25 years.
Twenty-Year Property	Municipal sewers; Property with ADR midpoint life of 25 years and more.
Property class	*Real property (real estate)*
27.5 Years	Residential rental property (does not include hotels and motels)
39 Years	Nonresidential real property

*The depreciation deduction for automobiles is limited to $2860 in the first tax year and is further reduced in subsequent years.

Half-Year Convention. Except for real property, a half-year convention is used. Under this convention all property is considered to be placed in service in the middle of the tax year, and a half year of depreciation is allowed in the first year. For each of the remaining years, one is allowed a full year of depreciation. If the property is disposed of prior to the end of the recovery period (property class life), a half year of depreciation is allowed in that year. If the property is held for the entire recovery period, a half year of depreciation is allowed for the year following the end of the recovery period (see Table A-4).

Table A-4. MACRS* Depreciation For Personal Property—Half-Year Convention

If the recovery year is:	*The applicable percentage for the class of property is:*			
	3-year class	5-year class	7-year class	10-year class
1	33.33	20.00	14.29	10.00
2	44.45	32.00	24.49	18.00
3	14.81†	19.20	17.49	14.40
4	7.41	11.52†	12.49	11.52
5		11.52	8.93†	9.22
6		5.76	8.92	7.37
7			8.93	6.55†
8			4.46	6.55
9				6.56
10				6.55
11				3.28

*In the *Fundamentals of Engineering Reference Handbook*, this table is called Modified ACRS Factors.

†Use straight-line depreciation for the year marked and all subsequent years.

Owing to the half-year convention, a general form of the double-declining-balance computation must be used to compute the year-by-year depreciation.

$$\text{DDB Depreciation in any year, } D_j = \frac{2}{n}(C - \text{Depreciation in years prior to } j)$$

Example 29

A \$5000 computer has an anticipated \$500 salvage value at the end of its five-year depreciable life. Compute the depreciation schedule for the machinery by a) Sum-Of-Years-Digits depreciation, and b) MACRS depreciation. Do the MACRS computation by hand, and then compare the results with the values from Table A-4.

Solution

a) Sum-of-Years-Digits depreciation

$$D_j = \frac{n-j+1}{\frac{n}{2}(n+1)}(C - S_n)$$

$$D_1 = \frac{5-1+1}{\frac{5}{2}(5+1)}(5000-500) = \$1500$$

$$D_2 = \frac{5-2+1}{\frac{5}{2}(5+1)}(5000-500) = \$1200$$

$$D_3 = \frac{5-3+1}{\frac{5}{2}(5+1)}(5000-500) = \quad 900$$

$$D_4 = \frac{5-4+1}{\frac{5}{2}(5+1)}(5000-500) = \quad 600$$

$$D_5 = \frac{5-5+1}{\frac{5}{2}(5+1)}(5000-500) = \quad 300$$

$4500

b) MACRS depreciation

Double declining balance with conversion to straight line. Five year property class. Half-year convention. Salvage value S_n is assumed to be zero for MACRS. Using the general DDB computation:

Year

1 (1/2 year) $D_1 = \frac{1}{2} \times \frac{2}{5}(5000-0) \quad = \$1000$

2 $D_2 = \frac{2}{5}(5000-1000) \quad = 1600$

3 $D_3 = \frac{2}{5}(5000-2600) \quad = 960$

4 $D_4 = \frac{2}{5}(5000-3560) \quad = 576$

5 $D_5 = \frac{2}{5}(5000-4136) \quad = 346$

6 (1/2 year) $D_6 = \frac{1}{2} \times \frac{2}{5}(5000-4482) = 104$

$4586

The computation must now be modified to convert to straight line depreciation at the point where the straight line depreciation will be larger. Using the alternate straight line computation:

$$D_5 = \frac{5000-4136-0}{1.5 \text{ years remaining}} = \$576$$

This is more than the $346 computed using DDB, hence switch to straight line for Year 5 and beyond.

$$D_6 \text{ (1/2 year)} = 1/2\ (576) = \$288$$

Answers:

Year	Depreciation SOYD	Depreciation MACRS
1	$1500	$1000
2	1200	1600
3	900	960
4	600	576
5	300	576
6	0	288
	$4500	$5000

The computed MACRS depreciation is identical with that obtained from Table A-4.

Tax Consequences

Income taxes represent another of the various kinds of disbursements encountered in an economic analysis. The starting point in an after-tax computation is the before-tax cash flow. Generally, the before-tax cash flow contains three types of entries:

1. Disbursements of money to purchase capital assets. These expenditures create no direct tax consequence for they are the exchange of one asset (money) for another (capital equipment).
2. Periodic receipts and/or disbursements representing operating income and/or expenses. These increase or decrease the year-by-year tax liability of the firm.
3. Receipts of money from the sale of capital assets, usually in the form of a salvage value when the equipment is removed. The tax consequences depend on the relationship between the book value (cost – depreciation taken) of the asset and its salvage value.

Situation	Tax Consequence
Salvage value > Book value	Capital gain on difference
Salvage value = Book value	No tax consequence
Salvage value < Book value	Capital loss on difference

After the before-tax cash flow, compute the depreciation schedule for any capital assets. Next, taxable income is the taxable component of the before-tax cash flow minus the depreciation. Then, the income tax is the taxable income times the appropriate tax rate. Finally, the after-tax cash flow is the before-tax cash flow adjusted for income taxes.

To organize these data, it is customary to arrange them in the form of a cash flow table, as follows:

Year	Before-tax cash flow	Depreciation	Taxable income	Income taxes	After-tax cash flow
0	•				•
1	•	•	•	•	•

Example 30

A corporation expects to receive $32,000 each year for 15 years from the sale of a product. There will be an initial investment of $150,000. Manufacturing and sales expenses will be $8067 per year. Assume straight line depreciation, a A-year useful life, and no salvage value. Use a 46% income tax rate. Determine the projected after-tax rate of return.

Solution

$$\text{Straight line depreciation, } D_j = \frac{C - S_n}{n} = \frac{150{,}000 - 0}{15} = \$10{,}000 \text{ per year}$$

Year	Before-tax cash flow	Depreciation	Taxable income	Income taxes	After-tax cash flow
0	−150,000				−150,000
1	+23,933	10,000	13,933	−6,409	+17,524
2	+23,933	10,000	13,933	−6,409	+17,524
•	•	•	•	•	•
•	•	•	•	•	•
•	•	•	•	•	•
15	+23,933	10,000	13,933	−6,409	+17,524

Take the after-tax cash flow and compute the rate of return at which PW of cost equals PW of benefits.

$$150{,}000 = 17{,}524\ (P/A,i\%,15)$$

$$(P/A,i\%,15) = \frac{150{,}000}{17{,}524} = 8.559$$

From the compound interest tables, the after-tax rate of return is $i = 8\%$.

Inflation

Inflation is characterized by rising prices for goods and services, while deflation produces a fall in prices. An inflationary trend makes future dollars have less purchasing power than present dollars. This helps long-term borrowers of money for they may repay a loan of present dollars in the future with dollars of reduced buying power. The help to borrowers is at the expense of lenders. Deflation has the opposite effect. Money borrowed at one point in time, followed by a deflationary period subjects the borrower to loan repayment with dollars of greater purchasing power than those he borrowed. This is to the lenders' advantage at the expense of borrowers.

Price changes occur in a variety of ways. One method of stating a price change is a uniform rate of price change per year.

f = General inflation rate per interest period
i = Effective interest rate per interest period

The following situation will illustrate the computations. A mortgage will be repaid in three equal payments of $5000 at the end of Years 1, 2, and 3. If the annual inflation rate, f, is 8% during this period, and the investor wishes a 12% annual interest rate (i), what is the maximum amount he would be willing to pay for the mortgage?

The computation is a two-step process. First, the three future payments must be converted into dollars with the same purchasing power as today's (Year 0) dollars.

Year	Actual Cash Flow		Multiplied by		Cash flow adjusted to today's (Yr. 0) dollars
0	—		—		—
1	+5000	×	$(1+0.08)^{-1}$	=	+4630
2	+5000	×	$(1+0.08)^{-2}$	=	+4286
3	+5000	×	$(1+0.08)^{-3}$	=	+3969

The general form of the adjusting multiplier is

$$(1+f)^{-n} \text{ which equals } (P/F,f,n)$$

Now that the problem has been converted to dollars of the same purchasing power (today's dollars in this example), we can proceed to compute the present worth of the future payments.

Year	Adjusted cash flow		Multiplied by		Present worth
0	—		—		—
1	+4630	×	$(1+0.12)^{-1}$	=	+4134
2	+4286	×	$(1+0.12)^{-2}$	=	+3417
3	+3969	×	$(1+0.12)^{-3}$	=	+2825
					$10,376

The general form of the discounting multiplier is

$$(1+i)^{-n} \text{ which equals } (P/F, i\%, n)$$

Alternate Solution

Instead of doing the inflation and interest rate computations separately, one can compute a combined equivalent interest rate, d.

$$d = (1+f)(1+i) - 1 = i + f + i(f)$$

For this cash flow, $d = 0.12 + 0.08 + 0.12(0.08) = 0.2096$. Since we do not have 20.96% interest tables, the problem has to be calculated using present worth equations.

$$\begin{aligned} \text{PW} &= 5000(1+0.2096)^{-1} + 5000(1+0.2096)^{-2} + 5000(1+0.2096)^{-3} \\ &= 4134 + 3417 + 2825 = \$10{,}376 \end{aligned}$$

Example 31

One economist has predicted that there will be a 7% per year inflation of prices during the next ten years. If this proves to be correct, an item that presently sells for $10 would sell for what price ten years hence?

Solution

$f = 7\%, \quad P = \$10$

$F = ? \quad n = 10$ years

Here the computation is to find the future worth F, rather than the present worth, P.

$$F = P(1+f)^{10} = 10(1+0.07)^{10} = \$19.67$$

Effect of Inflation on a Rate of Return

The effect of inflation on the computed rate of return for an investment depends on how future benefits respond to the inflation. If benefits produce constant dollars, which are not increased by inflation, the effect of inflation is to reduce the before-tax rate of return on the investment. If, on the other hand, the dollar benefits increase to keep up with the inflation, the before-tax rate of return will not be adversely affected by the inflation.

This is not true when an after-tax analysis is made. Even if the future benefits increase to match the inflation rate, the allowable depreciation schedule does not increase. The result will be increased taxable income and income tax payments. This reduces the available after-tax benefits and, therefore, the after-tax rate of return.

Example 32

A man bought a 5% tax-free municipal bond. It cost $1000 and will pay $50 interest each year for 20 years. The bond will mature at the end of 20 years and return the original $1000. If there is 2% annual inflation during this period, what rate of return will the investor receive after considering the effect of inflation?

Solution

$d = 0.05, \quad i = \text{unknown}, \quad j = 0.02$

$d = i + j + i(j)$

$0.05 = i + 0.02 + 0.02i$

$1.02i = 0.03, \quad i = 0.294 = 2.94\%$

Risk Analysis

Probability

Probability can be considered to be the long-run relative frequency of occurrence of an outcome. There are just two possible outcomes from flipping a coin (a Head or a Tail). If, for example, a coin is flipped over and over, we can expect in the long run that half the time Heads will appear and half the time Tails. We would say the probability of flipping a Head is 0.50 and of flipping a Tail is 0.50. Since the probabilities are defined so that the sum of probabilities for all possible outcomes is 1, the situation is

Probability of flipping a Head = 0.50
Probability of flipping a Tail = 0.50
Sum of all possible outcomes = 1.00

Example 33

If one were to roll one die (that is, one-half of a pair of dice), what is the probability that either a 1 or a 6 would result?

Solution

Since a die is a perfect six-sided cube, the probability of any side appearing is 1/6.

Probability of rolling a $1 = P(1) = 1/6$
$2 = P(2) = 1/6$
$3 = P(3) = 1/6$
$4 = P(4) = 1/6$
$5 = P(5) = 1/6$
$6 = P(6) = 1/6$

Sum of all possible outcomes = 6/6 = 1. The probability of rolling a 1 or a 6 = 1/6 + 1/6 = 1/3.

In the examples, the probability of each outcome was the same. This need not be the case.

Example 34

In the game of Blackjack, a perfect hand is a Ten or a facecard plus an Ace. What is the probability of being dealt a Ten or a facecard from a newly shuffled deck of 52 cards? What is the probability of being dealt an Ace in this same situation?

Solution

The three outcomes being examined are to be dealt a Ten or a facecard, an Ace, or some other card. Every card in the deck represents one of these three possible outcomes. There are 4 Aces; 16 Tens, Jacks, Queens, and Kings; and 32 other cards.

The probability of being dealt a Ten or a facecard = 16/52 = 0.31
The probability of being dealt an Ace = 4/52 = 0.08
The probability of being dealt some other card = 32/52 = 0.61
1.00

Risk

The term risk has a special meaning when it is used in statistics. It is defined as a situation where there are two or more possible outcomes and the probability associated with each outcome is known. In the two previous examples there is a risk situation. We could not know in advance what playing card would be dealt or what number would be rolled by the die. However, since the various probabilities could be computed, our definition of risk has been satisfied. Probability and risk are not restricted to gambling games. For example, in a particular engineering course, a student has computed the probability for each of the letter grades he might receive as follows:

Grade	Grade point	Probability P(grade)
A	4.0	0.10
B	3.0	0.30
C	2.0	0.25
D	1.0	0.20
F	0	0.15
		1.00

From the table we see that the grade with the highest probability is B. This, therefore, is the most likely grade. We also see that there is a substantial probability that some grade other than B will be received. And the probabilities indicate that if a B is not received, the grade will probably be something less than a B. But in saying the most likely grade is a B, other outcomes are ignored. In the next section we will show that a composite statistic may be computed using all the data.

Expected Value

In the last example the most likely grade of B in an engineering class had a probability of 0.30. That is not a very high probability. In some other course, say a math class, we might estimate a probability of 0.65 of obtaining a B, again making the B the most likely grade. While a B is most likely in both classes, it is more certain in the math class.

We can compute a weighted mean to give a better understanding of the total situation as represented by various possible outcomes. When the probabilities are used as the weighting factors, the result is called the *expected value* and is written:

$$\text{Expected value} = \text{Outcome}_A \times P(\text{A}) + \text{Outcome}_B \times P(\text{B}) + \ldots$$

Example 35

An engineer wishes to determine the risk of fire loss for her $200,000 home. From a fire rating bureau she obtains the following data:

Outcome	Probability
No fire loss	0.986 in any year
$10,000 fire loss	0.010
40,000 fire loss	0.003
200,000 fire loss	0.001

Compute the expected fire loss in any year.

Solution

Expected fire loss = 10,000 (0.010) + 40,000 (0.003) + 200,000 (0.001) = $420

Reference

Engineering Economic Analysis, Donald G. Newnan. Fifth edition, 1995. Engineering Press, Inc. (P.O. Box 1, San Jose, CA 95103-0001. Tel: 800-800-1651).

Problems and Solutions

A-1. A loan was made $2^1/_2$ years ago at 8% simple annual interest. The principal amount of the loan has just been repaid along with \$600 of interest. The principal amount of the loan was closest to

(a) \$300
(b) \$3000
(c) \$4000
(d) \$5000
(e) \$7500

Solution

$$F = P + Pin$$

$$600 + P = P + P(0.08)(2.50)$$

$$P = [600]/[0.08(2.50)] = \$3000$$

The answer is (b).

A-2. A \$1000 loan was made at 10% simple annual interest. It will take how many years for the amount of the loan and interest to equal \$1700?

(a) 6 years
(b) 7 years
(c) 8 years
(d) 9 years
(e) 10 years

Solution

$$F = P + Pin$$

$$1700 = 1000 + 1000(0.10)(n)$$

$$n = [700]/[1000(0.10)] = 7 \text{ years}$$

The answer is (b).

A-3. A retirement fund earns 8% interest, compounded quarterly. If \$400 is deposited every three months for 25 years, the amount in the fund at the end of 25 years is nearest to

(a) \$50,000
(b) \$75,000
(c) \$100,000
(d) \$125,000
(e) \$150,000

Solution

$$F = A(F/A,i\%,n) = 400(F/A,2\%,100)$$

$$= 400(312.23) = \$124,890$$

The answer is (d).

A-4. For some interest rate i, and some number of interest periods n, the uniform series capital recovery factor is 0.2091 and the sinking fund factor is 0.1941. The interest rate i must be closest to

(a) $1\frac{1}{2}\%$
(b) 2%
(c) 3%
(d) 4%
(e) 5%

Solution

The relationship between the capital recovery factor and the sinking fund factor is $(A/P,i\%,n) = (A/F,i\%,n) + i$. Substituting the values in the problem

$$0.2091 = 0.1941 + i$$
$$i = 0.2091 - 0.1941 = 0.015 = 1\tfrac{1}{2}\%$$

The answer is (a).

A-5. The repair costs for some handheld equipment is estimated to be $120 the first year, increasing by $30 per year in subsequent years. The amount a person will need to deposit into a bank account, paying 4% interest, to provide for the repair costs for the next five years is nearest to

(a) $500
(b) $600
(c) $700
(d) $800
(e) $900

Solution

$$\begin{aligned} P &= A(P/A,i\%,n) + G(P/G,i\%,n) \\ &= 120(P/A,4\%,5) + 30(P/G,4\%,5) \\ &= 120(4.452) + 30(8.555) = \$791 \end{aligned}$$

The answer is (d).

A-6. An "annuity" is defined as the

(a) earned interest due at the end of each interest period
(b) cost of producing a product or rendering a service
(c) total annual overhead assigned to a unit of production
(d) amount of interest earned by a unit of principal in a unit of time
(e) series of equal payments occurring at equal periods of time

Solution

The answer is (e).

A-7. One thousand dollars is borrowed for one year at an interest rate of 1% per month. If this same sum of money is borrowed for the same period at an interest rate of 12% per year, the saving in interest charges is closest to

(a) \$0
(b) \$3
(c) \$5
(d) \$7
(e) \$14

Solution

At i = 1%/month: $F = 1000(1 + 0.01)^{12} = \$1126.83$

At i = 12%/year: $F = 1000(1 + 0.12)^{1} = 1120.00$

Saving in interest charges = 1126.83 – 1120.00 = \$6.83

The answer is (d).

A-8. How much should a person invest in a fund that will pay 9%, compounded continuously, if he wishes to have \$10,000 in the fund at the end of 10 years? The amount is nearest to

(a) \$4000
(b) \$5000
(c) \$6000
(d) \$7000
(e) \$8000

Solution

$$P = Fe^{-rn} = 10{,}000e^{-0.09(10)} = 4066$$

The answer is (a).

A-9. A store charges $1\frac{1}{2}$ % interest per month on credit purchases. This is equivalent to a nominal annual interest rate of

(a) 1.5%
(b) 15.0%
(c) 18.0%
(d) 19.6%
(e) 21.0%

Solution

The nominal interest rate is the annual interest rate ignoring the effect of any compounding. Nominal interest rate = $1\frac{1}{2}\% \times 12 = 18\%$. The answer is (c).

A-10. A small company borrowed \$10,000 to expand its business. The entire principal of \$10,000 will be repaid in two years, but quarterly interest of \$330 must be paid every three months. The nominal annual interest rate the company is paying is closest to

(a) 3.3%
(b) 5.0%
(c) 6.6%
(d) 10.0%
(e) 13.2%

Solution

The interest paid per year = 330 × 4 = 1320. The nominal annual interest rate = 1320/10,000 = 0.132 = 13.2%. The answer is (e).

A-11. A store policy is to charge 3% interest every two months on the unpaid balance in charge accounts. The effective interest rate is closest to

(a) 6%
(b) 12%
(c) 15%
(d) 18%
(e) 19%

Solution

$i_{\text{eff}} = (1 + r/m)^m - 1 = (1 + 0.03)^6 - 1 = 0.194 = 19.4\%$

The answer is (e).

A-12. The effective interest rate is 19.56%. If there are 12 compounding periods per year, the nominal interest rate is closest to

(a) 1.5%
(b) 4.5%
(c) 9.0%
(d) 18.0%
(e) 19.6%

Solution

$i_{\text{eff}} = (1 + r/m)^m - 1$

$r/m = (1 + i_{\text{eff}})^{1/m} - 1 = (1 + 0.1956)^{1/12} - 1 = 0.015$

$r = 0.015(m) = 0.015 \times 12 = 0.18 = 18\%$

The answer is (d).

A-13. A deposit of $300 was made one year ago into an account paying monthly interest. If the account now has $320.52, the effective annual interest rate is closest to

(a) 7%
(b) 10%
(c) 12%
(d) 15%
(e) 18%

Solution

$i_{\text{eff}} = 20.52/300 = 0.0684 = 6.84\%$

The answer is (a).

A-14. In a situation where the effective interest rate per year is 12%, based on monthly compounding, the nominal interest rate per year is closest to

(a) 8.5%
(b) 9.3%
(c) 10.0%
(d) 11.4%
(e) 12.0%

Solution

$i_{eff} = (1 + r/m)^m - 1$

$0.12 = (1 + r/12)^{12} - 1$

$(1.12)^{1/12} = (1 + r/12)$

$1.00949 = (1 + r/12)$

$r = 0.00949 \times 12 = 0.1138 = 11.38\%$

The answer is (d).

A-15. If 10% nominal annual interest is compounded daily, the effective annual interest rate is nearest to

(a) 10.00%
(b) 10.38%
(c) 10.50%
(d) 10.75%
(e) 18.00%

Solution

$i_{eff} = (1 + r/m)^m - 1 = (1 + 0.10/365)^{365} - 1 = 0.1052 = 10.52\%$

The answer is (c).

A-16. If 10% nominal annual interest is compounded continuously, the effective annual interest rate is nearest to

(a) 10.00%
(b) 10.38%
(c) 10.50%
(d) 10.75%
(e) 18.00%

Solution

$i_{eff} = e^r - 1$

where r = nominal annual interest rate

$i_{eff} = e^{0.10} - 1 = 0.10517 = 10.52\%$

The answer is (c).

A-17. If the quarterly effective interest rate is $5^1/_2\%$ with continuous compounding, the nominal interest rate is nearest to

(a) 5.5%
(b) 11.0%
(c) 16.5%
(d) 21.4%
(e) 22.0%

Solution

For 3 months: $i_{eff} = e^r - 1$; $0.055 = e^r - 1$

The rate per quarter year is $r = \log_e(1.055) = 0.05354$; $r = 4 \times 0.05354 = 0.214 = 21.4\%$ per year

The answer is (d).

A-18. A continuously compounded loan has what effective interest rate if the nominal interest rate is 25%? Select one of the five choices.

(a) $e^{1.25}$
(b) $e^{0.25}$
(c) $\log_e(1.25)$
(d) $\log_e(0.25)$
(e) $e^{0.25} - 1$

Solution

$i_{eff} = e^r - 1 = e^{0.25} - 1$

The answer is (e).

A-19. A continuously compounded loan has what *nominal interest rate* if the *effective interest rate* is 25%? Select one of the five choices.

(a) $e^{1.25}$
(b) $e^{0.25}$
(c) $\log_e(1.25)$
(d) $\log_e(0.25)$
(e) $\log_{10}(1.25)$

Solution

$i_{eff} = e^r - 1 = 0.25 \quad e^r = 1.25$

$\log_e(e^r) = \log_e(1.25)$

$r = \log_e(1.25)$

The answer is (c).

A-20. An individual wishes to deposit a certain quantity of money now so that he will have \$500 at the end of five years. With interest at 4% per year, compounded semiannually, the amount of the deposit is nearest to

(a) \$340
(b) \$400
(c) \$410
(d) \$416
(e) \$608

Solution

$P = F(P/F,i\%,n) = 500(P/F,2\%,10) = 500(0.8203) = \$410$

The answer is (c).

A-21. A steam boiler is purchased on the basis of guaranteed performance. A test indicates that the operating cost will be \$300 more per year than the manufacturer guaranteed. If the expected life of the boiler is 20 years, and money is worth 8%, the amount the purchaser should deduct from the purchase price to compensate for the extra operating cost is nearest to

(a) $2950 (d) $5520

(b) $3320 (e) $6000

(c) $4100

Solution

$P = 300(P/A,8\%,20) = 300(9.818) = \$2945$

The answer is (a).

A-22. A consulting engineer bought a fax machine. There will be no maintenance cost the first year as it was sold with one year's free maintenance. In the second year the maintenance is estimated at $20. In subsequent years the maintenance cost will increase $20 per year (that is, 3rd year maintenance will be $40, 4th year maintenance will be $60, and so forth). The amount that must be set aside now at 6% interest to pay the maintenance costs on the fax machine for the first six years of ownership is nearest to

(a) $101 (d) $284

(b) $164 (e) $300

(c) $229

Solution

Using single payment present worth factors:

$P = 20(P/F,6\%,2) + 40(P/F,6\%,3) + 60(P/F,6\%,4) + 80(P/F,6\%,5) + 100(P/F,6\%,6) = \$229$

Alternate solution using the gradient present worth factor:

$P = 20(P/G,6\%,6) = 20(11.459) = \$229$

The answer is (c).

A-23. An investor is considering buying a 20-year corporate bond. The bond has a face value of $1000 and pays 6% interest per year in two semiannual payments. Thus the purchaser of the bond will receive $30 every six months, and in addition he will receive $1000 at the end of 20 years, along with the last $30 interest payment. If the investor believes he should receive 8% annual interest, compounded semiannually, the amount he is willing to pay for the bond value is closest to

(a) $500 (d) $800

(b) $600 (e) $900

(c) $700

Solution

$PW = 30(P/A,4\%,40) + 1000(P/F,4\%,40) = 30(19.793) + 1000(0.2083) = \$802$

The answer is (d).

A-24. Annual maintenance costs for a particular section of highway pavement are $2000. The placement of a new surface would reduce the annual maintenance cost to $500 per year for the first five years and to $1000 per year for the next five years. The annual maintenance after ten years would again be $2000. If maintenance costs are the only saving, the maximum investment that can be justified for the new surface, with interest at 4%, is closest to

(a) $5,500
(b) $7,170
(c) $10,000
(d) $10,340
(e) $12,500

Solution

Benefits are $1500 per year for the first five years and $1000 per year for the subsequent five years.

As Fig. A-24 indicates, the benefits may be considered as $1000 per year for ten years, plus an additional $500 benefit in each of the first five years.

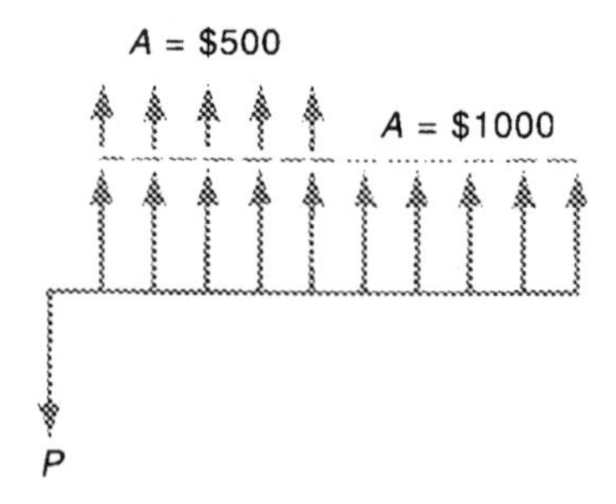

Fig. A-24

Maximum investment = Present worth of benefits
= 1000(*P*/*A*,4%,10) + 500(*P*/*A*,4%,5)
= 1000((8.111) + 500(4.452) = $10,337

The answer is (d).

A-25. A project has an initial cost of $10,000, uniform annual benefits of $2400, and a salvage value of $3000 at the end of its 10-year useful life. At 12% interest the net present worth of the project is closest to

(a) $2500
(b) $3500
(c) $4500
(d) $5500
(e) $6500

Solution

NPW = PW of benefits – PW of cost
= 2400(*P*/*A*,12%,10) + 3000(*P*/*F*,12%,10) – 10,000 = $4526

The answer is (c).

A-26. A person borrows $5000 at an interest rate of 18%, compounded monthly. Monthly payments of $167.10 are agreed upon. The length of the loan is closest to

(a) 12 months
(b) 20 months
(c) 24 months
(d) 30 months
(e) 40 months

Solution

PW of benefits = PW of cost
5000 = 167.10(*P*/*A*,1.5%,*n*)
(*P*/*A*,1.5%,*n*) = 5000/167.10 = 29.92

From the $1\frac{1}{2}$% interest table, $n = 40$. The answer is (e).

A-27. A machine costing $2000 to buy and $300 per year to operate will save labor expenses of $650 per year for eight years. The machine will be purchased if its salvage value at the end of eight years is sufficiently large to make the investment economically attractive. If an interest rate of 10% is used, the minimum salvage value must be closest to

(a) $100
(b) $200
(c) $300
(d) $400
(e) $500

Solution

$$\begin{aligned} \text{NPW} &= \text{PW of benefits} - \text{PW of cost} = 0 \\ &= (650 - 300)(P/A,10\%,8) + S_8(P/F,10\%,8) - 2000 = 0 \\ &= 350(5.335) + S_8(0.4665) - 2000 = 0 \end{aligned}$$

$$S_8 = 132.75/0.4665 = \$285$$

The answer is (c).

A-28. The amount of money deposited 50 years ago at 8% interest that would now provide a perpetual payment of $10,000 per year is nearest to

(a) $3,000
(b) $8,000
(c) $50,000
(d) $70,000
(e) $90,000

Solution

The amount of money needed now to begin the perpetual payments is $P' = A/i = 10{,}000/0.08 = 125{,}000$. From this we can compute the amount of money, P, that would need to have been deposited 50 years ago:

$$P = 125{,}000(P/F,8\%,50) = 125{,}000(0.0213) = \$2663$$

The answer is (a).

A-29. An industrial firm must pay a local jurisdiction the cost to expand its sewage treatment plant. In addition, the firm must pay $12,000 annually toward the plant operating costs. The industrial firm will pay sufficient money into a fund, that earns 5% per year, to pay its share of the plant operating costs forever. The amount to be paid to the fund is nearest to

(a) $15,000
(b) $30,000
(c) $60,000
(d) $120,000
(e) $240,000

Solution

$$P = A/i = 12000/0.05 = \$240{,}000$$

The answer is (e).

A-30. At an interest rate of 2% per month, money will double in value in how many months?

(a) 20 months
(b) 22 months
(c) 24 months
(d) 30 months
(e) 35 months

Solution

$2 = 1(F/P,i\%,n)$

$(F/P,2\%,n) = 2$

From the 2% interest table, n = about 35 months. The answer is (e).

A-31. A woman deposited $10,000 into an account at her credit union. The money was left on deposit for 80 months. During the first 50 months the woman earned 12% interest, compounded monthly. The credit union then changed its interest policy so that the woman earned 8% interest compounded quarterly during the next 30 months. The amount of money in the account at the end of 80 months is nearest to

(a) $10,000
(b) $12,500
(c) $15,000
(d) $17,500
(e) $20,000

Solution

At end of 50 months

$$F = 10{,}000(F/P,1\%,50) = 10{,}000(1.645) = \$16{,}450$$

At end of 80 months

$$F = 16{,}450(F/P,2\%,10) = 16{,}450(1.219) = \$20{,}053$$

The answer is (e).

A-32. An engineer deposited $200 quarterly in her savings account for three years at 6% interest, compounded quarterly. Then for five years she made no deposits or withdrawals. The amount in the account after eight years is closest to

(a) $1200
(b) $1800
(c) $2400
(d) $3000
(e) $3600

Solution

$$\text{FW} = 200(F/A,1.5\%,12)(F/P,1.5\%,20)$$
$$= 200(13.041)(1.347) = \$3513$$

The answer is (e).

A-33. A sum of money, Q, will be received six years from now. At 6% annual interest the present worth now of Q is \$60. At this same interest rate the value of Q ten years from now is closest to

(a) \$60
(b) \$77
(c) \$90
(d) \$107
(e) \$120

Solution

The present sum $P = 60$ is equivalent to Q six years hence at 6% interest. The future sum F may be calculated by either of two methods:

$$F = Q(F/P,6\%,4) \text{ and } Q = 60\,(F/P,6\%,6) \quad (1)$$

$$F = P(F/P,6\%,10) \quad (2)$$

Since P is known, the second equation may be solved directly.

$$F = P(F/P,6\%,10) = 60(1.791) = \$107$$

The answer is (d).

A-34. If \$200 is deposited in a savings account at the beginning of each of 15 years and the account earns interest at 6%, compounded annually, the value of the account at the end of 15 years will be most nearly

(a) \$4500
(b) \$4700
(c) \$4900
(d) \$5100
(e) \$5300

Solution

$$F' = A(F/A,i\%,n) = 200(F/A,6\%,15) = 200(23.276) = \$4655.20$$

$$F = F'(F/P,i\%,n) = 4655.20(F/P,6\%,1) = 4655.20(1.06) = \$4935$$

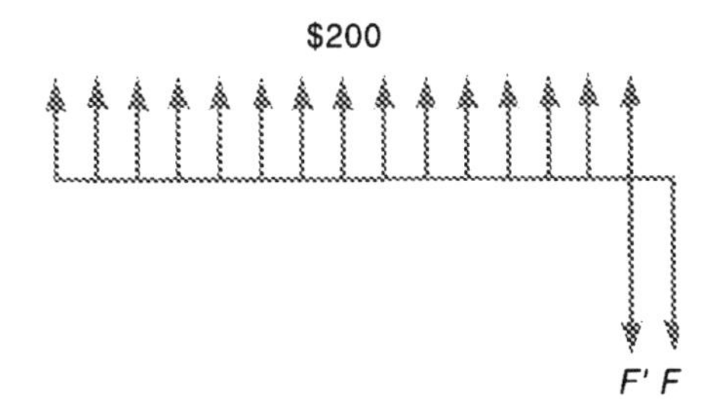

Fig. A-34

The answer is (c).

A-35. The maintenance expense on a piece of machinery is estimated as follows:

Year	1	2	3	4
Maintenance	\$150	\$300	\$450	\$600

If interest is 8%, the equivalent uniform annual maintenance cost is closest to

(a) \$250 (d) \$400

(b) \$300 (e) \$450

(c) \$350

Solution

$$\text{EUAC} = 150 + 150(A/G,8\%,4) = 150 + 150(1.404) = \$361$$

The answer is (c).

A-36. A payment of \$12,000 six years from now is equivalent, at 10% interest, to an annual payment for eight years starting at the end of this year. The annual payment is closest to

(a) \$1000 (d) \$1600

(b) \$1200 (e) \$1800

(c) \$1400

Solution

$$\begin{aligned}\text{Annual payment} &= 12{,}000(P/F,10\%,6)(A/P,10\%,8)\\ &= 12{,}000(0.5645)(0.1874) = \$1269\end{aligned}$$

The answer is (b).

A-37. A manufacturer purchased \$15,000 worth of equipment with a useful life of six years and a \$2000 salvage value at the end of the six years. Assuming a 12% interest rate, the equivalent uniform annual cost is nearest to

(a) \$1500 (d) \$4500

(b) \$2500 (e) \$5500

(c) \$3500

Solution

$$\begin{aligned}\text{EUAC} &= 15{,}000(A/P,12\%,6) - 2000(A/F,12\%,6)\\ &= 15{,}000(0.2432) - 2000(0.1232) = \$3402\end{aligned}$$

The answer is (c).

A-38. Consider a machine as follows:

Initial cost: \$80,000

End-of-useful life salvage value: \$20,000

Annual operating cost: \$18,000

Useful life: 20 years

Based on 10% interest, the equivalent uniform annual cost for the machine is closest to

(a) \$21,000
(b) \$23,000
(c) \$25,000
(d) \$27,000
(e) \$29,000

Solution

$$\begin{aligned} \text{EUAC} &= 80{,}000\,(A/P,10\%,20) - 20{,}000\,(A/F,10\%,20) + \text{annual operating cost} \\ &= 80{,}000\,(0.1175) - 20{,}000\,(0.0175) + 18{,}000 \\ &= 9400 - 350 + 18{,}000 = \$27{,}050 \end{aligned}$$

The answer is (d).

A-39. Consider a machine as follows:

Initial cost: \$80,000

Annual operating cost: \$18,000

Useful life: 20 years

What must be the salvage value of the machine at the end of 20 years for the machine to have an equivalent uniform annual cost of \$27,000? Assume a 10% interest rate. The salvage value is closest to

(a) \$10,000
(b) \$20,000
(c) \$30,000
(d) \$40,000
(e) \$50,000

Solution

$$\begin{aligned} \text{EUAC} &= \text{EUAB} \\ 27{,}000 &= 80{,}000(A/P,10\%,20) + 18{,}000 - S(A/F,10\%,20) \\ &= 80{,}000(0.1175) + 18{,}000 - S(0.0175) \end{aligned}$$

$$S = (27{,}400 - 27{,}000)/0.0175 = \$22{,}857$$

The answer is (b).

A-40. Twenty-five thousand dollars is deposited in a savings account that pays 5% interest, compounded semiannually. Equal annual withdrawals are to be made from the account beginning one year from now and continuing forever. The maximum amount of the equal annual withdrawals is closest to

(a) \$625
(b) \$1000
(c) \$1250
(d) \$1265
(e) \$1365

Solution

The general equation for an infinite life, $P = A/i$, must be used to solve the problem.

$i_{eff} = (1 + 0.025)^2 - 1 = 0.050625$

The maximum annual withdrawal will be $A = Pi = 25{,}000(0.050625) = \$1266$

The answer is (d).

A-41. An investor is considering the investment of $10,000 in a piece of land. The property taxes are $100 per year. The lowest selling price the investor must receive if she wishes to earn a 10% interest rate after keeping the land for 10 years is

(a) $20,000
(b) $21,000
(c) $23,000
(d) $25,000
(e) $27,000

Solution

$$\text{Minimum sale price} = 10{,}000(F/P,10\%,10) + 100(F/A,10\%,10)$$
$$= 10{,}000(2.594) + 100(15.937) = \$27{,}530$$

The answer is (e).

A-42. The rate of return for a $10,000 investment that will yield $1000 per year for 20 years is closest to

(a) 1%
(b) 4%
(c) 8%
(d) 12%
(e) 18%

Solution

$NPW = 1000(P/A,i\%,20) - 10{,}000 = 0$

$(P/A,i\%,20) = 10{,}000/1000 = 10$

From interest tables: $6\% < i < 8\%$. The answer is (c).

A-43. An engineer invested $10,000 in a company. In return he received $600 per year for six years and his $10,000 investment back at the end of the six years. His rate of return on the investment was closest to

(a) 6%
(b) 10%
(c) 12%
(d) 15%
(e) 18%

Solution

The rate of return was = 600/10,000 = 0.06 = 6%

The answer is (a).

A-44. An engineer made ten annual end-of-year purchases of $1000 of common stock. At the end of the tenth year, just after the last purchase, the engineer sold all the stock for $12,000. The rate of return received on the investment is closest to

(a) 2%
(b) 4%
(c) 8%
(d) 10%
(e) 12%

Solution

$$F = A(F/A,i\%,n)$$

$$12{,}000 = 1000(F/A,i\%,10)$$

$$(F/A,i\%,10) = 12{,}000/1000 = 12$$

In the 4% interest table: $(F/A,4\%,10) = 12.006$, so $i = 4\%$. The answer is (b).

A-45. A company is considering buying a new piece of machinery.

Initial cost: $80,000

End-of-useful life salvage value: $20,000

Annual operating cost: $18,000

Useful life: 20 years

The machine will produce an annual saving in material of $25,700. What is the before-tax rate of return if the machine is installed? The rate of return is closest to

(a) 6%
(b) 8%
(c) 10%
(d) 15%
(e) 20%

Solution

PW of cost = PW of benefits

$$80{,}000 = (25{,}700 - 18{,}000)(P/A,i\%,20) + 20{,}000(P/F,i\%,20)$$

Try $i = 8\%$

$$80{,}000 = 7700(9.818) + 20{,}000(0.2145) = 79{,}889$$

Therefore, the rate of return is very close to 8%. The answer is (b).

A-46. Consider the following situation: Invest $100 now and receive two payments of $102.15—one at the end of Year 3, and one at the end of Year 6. The rate of return is nearest to

(a) 6%
(b) 8%
(c) 10%
(d) 12%
(e) 18%

Solution

PW of cost = PW of benefits

$$100 = 102.15(P/F,i\%,3) + 102.15(P/F,i\%,6)$$

Solve by trial and error:

Try $i = 12\%$

$$100 = 102.15(0.7118) + 102.15(0.5066) = 124.46$$

The PW of benefits exceeds the PW of cost. This indicates that the interest rate i is too low.

Try $i = 18\%$

$$100 = 102.15(0.6086) + 102.15(0.3704) = 100.00$$

Therefore, the rate of return is 18%. The answer is (e).

A-47. Two mutually exclusive alternatives are being considered:

Year	A	B
0	–$2500	–$6000
1	+746	+1664
2	+746	+1664
3	+746	+1664
4	+746	+1664
5	+746	+1664

The rate of return on the difference between the alternatives is closest to

(a) 6% (d) 12%
(b) 8% (e) 15%
(c) 10%

Solution

The difference between the alternatives:

Incremental cost = 6000 – 2500 = $3500

Incremental annual benefit = 1664 – 746 = $918

PW of cost = PW of benefits

$$3500 = 918(P/A,i\%,5)$$

$$(P/A,i\%,5) = 3500/918 = 3.81$$

From the interest tables, i is very close to 10%. The answer is (c).

A-48. A project will cost $50,000. The benefits at the end of the first year are estimated to be $10,000, increasing $1000 per year in subsequent years. Assuming a 12% interest rate, no salvage value, and an eight-year analysis period, the Benefit-Cost ratio is closest to

(a) 0.78 (d) 1.45
(b) 1.00 (e) 1.60
(c) 1.28

Solution

$$B/C = \frac{\text{PW of benefits}}{\text{PW of cost}} = \frac{10{,}000(P/A,12\%,8) + 1000(P/G,12\%,8)}{50{,}000}$$

$$= \frac{10{,}000(4.968) + 1000(14.471)}{50{,}000} = 1.28$$

The answer is (c).

A-49. Two alternatives are being considered.

	A	B
Initial cost:	\$500	\$800
Uniform annual benefit:	\$140	\$200
Useful life, years:	8	8

The Benefit-Cost ratio of the difference between the alternatives, based on a 12% interest rate, is closest to

(a) 0.60
(b) 0.80
(c) 1.00
(d) 1.20
(e) 1.40

Solution

$$B/C = \frac{\text{PW of benefits}}{\text{PW of cost}} = \frac{60(P/A,12\%,8)}{300} = \frac{60(4.968)}{300} = 0.99$$

Alternate Solution:

$$B/C = \frac{\text{EUAB}}{\text{EUAC}} = \frac{60}{300(A/P,12\%,8)} = \frac{60}{300(0.2013)} = 0.99$$

The answer is (c).

A-50. An engineer will invest in a mining project if the Benefit-Cost ratio is greater than one, based on an 18% interest rate. The project cost is \$57,000. The net annual return is estimated at \$14,000 for each of the next eight years. At the end of eight years the mining project will be worthless. The Benefit-Cost ratio is closest to

(a) 1.00
(b) 1.05
(c) 1.21
(d) 1.57
(e) 1.96

Solution

$$B/C = \frac{\text{PW of benefits}}{\text{PW of cost}} = \frac{14{,}000(P/A,18\%,8)}{57{,}000} = \frac{14{,}000(4.078)}{57{,}000} = 1.00$$

The answer is (a).

A-51. A city has retained your firm to do a Benefit-Cost analysis of the following project:

Project cost: \$60,000,000

Gross income: \$20,000,000 per year

Operating costs: \$5,500,000 per year

Salvage value after 10 years: None

The project life is ten years. Use 8% interest in the analysis. The computed Benefit-Cost ratio is closest to

(a) 0.80 (d) 1.40
(b) 1.00 (e) 1.60
(c) 1.20

Solution

$$B/C = \frac{EUAB}{EUAC} = \frac{20{,}000{,}000 - 5{,}500{,}000}{60{,}000{,}000(A/P,8\%,10)} = 1.62$$

The answer is (e).

A-52. A piece of property is purchased for \$10,000 and yields a \$1000 yearly profit. If the property is sold after five years, the minimum price to break even, with interest at 6%, is closest to

(a) \$5000 (d) \$8300
(b) \$6500 (e) \$9700
(c) \$7700

Solution

$$F = 10{,}000(F/P,6\%,5) - 1000(F/A,6\%,5)$$
$$= 10{,}000(1.338) - 1000(5.637) = \$7743$$

The answer is (c).

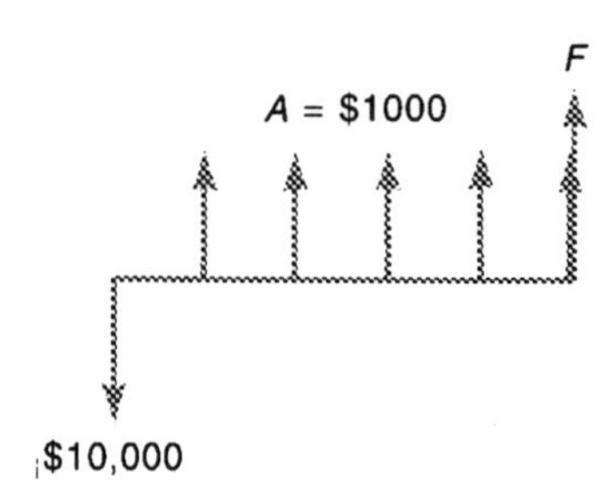

Fig. A-52

A-53. Given two machines:

	A	B
Initial cost:	\$55,000	\$75,000
Total annual costs:	\$16,200	\$12,450

With interest at 10% per year, at what service life do these two machines have the same equivalent uniform annual cost? The service life is closest to

(a) 4 years (d) 7 years
(b) 5 years (e) 8 years
(c) 6 years

Solution

$$\text{PW of cost}_A = \text{PW of cost}_B$$
$$55{,}000 + 16{,}200(P/A,10\%,n) = 75{,}000 + 12{,}450(P/A,10\%,n)$$
$$(P/A,10\%,n) = (75{,}000 - 55{,}000)/(16{,}200 - 12{,}450) = 5.33$$

From the 10% interest tables, $n = 8$ years. The answer is (e).

A-54. A machine part that is operating in a corrosive atmosphere is made of low-carbon steel. It costs $350 installed, and lasts six years. If the part is treated for corrosion resistance it will cost $700 installed. How long must the treated part last to be as economic as the untreated part, if money is worth 6%?

(a) 8 years
(b) 11 years
(c) 15 years
(d) 17 years
(e) 20 years

Solution

$$\text{EUAC}_{\text{untreated}} = \text{EUAC}_{\text{treated}}$$

$$350(A/P,6\%,6) = 700(A/P,6\%,n)$$

$$350(0.2034) = 700(A/P,6\%,n)$$

$$(A/P,6\%,n) = 71.19/700 = 0.1017$$

From the 6% interest table, n = 15+ years. The answer is (c).

A-55. A firm has determined the two best paints for its machinery are Tuff-Coat at $45 per gallon and Quick at $22 per gallon. The Quick paint is expected to prevent rust for five years. Both paints take $40 of labor per gallon to apply, and both cover the same area. If a 12% interest rate is used, how long must the Tuff-Coat paint prevent rust to justify its use?

(a) 5 years
(b) 6 years
(c) 7 years
(d) 8 years
(e) 9 years

Solution

$$\text{EUAC}_{\text{T-C}} = \text{EUAC}_{\text{Quick}}$$

$$(45 + 40)(A/P,12\%,n) = (22 + 40)(A/P,12\%,5)$$

$$(\text{A/P},12\%,n) = 17.20/85 = 0.202$$

From the 12% interest table, n = 8. The answer is (d).

A-56. Two alternatives are being considered:

	A	B
Cost:	$1000	$2000
Useful life in years:	10	10
End-of-useful-life salvage value:	100	400

The net annual benefit of *A* is \$150. If interest is 8%, what must be the net annual benefit of *B* for the two alternatives to be equally desirable? The net annual benefit of *B* must be closest to

(a) \$150
(b) \$200
(c) \$225
(d) \$275
(e) \$325

Solution

At breakeven,

$$NPW_A = NPW_B$$

$$150(P/A,8\%,10) + 100(P/F,8\%,10) - 1000 = NAB(P/A,8\%,10) + 400(P/F,8\%,10) - 2000$$

$$52.82 = 6.71(NAB) - 1814.72$$

Net annual benefit (NAB) = (1814.72 + 52.82)/6.71 = \$278. The answer is (d).

A-57. Which one of the following is *NOT* a method of depreciating plant equipment for accounting and engineering economics purposes?

(a) Double-entry method
(b) Modified accelerated cost recovery system
(c) Sum-of-years-digits method
(d) Straight-line method
(e) Sinking-fund method

Solution

"Double entry" probably is a reference to double entry accounting. It is not a method of depreciation. The answer is (a).

A-58. A machine costs \$80,000, has a 20-year useful life, and an estimated \$20,000 end-of-useful-life salvage value. Assuming Sum-Of-Years-Digits depreciation, the book value of the machine after two years is closest to

(a) \$21,000
(b) \$42,000
(c) \$59,000
(d) \$69,000
(e) \$79,000

Solution

Sum-Of-Years-Digits depreciation:

$$D_j = \frac{n-j+1}{\frac{n}{2}(n+1)}(C - S_n)$$

$$D_1 = \frac{20-1+1}{\frac{20}{2}(20+1)}(80{,}000 - 20{,}000) = 5714$$

$$D_2 = \frac{20-2+1}{\frac{20}{2}(20+1)}(80{,}000 - 20{,}000) = 5429$$

Total: \$11,143

$$\text{Book value} = \text{Cost} - \text{Depreciation to date}$$
$$= 80{,}000 - 11{,}143 = \$68{,}857$$

The answer is (d).

A-59. A machine costs $100,000. After its 25-year useful life, its estimated salvage value is $5,000. Based on Double-Declining-Balance depreciation, what will be the book value of the machine at the end of three years? The book value is closest to

(a) $16,000
(b) $22,000
(c) $58,000
(d) $78,000
(e) $83,000

Solution

Double-Declining-Balance depreciation:

$$BV_j = C\left(1 - \frac{2}{n}\right)^j$$

$$BV_3 = 100{,}000\left(1 - \frac{2}{25}\right)^3 = \$77{,}869$$

The answer is (d).

Questions A-60 to A-63

Special tools for the manufacture of finished plastic products cost $15,000 and have an estimated $1000 salvage value at the end of an estimated three-year useful life.

A-60. The third-year straight line depreciation is closest to

(a) $3000
(b) $3500
(c) $4000
(d) $4500
(e) $5000

Solution

$D_3 = (C - S)/n = (15{,}000 - 1000)/3 = \$4666$

The answer is (d).

A-61. The first year Modified-Accelerated-Cost-Recovery-System (MACRS) depreciation is closest to

(a) $3000
(b) $3500
(c) $4000
(d) $4500
(e) $5000

Solution

The half-year convention applies here; Double-declining balance must be used with an assumed salvage value of zero. In general,

$$D_j = \frac{2C}{n}\left(1-\frac{2}{n}\right)^{j-1}$$

For a half-year in Year 1:

$$D_1 = \frac{1}{2}\times\frac{2\times 15,000}{3}\left(1-\frac{2}{3}\right)^{1-1} = \$5000$$

The answer is (e).

A-62. The second-year Sum-Of-Years-Digits (SOYD) depreciation is closest to

(a) \$3000
(b) \$3500
(c) \$4000
(d) \$4500
(e) \$5000

Solution

$$D_j = \frac{n-j+1}{\frac{n}{2}(n+1)}(C-S_n)$$

$$D_2 = \frac{3-2+1}{\frac{3}{2}(3+1)}(15,000-1000) = \$4667$$

The answer is (d).

A-63. The second-year sinking-fund depreciation, based on 8% interest, is nearest to

(a) \$3000
(b) \$3500
(c) \$4000
(d) \$4500
(e) \$5000

Solution

$$\begin{aligned} D_2 &= (15,000 - 1000)(A/F,8\%,3)(F/P,8\%,1) \\ &= 14,000(0.3080)(1.08) = \$4657 \end{aligned}$$

The answer is (d).

A-64. An individual, who has a 28% incremental income tax rate, is considering purchasing a \$1000 taxable corporation bond. As the bondholder he will receive \$100 a year in interest and his \$1000 back when the bond becomes due in six years. This individual's after-tax rate of return from the bond is nearest to

(a) 6%
(b) 7%
(c) 8%
(d) 9%
(e) 10%

Solution

Twenty-eight percent of the $100 interest income must be paid in taxes. The balance of $72 is the after-tax income. Thus the after-tax rate of return = 72/1000 = 0.072 = 7.2%. The answer is (b).

A-65. A $20,000 investment in equipment will produce $6000 of net annual benefits for the next eight years. The equipment will be depreciated by straight-line depreciation over its eight year useful life. The equipment has no salvage value. Assuming a 34% income tax rate, the after-tax rate of return for this investment is closest to

(a) 6% (d) 12%
(b) 8% (e) 18%
(c) 10%

Solution

Year	Before-tax cash flow	SL deprec	Taxable income	34% Income taxes	After-tax cash flow
0	–$20,000				–$20,000
1-8	+6000	2500	3500	1190	+4810

$$D_j = (P - S)/n = \frac{20{,}000 - 0}{8} = 2500$$

$$\text{PW of cost} = \text{PW of benefits}$$

$$20{,}000 = 4810\ (P/A,i\%,8)$$

$$(P/A,i\%,8) = 20{,}000/4810 = 4.158$$

From interest tables, i = 18%. The answer is (e).

A-66. An individual bought a one-year savings certificate for $10,000, and it pays 6%. He has a taxable income that puts him at the 28% incremental income tax rate. His after-tax rate of return on this investment is closest to

(a) 2% (d) 5%
(b) 3% (e) 6%
(c) 4%

Solution

Additional taxable income = 0.06(10,000) = 600

Additional income tax = 0.28(600) = 168

After-tax income = 600 – 168 = 432

After-tax rate of return = 432/10,000 = 0.043 = 4.3%

Alternate Solution

After-tax rate of return = (1 – Incremental tax rate)(Before-tax rate of return)

$$= (1 - 0.28)(0.06) = 0.043 = 4.3\%$$

The answer is (c).

A-67. A tool costing $300 has no salvage value. Its resulting before-tax cash flow is shown in the following partially completed cash flow table.

Year	Before-tax cash flow	Effect on SOYD Deprec	Effect on taxable income	Income taxes	After-tax cash flow
0	–$300				
1	+100				
2	+150				
3	+200				

The tool is to be depreciated over three years using Sum-Of-Years-Digits depreciation. The income tax rate is 50%. The after-tax rate of return is nearest to

(a) 8%
(b) 10%
(c) 12%
(d) 15%
(e) 20%

Solution

Year	Before-tax cash flow	Effect on SOYD Deprec	Effect on taxable income	50% Income taxes	After-tax cash flow
0	–$300				–$300
1	+100	–150	–50	+25	+125
2	+150	–100	+50	–25	+125
3	+200	–50	+150	–75	+125

For the after-tax cash flow:

$$\text{PW of cost} = \text{PW of benefits}$$

$$300 = 125(P/A,i\%,3)$$

$$(P/A,i\%,3) = 300/125 = 2.40$$

From the interest tables, we find that i (the after-tax rate of return) is close to 12%. The answer is (c).

A-68. An engineer is considering the purchase of an annuity that will pay $1000 per year for ten years. The engineer feels he should obtain a 5% rate of return on the annuity after considering the effect of an estimated 6% inflation per year. The amount he would be willing to pay to purchase the annuity is closest to

(a) \$1500
(b) \$3000
(c) \$4500
(d) \$6000
(e) \$7500

Solution

$d = i + f + if = 0.05 + 0.06 + 0.05(0.06) = 0.113 = 11.3\%$

$$P = A(P/A, 11.3\%, 10) = 1000\left[\frac{(1+0.113)^{10}-1}{0.113(1+0.113)^{10}}\right] = 1000\left[\frac{1.9171}{0.3296}\right] = \$5816$$

The answer is (d).

A-69. An automobile costs \$20,000 today. You can earn 12% tax free on an "auto purchase account." If you expect the cost of the auto to increase by 10% per year, the amount you would need to deposit in the account to provide for the purchase of the auto five years from now is closest to

(a) \$12,000
(b) \$14,000
(c) \$16,000
(d) \$18,000
(e) \$20,000

Solution

$$\begin{aligned}\text{Cost of auto 5 years hence } (F) &= P(1 + \text{inflation rate})^n \\ &= 20{,}000(1 + 0.10)^5 = 32{,}210\end{aligned}$$

Amount to deposit now to have \$32,210 available 5 years hence:

$P = F(P/F,i\%,n) = 32{,}210(P/F,12\%,5) = 32{,}210(0.5674) = \$18{,}276$

The answer is (d).

A-70. An engineer purchases a building lot for \$40,000 cash and plans to sell it after five years. If he wants an 18% before-tax rate of return, after taking the 6% annual inflation rate into account, the selling price must be nearest to

(a) \$55,000
(b) \$65,000
(c) \$75,000
(d) \$100,000
(e) \$125,000

Solution

$$\begin{aligned}\text{Selling price } (F) &= 40{,}000(F/P,18\%,5)(F/P,6\%,5) \\ &= 40{,}000(2.288)(1.338) = \$122{,}500\end{aligned}$$

The answer is (e).

A-71. A piece of equipment with a list price of \$450 can actually be purchased for either \$400 cash or \$50 immediately plus four additional annual payments of \$115.25. All values are in dollars of current purchasing power. If the typical customer considered a 5% interest rate appropriate, the inflation rate at which the two purchase alternatives are equivalent is nearest to

(a) 5%
(b) 6%
(c) 8%
(d) 10%
(e) 12%

Solution

$$\text{PW of cash purchase} = \text{PW of installment purchase}$$

$$400 = 50 + 115.25(P/A,d\%,4)$$

$$(P/A,d\%,4) = 350/115.25 = 3.037$$

From the interest tables, $d = 12\%$.

$$d = i + f + i(f)$$

$$0.12 = 0.05 + f + 0.05f$$

$f = 0.07/1.05 = 0.0667 = 6.67\%$

The answer is (b).

A-72. A man wants to determine whether to invest \$1000 in a friend's speculative venture. He will do so if he thinks he can get his money back. The probabilities of the various outcomes at the end of one year are:

Result	Probability
\$2000 (double his money)	0.3
1500	0.1
1000	0.2
500	0.3
0 (lose everything)	0.1

His expected outcome if he invests the \$1000 is closest to

(a) \$800
(b) \$900
(c) \$1000
(d) \$1100
(e) \$1200

Solution

The expected income = 0.3(2000) + 0.1(1500) + 0.2(1000) + 0.3(500) + 0.1(0) = \$1100. The answer is (d).

A-73. The amount you would be willing to pay for an insurance policy protecting you against a one in twenty chance of losing $10,000 three years from now, if interest is 10%, is closest to

(a) $175 (d) $1500

(b) $350 (e) $2000

(c) $1000

Solution PW of benefit = (10,000/20)(P/F,10%,3)

= 500(0.7513) = $376

The answer is (b).

½% Compound Interest Factors ½%

	Single Payment		Uniform Payment Series				Uniform Gradient		
	Compound Amount Factor	Present Worth Factor	Sinking Fund Factor	Capital Recovery Factor	Compound Amount Factor	Present Worth Factor	Gradient Uniform Series	Gradient Present Worth	
n	Find *F* Given *P* *F/P*	Find *P* Given *F* *P/F*	Find *A* Given *F* *A/F*	Find *A* Given *P* *A/P*	Find *F* Given *A* *F/A*	Find *P* Given *A* *P/A*	Find *A* Given *G* *A/G*	Find *P* Given *G* *P/G*	*n*
1	1.005	.9950	1.0000	1.0050	1.000	0.995	0	0	**1**
2	1.010	.9901	.4988	.5038	2.005	1.985	0.499	0.991	**2**
3	1.015	.9851	.3317	.3367	3.015	2.970	0.996	2.959	**3**
4	1.020	.9802	.2481	.2531	4.030	3.951	1.494	5.903	**4**
5	1.025	.9754	.1980	.2030	5.050	4.926	1.990	9.803	**5**
6	1.030	.9705	.1646	.1696	6.076	5.896	2.486	14.660	**6**
7	1.036	.9657	.1407	.1457	7.106	6.862	2.980	20.448	**7**
8	1.041	.9609	.1228	.1278	8.141	7.823	3.474	27.178	**8**
9	1.046	.9561	.1089	.1139	9.182	8.779	3.967	34.825	**9**
10	1.051	.9513	.0978	.1028	10.228	9.730	4.459	43.389	**10**
11	1.056	.9466	.0887	.0937	11.279	10.677	4.950	52.855	**11**
12	1.062	.9419	.0811	.0861	12.336	11.619	5.441	63.218	**12**
13	1.067	.9372	.0746	.0796	13.397	12.556	5.931	74.465	**13**
14	1.072	.9326	.0691	.0741	14.464	13.489	6.419	86.590	**14**
15	1.078	.9279	.0644	.0694	15.537	14.417	6.907	99.574	**15**
16	1.083	.9233	.0602	.0652	16.614	15.340	7.394	113.427	**16**
17	1.088	.9187	.0565	.0615	17.697	16.259	7.880	128.125	**17**
18	1.094	.9141	.0532	.0582	18.786	17.173	8.366	143.668	**18**
19	1.099	.9096	.0503	.0553	19.880	18.082	8.850	160.037	**19**
20	1.105	.9051	.0477	.0527	20.979	18.987	9.334	177.237	**20**
21	1.110	.9006	.0453	.0503	22.084	19.888	9.817	195.245	**21**
22	1.116	.8961	.0431	.0481	23.194	20.784	10.300	214.070	**22**
23	1.122	.8916	.0411	.0461	24.310	21.676	10.781	233.680	**23**
24	1.127	.8872	.0393	.0443	25.432	22.563	11.261	254.088	**24**
25	1.133	.8828	.0377	.0427	26.559	23.446	11.741	275.273	**25**
26	1.138	.8784	.0361	.0411	27.692	24.324	12.220	297.233	**26**
27	1.144	.8740	.0347	.0397	28.830	25.198	12.698	319.955	**27**
28	1.150	.8697	.0334	.0384	29.975	26.068	13.175	343.439	**28**
29	1.156	.8653	.0321	.0371	31.124	26.933	13.651	367.672	**29**
30	1.161	.8610	.0310	.0360	32.280	27.794	14.127	392.640	**30**
36	1.197	.8356	.0254	.0304	39.336	32.871	16.962	557.564	**36**
40	1.221	.8191	.0226	.0276	44.159	36.172	18.836	681.341	**40**
48	1.270	.7871	.0185	.0235	54.098	42.580	22.544	959.928	**48**
50	1.283	.7793	.0177	.0227	56.645	44.143	23.463	1 035.70	**50**
52	1.296	.7716	.0169	.0219	59.218	45.690	24.378	1 113.82	**52**
60	1.349	.7414	.0143	.0193	69.770	51.726	28.007	1 448.65	**60**
70	1.418	.7053	.0120	.0170	83.566	58.939	32.468	1 913.65	**70**
72	1.432	.6983	.0116	.0166	86.409	60.340	33.351	2 012.35	**72**
80	1.490	.6710	.0102	.0152	98.068	65.802	36.848	2 424.65	**80**
84	1.520	.6577	.00961	.0146	104.074	68.453	38.576	2 640.67	**84**
90	1.567	.6383	.00883	.0138	113.311	72.331	41.145	2 976.08	**90**
96	1.614	.6195	.00814	.0131	122.829	76.095	43.685	3 324.19	**96**
100	1.647	.6073	.00773	.0127	129.334	78.543	45.361	3 562.80	**100**
104	1.680	.5953	.00735	.0124	135.970	80.942	47.025	3 806.29	**104**
120	1.819	.5496	.00610	.0111	163.880	90.074	53.551	4 823.52	**120**
240	3.310	.3021	.00216	.00716	462.041	139.581	96.113	13 415.56	**240**
360	6.023	.1660	.00100	.00600	1 004.5	166.792	128.324	21 403.32	**360**
480	10.957	.0913	.00050	.00550	1 991.5	181.748	151.795	27 588.37	**480**

1% Compound Interest Factors 1%

	Single Payment		Uniform Payment Series				Uniform Gradient		
	Compound Amount Factor	Present Worth Factor	Sinking Fund Factor	Capital Recovery Factor	Compound Amount Factor	Present Worth Factor	Gradient Uniform Series	Gradient Present Worth	
n	Find *F* Given *P* *F/P*	Find *P* Given *F* *P/F*	Find *A* Given *F* *A/F*	Find *A* Given *P* *A/P*	Find *F* Given *A* *F/A*	Find *P* Given *A* *P/A*	Find *A* Given *G* *A/G*	Find *P* Given *G* *P/G*	*n*
1	1.010	.9901	1.0000	1.0100	1.000	0.990	0	0	**1**
2	1.020	.9803	.4975	.5075	2.010	1.970	0.498	0.980	**2**
3	1.030	.9706	.3300	.3400	3.030	2.941	0.993	2.921	**3**
4	1.041	.9610	.2463	.2563	4.060	3.902	1.488	5.804	**4**
5	1.051	.9515	.1960	.2060	5.101	4.853	1.980	9.610	**5**
6	1.062	.9420	.1625	.1725	6.152	5.795	2.471	14.320	**6**
7	1.072	.9327	.1386	.1486	7.214	6.728	2.960	19.917	**7**
8	1.083	.9235	.1207	.1307	8.286	7.652	3.448	26.381	**8**
9	1.094	.9143	.1067	.1167	9.369	8.566	3.934	33.695	**9**
10	1.105	.9053	.0956	.1056	10.462	9.471	4.418	41.843	**10**
11	1.116	.8963	.0865	.0965	11.567	10.368	4.900	50.806	**11**
12	1.127	.8874	.0788	.0888	12.682	11.255	5.381	60.568	**12**
13	1.138	.8787	.0724	.0824	13.809	12.134	5.861	71.112	**13**
14	1.149	.8700	.0669	.0769	14.947	13.004	6.338	82.422	**14**
15	1.161	.8613	.0621	.0721	16.097	13.865	6.814	94.481	**15**
16	1.173	.8528	.0579	.0679	17.258	14.718	7.289	107.273	**16**
17	1.184	.8444	.0543	.0643	18.430	15.562	7.761	120.783	**17**
18	1.196	.8360	.0510	.0610	19.615	16.398	8.232	134.995	**18**
19	1.208	.8277	.0481	.0581	20.811	17.226	8.702	149.895	**19**
20	1.220	.8195	.0454	.0554	22.019	18.046	9.169	165.465	**20**
21	1.232	.8114	.0430	.0530	23.239	18.857	9.635	181.694	**21**
22	1.245	.8034	.0409	.0509	24.472	19.660	10.100	198.565	**22**
23	1.257	.7954	.0389	.0489	25.716	20.456	10.563	216.065	**23**
24	1.270	.7876	.0371	.0471	26.973	21.243	11.024	234.179	**24**
25	1.282	.7798	.0354	.0454	28.243	22.023	11.483	252.892	**25**
26	1.295	.7720	.0339	.0439	29.526	22.795	11.941	272.195	**26**
27	1.308	.7644	.0324	.0424	30.821	23.560	12.397	292.069	**27**
28	1.321	.7568	.0311	.0411	32.129	24.316	12.852	312.504	**28**
29	1.335	.7493	.0299	.0399	33.450	25.066	13.304	333.486	**29**
30	1.348	.7419	.0287	.0387	34.785	25.808	13.756	355.001	**30**
36	1.431	.6989	.0232	.0332	43.077	30.107	16.428	494.620	**36**
40	1.489	.6717	.0205	.0305	48.886	32.835	18.178	596.854	**40**
48	1.612	.6203	.0163	.0263	61.223	37.974	21.598	820.144	**48**
50	1.645	.6080	.0155	.0255	64.463	39.196	22.436	879.417	**50**
52	1.678	.5961	.0148	.0248	67.769	40.394	23.269	939.916	**52**
60	1.817	.5504	.0122	.0222	81.670	44.955	26.533	1 192.80	**60**
70	2.007	.4983	.00993	.0199	100.676	50.168	30.470	1 528.64	**70**
72	2.047	.4885	.00955	.0196	104.710	51.150	31.239	1 597.86	**72**
80	2.217	.4511	.00822	.0182	121.671	54.888	34.249	1 879.87	**80**
84	2.307	.4335	.00765	.0177	130.672	56.648	35.717	2 023.31	**84**
90	2.449	.4084	.00690	.0169	144.863	59.161	37.872	2 240.56	**90**
96	2.599	.3847	.00625	.0163	159.927	61.528	39.973	2 459.42	**96**
100	2.705	.3697	.00587	.0159	170.481	63.029	41.343	2 605.77	**100**
104	2.815	.3553	.00551	.0155	181.464	64.471	42.688	2 752.17	**104**
120	3.300	.3030	.00435	.0143	230.039	69.701	47.835	3 334.11	**120**
240	10.893	.0918	.00101	.0110	989.254	90.819	75.739	6 878.59	**240**
360	35.950	.0278	.00029	.0103	3 495.0	97.218	89.699	8 720.43	**360**
480	118.648	.00843	.00008	.0101	11 764.8	99.157	95.920	9 511.15	**480**

1½% Compound Interest Factors 1½%

n	Single Payment: Compound Amount Factor, Find F Given P, F/P	Single Payment: Present Worth Factor, Find P Given F, P/F	Uniform Payment Series: Sinking Fund Factor, Find A Given F, A/F	Uniform Payment Series: Capital Recovery Factor, Find A Given P, A/P	Uniform Payment Series: Compound Amount Factor, Find F Given A, F/A	Uniform Payment Series: Present Worth Factor, Find P Given A, P/A	Uniform Gradient: Gradient Uniform Series, Find A Given G, A/G	Uniform Gradient: Gradient Present Worth, Find P Given G, P/G	n
1	1.015	.9852	1.0000	1.0150	1.000	0.985	0	0	1
2	1.030	.9707	.4963	.5113	2.015	1.956	0.496	0.970	2
3	1.046	.9563	.3284	.3434	3.045	2.912	0.990	2.883	3
4	1.061	.9422	.2444	.2594	4.091	3.854	1.481	5.709	4
5	1.077	.9283	.1941	.2091	5.152	4.783	1.970	9.422	5
6	1.093	.9145	.1605	.1755	6.230	5.697	2.456	13.994	6
7	1.110	.9010	.1366	.1516	7.323	6.598	2.940	19.400	7
8	1.126	.8877	.1186	.1336	8.433	7.486	3.422	25.614	8
9	1.143	.8746	.1046	.1196	9.559	8.360	3.901	32.610	9
10	1.161	.8617	.0934	.1084	10.703	9.222	4.377	40.365	10
11	1.178	.8489	.0843	.0993	11.863	10.071	4.851	48.855	11
12	1.196	.8364	.0767	.0917	13.041	10.907	5.322	58.054	12
13	1.214	.8240	.0702	.0852	14.237	11.731	5.791	67.943	13
14	1.232	.8118	.0647	.0797	15.450	12.543	6.258	78.496	14
15	1.250	.7999	.0599	.0749	16.682	13.343	6.722	89.694	15
16	1.269	.7880	.0558	.0708	17.932	14.131	7.184	101.514	16
17	1.288	.7764	.0521	.0671	19.201	14.908	7.643	113.937	17
18	1.307	.7649	.0488	.0638	20.489	15.673	8.100	126.940	18
19	1.327	.7536	.0459	.0609	21.797	16.426	8.554	140.505	19
20	1.347	.7425	.0432	.0582	23.124	17.169	9.005	154.611	20
21	1.367	.7315	.0409	.0559	24.470	17.900	9.455	169.241	21
22	1.388	.7207	.0387	.0537	25.837	18.621	9.902	184.375	22
23	1.408	.7100	.0367	.0517	27.225	19.331	10.346	199.996	23
24	1.430	.6995	.0349	.0499	28.633	20.030	10.788	216.085	24
25	1.451	.6892	.0333	.0483	30.063	20.720	11.227	232.626	25
26	1.473	.6790	.0317	.0467	31.514	21.399	11.664	249.601	26
27	1.495	.6690	.0303	.0453	32.987	22.068	12.099	266.995	27
28	1.517	.6591	.0290	.0440	34.481	22.727	12.531	284.790	28
29	1.540	.6494	.0278	.0428	35.999	23.376	12.961	302.972	29
30	1.563	.6398	.0266	.0416	37.539	24.016	13.388	321.525	30
36	1.709	.5851	.0212	.0362	47.276	27.661	15.901	439.823	36
40	1.814	.5513	.0184	.0334	54.268	29.916	17.528	524.349	40
48	2.043	.4894	.0144	.0294	69.565	34.042	20.666	703.537	48
50	2.105	.4750	.0136	.0286	73.682	35.000	21.428	749.955	50
52	2.169	.4611	.0128	.0278	77.925	35.929	22.179	796.868	52
60	2.443	.4093	.0104	.0254	96.214	39.380	25.093	988.157	60
70	2.835	.3527	.00817	.0232	122.363	43.155	28.529	1 231.15	70
72	2.921	.3423	.00781	.0228	128.076	43.845	29.189	1 279.78	72
80	3.291	.3039	.00655	.0215	152.710	46.407	31.742	1 473.06	80
84	3.493	.2863	.00602	.0210	166.172	47.579	32.967	1 568.50	84
90	3.819	.2619	.00532	.0203	187.929	49.210	34.740	1 709.53	90
96	4.176	.2395	.00472	.0197	211.719	50.702	36.438	1 847.46	96
100	4.432	.2256	.00437	.0194	228.802	51.625	37.529	1 937.43	100
104	4.704	.2126	.00405	.0190	246.932	52.494	38.589	2 025.69	104
120	5.969	.1675	.00302	.0180	331.286	55.498	42.518	2 359.69	120
240	35.632	.0281	.00043	.0154	2 308.8	64.796	59.737	3 870.68	240
360	212.700	.00470	.00007	.0151	14 113.3	66.353	64.966	4 310.71	360
480	1 269.7	.00079	.00001	.0150	84 577.8	66.614	66.288	4 415.74	480

2% Compound Interest Factors 2%

n	Single Payment: Compound Amount Factor, Find F Given P, F/P	Single Payment: Present Worth Factor, Find P Given F, P/F	Uniform Payment Series: Sinking Fund Factor, Find A Given F, A/F	Uniform Payment Series: Capital Recovery Factor, Find A Given P, A/P	Uniform Payment Series: Compound Amount Factor, Find F Given A, F/A	Uniform Payment Series: Present Worth Factor, Find P Given A, P/A	Uniform Gradient: Gradient Uniform Series, Find A Given G, A/G	Uniform Gradient: Gradient Present Worth, Find P Given G, P/G	n
1	1.020	.9804	1.0000	1.0200	1.000	0.980	0	0	1
2	1.040	.9612	.4951	.5151	2.020	1.942	0.495	0.961	2
3	1.061	.9423	.3268	.3468	3.060	2.884	0.987	2.846	3
4	1.082	.9238	.2426	.2626	4.122	3.808	1.475	5.617	4
5	1.104	.9057	.1922	.2122	5.204	4.713	1.960	9.240	5
6	1.126	.8880	.1585	.1785	6.308	5.601	2.442	13.679	6
7	1.149	.8706	.1345	.1545	7.434	6.472	2.921	18.903	7
8	1.172	.8535	.1165	.1365	8.583	7.325	3.396	24.877	8
9	1.195	.8368	.1025	.1225	9.755	8.162	3.868	31.571	9
10	1.219	.8203	.0913	.1113	10.950	8.983	4.337	38.954	10
11	1.243	.8043	.0822	.1022	12.169	9.787	4.802	46.996	11
12	1.268	.7885	.0746	.0946	13.412	10.575	5.264	55.669	12
13	1.294	.7730	.0681	.0881	14.680	11.348	5.723	64.946	13
14	1.319	.7579	.0626	.0826	15.974	12.106	6.178	74.798	14
15	1.346	.7430	.0578	.0778	17.293	12.849	6.631	85.200	15
16	1.373	.7284	.0537	.0737	18.639	13.578	7.080	96.127	16
17	1.400	.7142	.0500	.0700	20.012	14.292	7.526	107.553	17
18	1.428	.7002	.0467	.0667	21.412	14.992	7.968	119.456	18
19	1.457	.6864	.0438	.0638	22.840	15.678	8.407	131.812	19
20	1.486	.6730	.0412	.0612	24.297	16.351	8.843	144.598	20
21	1.516	.6598	.0388	.0588	25.783	17.011	9.276	157.793	21
22	1.546	.6468	.0366	.0566	27.299	17.658	9.705	171.377	22
23	1.577	.6342	.0347	.0547	28.845	18.292	10.132	185.328	23
24	1.608	.6217	.0329	.0529	30.422	18.914	10.555	199.628	24
25	1.641	.6095	.0312	.0512	32.030	19.523	10.974	214.256	25
26	1.673	.5976	.0297	.0497	33.671	20.121	11.391	229.196	26
27	1.707	.5859	.0283	.0483	35.344	20.707	11.804	244.428	27
28	1.741	.5744	.0270	.0470	37.051	21.281	12.214	259.936	28
29	1.776	.5631	.0258	.0458	38.792	21.844	12.621	275.703	29
30	1.811	.5521	.0247	.0447	40.568	22.396	13.025	291.713	30
36	2.040	.4902	.0192	.0392	51.994	25.489	15.381	392.036	36
40	2.208	.4529	.0166	.0366	60.402	27.355	16.888	461.989	40
48	2.587	.3865	.0126	.0326	79.353	30.673	19.755	605.961	48
50	2.692	.3715	.0118	.0318	84.579	31.424	20.442	642.355	50
52	2.800	.3571	.0111	.0311	90.016	32.145	21.116	678.779	52
60	3.281	.3048	.00877	.0288	114.051	34.761	23.696	823.692	60
70	4.000	.2500	.00667	.0267	149.977	37.499	26.663	999.829	70
72	4.161	.2403	.00633	.0263	158.056	37.984	27.223	1 034.050	72
80	4.875	.2051	.00516	.0252	193.771	39.744	29.357	1 166.781	80
84	5.277	.1895	.00468	.0247	213.865	40.525	30.361	1 230.413	84
90	5.943	.1683	.00405	.0240	247.155	41.587	31.793	1 322.164	90
96	6.693	.1494	.00351	.0235	284.645	42.529	33.137	1 409.291	96
100	7.245	.1380	.00320	.0232	312.230	43.098	33.986	1 464.747	100
104	7.842	.1275	.00292	.0229	342.090	43.624	34.799	1 518.082	104
120	10.765	.0929	.00205	.0220	488.255	45.355	37.711	1 710.411	120
240	115.887	.00863	.00017	.0202	5 744.4	49.569	47.911	2 374.878	240
360	1 247.5	.00080	.00002	.0200	62 326.8	49.960	49.711	2 483.567	360
480	13 429.8	.00007		.0200	671 442.0	49.996	49.964	2 498.027	480

4% Compound Interest Factors 4%

n	F/P	P/F	A/F	A/P	F/A	P/A	A/G	P/G	n
1	1.040	.9615	1.0000	1.0400	1.000	0.962	0	0	1
2	1.082	.9246	.4902	.5302	2.040	1.886	0.490	0.925	2
3	1.125	.8890	.3203	.3603	3.122	2.775	0.974	2.702	3
4	1.170	.8548	.2355	.2755	4.246	3.630	1.451	5.267	4
5	1.217	.8219	.1846	.2246	5.416	4.452	1.922	8.555	5
6	1.265	.7903	.1508	.1908	6.633	5.242	2.386	12.506	6
7	1.316	.7599	.1266	.1666	7.898	6.002	2.843	17.066	7
8	1.369	.7307	.1085	.1485	9.214	6.733	3.294	22.180	8
9	1.423	.7026	.0945	.1345	10.583	7.435	3.739	27.801	9
10	1.480	.6756	.0833	.1233	12.006	8.111	4.177	33.881	10
11	1.539	.6496	.0741	.1141	13.486	8.760	4.609	40.377	11
12	1.601	.6246	.0666	.1066	15.026	9.385	5.034	47.248	12
13	1.665	.6006	.0601	.1001	16.627	9.986	5.453	54.454	13
14	1.732	.5775	.0547	.0947	18.292	10.563	5.866	61.962	14
15	1.801	.5553	.0499	.0899	20.024	11.118	6.272	69.735	15
16	1.873	.5339	.0458	.0858	21.825	11.652	6.672	77.744	16
17	1.948	.5134	.0422	.0822	23.697	12.166	7.066	85.958	17
18	2.026	.4936	.0390	.0790	25.645	12.659	7.453	94.350	18
19	2.107	.4746	.0361	.0761	27.671	13.134	7.834	102.893	19
20	2.191	.4564	.0336	.0736	29.778	13.590	8.209	111.564	20
21	2.279	.4388	.0313	.0713	31.969	14.029	8.578	120.341	21
22	2.370	.4220	.0292	.0692	34.248	14.451	8.941	129.202	22
23	2.465	.4057	.0273	.0673	36.618	14.857	9.297	138.128	23
24	2.563	.3901	.0256	.0656	39.083	15.247	9.648	147.101	24
25	2.666	.3751	.0240	.0640	41.646	15.622	9.993	156.104	25
26	2.772	.3607	.0226	.0626	44.312	15.983	10.331	165.121	26
27	2.883	.3468	.0212	.0612	47.084	16.330	10.664	174.138	27
28	2.999	.3335	.0200	.0600	49.968	16.663	10.991	183.142	28
29	3.119	.3207	.0189	.0589	52.966	16.984	11.312	192.120	29
30	3.243	.3083	.0178	.0578	56.085	17.292	11.627	201.062	30
31	3.373	.2965	.0169	.0569	59.328	17.588	11.937	209.955	31
32	3.508	.2851	.0159	.0559	62.701	17.874	12.241	218.792	32
33	3.648	.2741	.0151	.0551	66.209	18.148	12.540	227.563	33
34	3.794	.2636	.0143	.0543	69.858	18.411	12.832	236.260	34
35	3.946	.2534	.0136	.0536	73.652	18.665	13.120	244.876	35
40	4.801	.2083	.0105	.0505	95.025	19.793	14.476	286.530	40
45	5.841	.1712	.00826	.0483	121.029	20.720	15.705	325.402	45
50	7.107	.1407	.00655	.0466	152.667	21.482	16.812	361.163	50
55	8.646	.1157	.00523	.0452	191.159	22.109	17.807	393.689	55
60	10.520	.0951	.00420	.0442	237.990	22.623	18.697	422.996	60
65	12.799	.0781	.00339	.0434	294.968	23.047	19.491	449.201	65
70	15.572	.0642	.00275	.0427	364.290	23.395	20.196	472.479	70
75	18.945	.0528	.00223	.0422	448.630	23.680	20.821	493.041	75
80	23.050	.0434	.00181	.0418	551.244	23.915	21.372	511.116	80
85	28.044	.0357	.00148	.0415	676.089	24.109	21.857	526.938	85
90	34.119	.0293	.00121	.0412	827.981	24.267	22.283	540.737	90
95	41.511	.0241	.00099	.0410	1 012.8	24.398	22.655	552.730	95
100	50.505	.0198	.00081	.0408	1 237.6	24.505	22.980	563.125	100

6% Compound Interest Factors 6%

n	F/P	P/F	A/F	A/P	F/A	P/A	A/G	P/G	n
1	1.060	.9434	1.0000	1.0600	1.000	0.943	0	0	1
2	1.124	.8900	.4854	.5454	2.060	1.833	0.485	0.890	2
3	1.191	.8396	.3141	.3741	3.184	2.673	0.961	2.569	3
4	1.262	.7921	.2286	.2886	4.375	3.465	1.427	4.945	4
5	1.338	.7473	.1774	.2374	5.637	4.212	1.884	7.934	5
6	1.419	.7050	.1434	.2034	6.975	4.917	2.330	11.459	6
7	1.504	.6651	.1191	.1791	8.394	5.582	2.768	15.450	7
8	1.594	.6274	.1010	.1610	9.897	6.210	3.195	19.841	8
9	1.689	.5919	.0870	.1470	11.491	6.802	3.613	24.577	9
10	1.791	.5584	.0759	.1359	13.181	7.360	4.022	29.602	10
11	1.898	.5268	.0668	.1268	14.972	7.887	4.421	34.870	11
12	2.012	.4970	.0593	.1193	16.870	8.384	4.811	40.337	12
13	2.133	.4688	.0530	.1130	18.882	8.853	5.192	45.963	13
14	2.261	.4423	.0476	.1076	21.015	9.295	5.564	51.713	14
15	2.397	.4173	.0430	.1030	23.276	9.712	5.926	57.554	15
16	2.540	.3936	.0390	.0990	25.672	10.106	6.279	63.459	16
17	2.693	.3714	.0354	.0954	28.213	10.477	6.624	69.401	17
18	2.854	.3503	.0324	.0924	30.906	10.828	6.960	75.357	18
19	3.026	.3305	.0296	.0896	33.760	11.158	7.287	81.306	19
20	3.207	.3118	.0272	.0872	36.786	11.470	7.605	87.230	20
21	3.400	.2942	.0250	.0850	39.993	11.764	7.915	93.113	21
22	3.604	.2775	.0230	.0830	43.392	12.042	8.217	98.941	22
23	3.820	.2618	.0213	.0813	46.996	12.303	8.510	104.700	23
24	4.049	.2470	.0197	.0797	50.815	12.550	8.795	110.381	24
25	4.292	.2330	.0182	.0782	54.864	12.783	9.072	115.973	25
26	4.549	.2198	.0169	.0769	59.156	13.003	9.341	121.468	26
27	4.822	.2074	.0157	.0757	63.706	13.211	9.603	126.860	27
28	5.112	.1956	.0146	.0746	68.528	13.406	9.857	132.142	28
29	5.418	.1846	.0136	.0736	73.640	13.591	10.103	137.309	29
30	5.743	.1741	.0126	.0726	79.058	13.765	10.342	142.359	30
31	6.088	.1643	.0118	.0718	84.801	13.929	10.574	147.286	31
32	6.453	.1550	.0110	.0710	90.890	14.084	10.799	152.090	32
33	6.841	.1462	.0103	.0703	97.343	14.230	11.017	156.768	33
34	7.251	.1379	.00960	.0696	104.184	14.368	11.228	161.319	34
35	7.686	.1301	.00897	.0690	111.435	14.498	11.432	165.743	35
40	10.286	.0972	.00646	.0665	154.762	15.046	12.359	185.957	40
45	13.765	.0727	.00470	.0647	212.743	15.456	13.141	203.109	45
50	18.420	.0543	.00344	.0634	290.335	15.762	13.796	217.457	50
55	24.650	.0406	.00254	.0625	394.171	15.991	14.341	229.322	55
60	32.988	.0303	.00188	.0619	533.126	16.161	14.791	239.043	60
65	44.145	.0227	.00139	.0614	719.080	16.289	15.160	246.945	65
70	59.076	.0169	.00103	.0610	967.928	16.385	15.461	253.327	70
75	79.057	.0126	.00077	.0608	1 300.9	16.456	15.706	258.453	75
80	105.796	.00945	.00057	.0606	1 746.6	16.509	15.903	262.549	80
85	141.578	.00706	.00043	.0604	2 343.0	16.549	16.062	265.810	85
90	189.464	.00528	.00032	.0603	3 141.1	16.579	16.189	268.395	90
95	253.545	.00394	.00024	.0602	4 209.1	16.601	16.290	270.437	95
100	339.300	.00295	.00018	.0602	5 638.3	16.618	16.371	272.047	100

8% Compound Interest Factors 8%

n	Single Payment: Compound Amount Factor (Find F Given P, F/P)	Single Payment: Present Worth Factor (Find P Given F, P/F)	Uniform Payment Series: Sinking Fund Factor (Find A Given F, A/F)	Uniform Payment Series: Capital Recovery Factor (Find A Given P, A/P)	Uniform Payment Series: Compound Amount Factor (Find F Given A, F/A)	Uniform Payment Series: Present Worth Factor (Find P Given A, P/A)	Uniform Gradient: Gradient Uniform Series (Find A Given G, A/G)	Uniform Gradient: Gradient Present Worth (Find P Given G, P/G)	n
1	1.080	.9259	1.0000	1.0800	1.000	0.926	0	0	1
2	1.166	.8573	.4808	.5608	2.080	1.783	0.481	0.857	2
3	1.260	.7938	.3080	.3880	3.246	2.577	0.949	2.445	3
4	1.360	.7350	.2219	.3019	4.506	3.312	1.404	4.650	4
5	1.469	.6806	.1705	.2505	5.867	3.993	1.846	7.372	5
6	1.587	.6302	.1363	.2163	7.336	4.623	2.276	10.523	6
7	1.714	.5835	.1121	.1921	8.923	5.206	2.694	14.024	7
8	1.851	.5403	.0940	.1740	10.637	5.747	3.099	17.806	8
9	1.999	.5002	.0801	.1601	12.488	6.247	3.491	21.808	9
10	2.159	.4632	.0690	.1490	14.487	6.710	3.871	25.977	10
11	2.332	.4289	.0601	.1401	16.645	7.139	4.240	30.266	11
12	2.518	.3971	.0527	.1327	18.977	7.536	4.596	34.634	12
13	2.720	.3677	.0465	.1265	21.495	7.904	4.940	39.046	13
14	2.937	.3405	.0413	.1213	24.215	8.244	5.273	43.472	14
15	3.172	.3152	.0368	.1168	27.152	8.559	5.594	47.886	15
16	3.426	.2919	.0330	.1130	30.324	8.851	5.905	52.264	16
17	3.700	.2703	.0296	.1096	33.750	9.122	6.204	56.588	17
18	3.996	.2502	.0267	.1067	37.450	9.372	6.492	60.843	18
19	4.316	.2317	.0241	.1041	41.446	9.604	6.770	65.013	19
20	4.661	.2145	.0219	.1019	45.762	9.818	7.037	69.090	20
21	5.034	.1987	.0198	.0998	50.423	10.017	7.294	73.063	21
22	5.437	.1839	.0180	.0980	55.457	10.201	7.541	76.926	22
23	5.871	.1703	.0164	.0964	60.893	10.371	7.779	80.673	23
24	6.341	.1577	.0150	.0950	66.765	10.529	8.007	84.300	24
25	6.848	.1460	.0137	.0937	73.106	10.675	8.225	87.804	25
26	7.396	.1352	.0125	.0925	79.954	10.810	8.435	91.184	26
27	7.988	.1252	.0114	.0914	87.351	10.935	8.636	94.439	27
28	8.627	.1159	.0105	.0905	95.339	11.051	8.829	97.569	28
29	9.317	.1073	.00962	.0896	103.966	11.158	9.013	100.574	29
30	10.063	.0994	.00883	.0888	113.283	11.258	9.190	103.456	30
31	10.868	.0920	.00811	.0881	123.346	11.350	9.358	106.216	31
32	11.737	.0852	.00745	.0875	134.214	11.435	9.520	108.858	32
33	12.676	.0789	.00685	.0869	145.951	11.514	9.674	111.382	33
34	13.690	.0730	.00630	.0863	158.627	11.587	9.821	113.792	34
35	14.785	.0676	.00580	.0858	172.317	11.655	9.961	116.092	35
40	21.725	.0460	.00386	.0839	259.057	11.925	10.570	126.042	40
45	31.920	.0313	.00259	.0826	386.506	12.108	11.045	133.733	45
50	46.902	.0213	.00174	.0817	573.771	12.233	11.411	139.593	50
55	68.914	.0145	.00118	.0812	848.925	12.319	11.690	144.006	55
60	101.257	.00988	.00080	.0808	1 253.2	12.377	11.902	147.300	60
65	148.780	.00672	.00054	.0805	1 847.3	12.416	12.060	149.739	65
70	218.607	.00457	.00037	.0804	2 720.1	12.443	12.178	151.533	70
75	321.205	.00311	.00025	.0802	4 002.6	12.461	12.266	152.845	75
80	471.956	.00212	.00017	.0802	5 887.0	12.474	12.330	153.800	80
85	693.458	.00144	.00012	.0801	8 655.7	12.482	12.377	154.492	85
90	1 018.9	.00098	.00008	.0801	12 724.0	12.488	12.412	154.993	90
95	1 497.1	.00067	.00005	.0801	18 701.6	12.492	12.437	155.352	95
100	2 199.8	.00045	.00004	.0800	27 484.6	12.494	12.455	155.611	100

10% Compound Interest Factors 10%

n	Single Payment: Compound Amount Factor (Find F Given P, F/P)	Single Payment: Present Worth Factor (Find P Given F, P/F)	Uniform Payment Series: Sinking Fund Factor (Find A Given F, A/F)	Uniform Payment Series: Capital Recovery Factor (Find A Given P, A/P)	Uniform Payment Series: Compound Amount Factor (Find F Given A, F/A)	Uniform Payment Series: Present Worth Factor (Find P Given A, P/A)	Uniform Gradient: Gradient Uniform Series (Find A Given G, A/G)	Uniform Gradient: Gradient Present Worth (Find P Given G, P/G)	n
1	1.100	.9091	1.0000	1.1000	1.000	0.909	0	0	1
2	1.210	.8264	.4762	.5762	2.100	1.736	0.476	0.826	2
3	1.331	.7513	.3021	.4021	3.310	2.487	0.937	2.329	3
4	1.464	.6830	.2155	.3155	4.641	3.170	1.381	4.378	4
5	1.611	.6209	.1638	.2638	6.105	3.791	1.810	6.862	5
6	1.772	.5645	.1296	.2296	7.716	4.355	2.224	9.684	6
7	1.949	.5132	.1054	.2054	9.487	4.868	2.622	12.763	7
8	2.144	.4665	.0874	.1874	11.436	5.335	3.004	16.029	8
9	2.358	.4241	.0736	.1736	13.579	5.759	3.372	19.421	9
10	2.594	.3855	.0627	.1627	15.937	6.145	3.725	22.891	10
11	2.853	.3505	.0540	.1540	18.531	6.495	4.064	26.396	11
12	3.138	.3186	.0468	.1468	21.384	6.814	4.388	29.901	12
13	3.452	.2897	.0408	.1408	24.523	7.103	4.699	33.377	13
14	3.797	.2633	.0357	.1357	27.975	7.367	4.996	36.801	14
15	4.177	.2394	.0315	.1315	31.772	7.606	5.279	40.152	15
16	4.595	.2176	.0278	.1278	35.950	7.824	5.549	43.416	16
17	5.054	.1978	.0247	.1247	40.545	8.022	5.807	46.582	17
18	5.560	.1799	.0219	.1219	45.599	8.201	6.053	49.640	18
19	6.116	.1635	.0195	.1195	51.159	8.365	6.286	52.583	19
20	6.728	.1486	.0175	.1175	57.275	8.514	6.508	55.407	20
21	7.400	.1351	.0156	.1156	64.003	8.649	6.719	58.110	21
22	8.140	.1228	.0140	.1140	71.403	8.772	6.919	60.689	22
23	8.954	.1117	.0126	.1126	79.543	8.883	7.108	63.146	23
24	9.850	.1015	.0113	.1113	88.497	8.985	7.288	65.481	24
25	10.835	.0923	.0102	.1102	98.347	9.077	7.458	67.696	25
26	11.918	.0839	.00916	.1092	109.182	9.161	7.619	69.794	26
27	13.110	.0763	.00826	.1083	121.100	9.237	7.770	71.777	27
28	14.421	.0693	.00745	.1075	134.210	9.307	7.914	73.650	28
29	15.863	.0630	.00673	.1067	148.631	9.370	8.049	75.415	29
30	17.449	.0573	.00608	.1061	164.494	9.427	8.176	77.077	30
31	19.194	.0521	.00550	.1055	181.944	9.479	8.296	78.640	31
32	21.114	.0474	.00497	.1050	201.138	9.526	8.409	80.108	32
33	23.225	.0431	.00450	.1045	222.252	9.569	8.515	81.486	33
34	25.548	.0391	.00407	.1041	245.477	9.609	8.615	82.777	34
35	28.102	.0356	.00369	.1037	271.025	9.644	8.709	83.987	35
40	45.259	.0221	.00226	.1023	442.593	9.779	9.096	88.953	40
45	72.891	.0137	.00139	.1014	718.905	9.863	9.374	92.454	45
50	117.391	.00852	.00086	.1009	1 163.9	9.915	9.570	94.889	50
55	189.059	.00529	.00053	.1005	1 880.6	9.947	9.708	96.562	55
60	304.482	.00328	.00033	.1003	3 034.8	9.967	9.802	97.701	60
65	490.371	.00204	.00020	.1002	4 893.7	9.980	9.867	98.471	65
70	789.748	.00127	.00013	.1001	7 887.5	9.987	9.911	98.987	70
75	1 271.9	.00079	.00008	.1001	12 709.0	9.992	9.941	99.332	75
80	2 048.4	.00049	.00005	.1000	20 474.0	9.995	9.961	99.561	80
85	3 299.0	.00030	.00003	.1000	32 979.7	9.997	9.974	99.712	85
90	5 313.0	.00019	.00002	.1000	53 120.3	9.998	9.983	99.812	90
95	8 556.7	.00012	.00001	.1000	85 556.9	9.999	9.989	99.877	95
100	13 780.6	.00007	.00001	.1000	137 796.3	9.999	9.993	99.920	100

12% Compound Interest Factors 12%

n	F/P	P/F	A/F	A/P	F/A	P/A	A/G	P/G	n
1	1.120	.8929	1.0000	1.1200	1.000	0.893	0	0	1
2	1.254	.7972	.4717	.5917	2.120	1.690	0.472	0.797	2
3	1.405	.7118	.2963	.4163	3.374	2.402	0.925	2.221	3
4	1.574	.6355	.2092	.3292	4.779	3.037	1.359	4.127	4
5	1.762	.5674	.1574	.2774	6.353	3.605	1.775	6.397	5
6	1.974	.5066	.1232	.2432	8.115	4.111	2.172	8.930	6
7	2.211	.4523	.0991	.2191	10.089	4.564	2.551	11.644	7
8	2.476	.4039	.0813	.2013	12.300	4.968	2.913	14.471	8
9	2.773	.3606	.0677	.1877	14.776	5.328	3.257	17.356	9
10	3.106	.3220	.0570	.1770	17.549	5.650	3.585	20.254	10
11	3.479	.2875	.0484	.1684	20.655	5.938	3.895	23.129	11
12	3.896	.2567	.0414	.1614	24.133	6.194	4.190	25.952	12
13	4.363	.2292	.0357	.1557	28.029	6.424	4.468	28.702	13
14	4.887	.2046	.0309	.1509	32.393	6.628	4.732	31.362	14
15	5.474	.1827	.0268	.1468	37.280	6.811	4.980	33.920	15
16	6.130	.1631	.0234	.1434	42.753	6.974	5.215	36.367	16
17	6.866	.1456	.0205	.1405	48.884	7.120	5.435	38.697	17
18	7.690	.1300	.0179	.1379	55.750	7.250	5.643	40.908	18
19	8.613	.1161	.0158	.1358	63.440	7.366	5.838	42.998	19
20	9.646	.1037	.0139	.1339	72.052	7.469	6.020	44.968	20
21	10.804	.0926	.0122	.1322	81.699	7.562	6.191	46.819	21
22	12.100	.0826	.0108	.1308	92.503	7.645	6.351	48.554	22
23	13.552	.0738	.00956	.1296	104.603	7.718	6.501	50.178	23
24	15.179	.0659	.00846	.1285	118.155	7.784	6.641	51.693	24
25	17.000	.0588	.00750	.1275	133.334	7.843	6.771	53.105	25
26	19.040	.0525	.00665	.1267	150.334	7.896	6.892	54.418	26
27	21.325	.0469	.00590	.1259	169.374	7.943	7.005	55.637	27
28	23.884	.0419	.00524	.1252	190.699	7.984	7.110	56.767	28
29	26.750	.0374	.00466	.1247	214.583	8.022	7.207	57.814	29
30	29.960	.0334	.00414	.1241	241.333	8.055	7.297	58.782	30
31	33.555	.0298	.00369	.1237	271.293	8.085	7.381	59.676	31
32	37.582	.0266	.00328	.1233	304.848	8.112	7.459	60.501	32
33	42.092	.0238	.00292	.1229	342.429	8.135	7.530	61.261	33
34	47.143	.0212	.00260	.1226	384.521	8.157	7.596	61.961	34
35	52.800	.0189	.00232	.1223	431.663	8.176	7.658	62.605	35
40	93.051	.0107	.00130	.1213	767.091	8.244	7.899	65.116	40
45	163.988	.00610	.00074	.1207	1 358.2	8.283	8.057	66.734	45
50	289.002	.00346	.00042	.1204	2 400.0	8.304	8.160	67.762	50
55	509.321	.00196	.00024	.1202	4 236.0	8.317	8.225	68.408	55
60	897.597	.00111	.00013	.1201	7 471.6	8.324	8.266	68.810	60
65	1 581.9	.00063	.00008	.1201	13 173.9	8.328	8.292	69.058	65
70	2 787.8	.00036	.00004	.1200	23 223.3	8.330	8.308	69.210	70
75	4 913.1	.00020	.00002	.1200	40 933.8	8.332	8.318	69.303	75
80	8 658.5	.00012	.00001	.1200	72 145.7	8.332	8.324	69.359	80
85	15 259.2	.00007	.00001	.1200	127 151.7	8.333	8.328	69.393	85
90	26 891.9	.00004		.1200	224 091.1	8.333	8.330	69.414	90
95	47 392.8	.00002		.1200	394 931.4	8.333	8.331	69.426	95
100	83 522.3	.00001		.1200	696 010.5	8.333	8.332	69.434	100

18% Compound Interest Factors 18%

n	F/P	P/F	A/F	A/P	F/A	P/A	A/G	P/G	n
1	1.180	.8475	1.0000	1.1800	1.000	0.847	0	0	1
2	1.392	.7182	.4587	.6387	2.180	1.566	0.459	0.718	2
3	1.643	.6086	.2799	.4599	3.572	2.174	0.890	1.935	3
4	1.939	.5158	.1917	.3717	5.215	2.690	1.295	3.483	4
5	2.288	.4371	.1398	.3198	7.154	3.127	1.673	5.231	5
6	2.700	.3704	.1059	.2859	9.442	3.498	2.025	7.083	6
7	3.185	.3139	.0824	.2624	12.142	3.812	2.353	8.967	7
8	3.759	.2660	.0652	.2452	15.327	4.078	2.656	10.829	8
9	4.435	.2255	.0524	.2324	19.086	4.303	2.936	12.633	9
10	5.234	.1911	.0425	.2225	23.521	4.494	3.194	14.352	10
11	6.176	.1619	.0348	.2148	28.755	4.656	3.430	15.972	11
12	7.288	.1372	.0286	.2086	34.931	4.793	3.647	17.481	12
13	8.599	.1163	.0237	.2037	42.219	4.910	3.845	18.877	13
14	10.147	.0985	.0197	.1997	50.818	5.008	4.025	20.158	14
15	11.974	.0835	.0164	.1964	60.965	5.092	4.189	21.327	15
16	14.129	.0708	.0137	.1937	72.939	5.162	4.337	22.389	16
17	16.672	.0600	.0115	.1915	87.068	5.222	4.471	23.348	17
18	19.673	.0508	.00964	.1896	103.740	5.273	4.592	24.212	18
19	23.214	.0431	.00810	.1881	123.413	5.316	4.700	24.988	19
20	27.393	.0365	.00682	.1868	146.628	5.353	4.798	25.681	20
21	32.324	.0309	.00575	.1857	174.021	5.384	4.885	26.300	21
22	38.142	.0262	.00485	.1848	206.345	5.410	4.963	26.851	22
23	45.008	.0222	.00409	.1841	244.487	5.432	5.033	27.339	23
24	53.109	.0188	.00345	.1835	289.494	5.451	5.095	27.772	24
25	62.669	.0160	.00292	.1829	342.603	5.467	5.150	28.155	25
26	73.949	.0135	.00247	.1825	405.272	5.480	5.199	28.494	26
27	87.260	.0115	.00209	.1821	479.221	5.492	5.243	28.791	27
28	102.966	.00971	.00177	.1818	566.480	5.502	5.281	29.054	28
29	121.500	.00823	.00149	.1815	669.447	5.510	5.315	29.284	29
30	143.370	.00697	.00126	.1813	790.947	5.517	5.345	29.486	30
31	169.177	.00591	.00107	.1811	934.317	5.523	5.371	29.664	31
32	199.629	.00501	.00091	.1809	1 103.5	5.528	5.394	29.819	32
33	235.562	.00425	.00077	.1808	1 303.1	5.532	5.415	29.955	33
34	277.963	.00360	.00065	.1806	1 538.7	5.536	5.433	30.074	34
35	327.997	.00305	.00055	.1806	1 816.6	5.539	5.449	30.177	35
40	750.377	.00133	.00024	.1802	4 163.2	5.548	5.502	30.527	40
45	1 716.7	.00058	.00010	.1801	9 531.6	5.552	5.529	30.701	45
50	3 927.3	.00025	.00005	.1800	21 813.0	5.554	5.543	30.786	50
55	8 984.8	.00011	.00002	.1800	49 910.1	5.555	5.549	30.827	55
60	20 555.1	.00005	.00001	.1800	114 189.4	5.555	5.553	30.846	60
65	47 025.1	.00002		.1800	261 244.7	5.555	5.554	30.856	65
70	107 581.9	.00001		.1800	597 671.7	5.556	5.555	30.860	70
75	246 122.1				1 367 339.2	5.556	5.555	30.862	75
100	15 424 131.9				85 689 616.2	5.556	5.555	30.864	100